Neurophysiology II

Publisher's Note

The *International Review of Physiology* remains a major force in the education of established scientists and advanced students of physiology throughout the world. It continues to present accurate, timely, and thorough reviews of key topics by distinguished authors charged with the responsibility of selecting and critically analyzing new facts and concepts important to the progress of physiology from the mass of information in their respective fields.

Following the successful format established by the earlier volumes in this series, new volumes of the *International Review of Physiology* will concentrate on current developments in neurophysiology and cardiovascular, respiratory, gastrointestinal, endocrine, kidney and urinary tract, environmental, and reproductive physiology. New volumes on a given subject generally appear at two-year intervals, or according to the demand created by new developments in the field. The scope of the series is flexible, however, so that future volumes may cover areas not included earlier.

University Park Press is honored to continue publication of the *International Review of Physiology* under its sole sponsorship beginning with Volume 9. The following is a list of volumes published and currently in preparation for the series:

Volume 1: **CARDIOVASCULAR PHYSIOLOGY** (A. C. Guyton and C. E. Jones)
Volume 2: **RESPIRATORY PHYSIOLOGY** (J. G. Widdicombe)
Volume 3: **NEUROPHYSIOLOGY** (C. C. Hunt)
Volume 4: **GASTROINTESTINAL PHYSIOLOGY** (E. D. Jacobson and L. L. Shanbour)
Volume 5: **ENDOCRINE PHYSIOLOGY** (S. M. McCann)
Volume 6: **KIDNEY AND URINARY TRACT PHYSIOLOGY** (K. Thurau)
Volume 7: **ENVIRONMENTAL PHYSIOLOGY** (D. Robertshaw)
Volume 8: **REPRODUCTIVE PHYSIOLOGY** (R. O. Greep)
Volume 9: **CARDIOVASCULAR PHYSIOLOGY II** (A. C. Guyton and A. W. Cowley, Jr.)
Volume 10: **NEUROPHYSIOLOGY II** (R. Porter)
Volume 11: **KIDNEY AND URINARY TRACT PHYSIOLOGY II** (K. Thurau)
Volume 12: **GASTROINTESTINAL PHYSIOLOGY II** (R. K. Crane)

(Series numbers for the following volumes will be assigned in order of publication)

RESPIRATORY PHYSIOLOGY II (J. G. Widdicombe)
REPRODUCTIVE PHYSIOLOGY II (R. O. Greep)
ENDOCRINE PHYSIOLOGY II (S. M. McCann)
ENVIRONMENTAL PHYSIOLOGY II (D. Robertshaw)

Consultant Editor: Arthur C. Guyton, M.D., Department of Physiology and Biophysics, University of Mississippi Medical Center

INTERNATIONAL
REVIEW OF PHYSIOLOGY

Volume 10

Neurophysiology II

Edited by

Robert Porter, M.D.
Department of Physiology
Monash University
Clayton, Victoria, Australia

UNIVERSITY PARK PRESS
Baltimore • London • Tokyo

UNIVERSITY PARK PRESS
International Publishers in Science and Medicine
Chamber of Commerce Building
Baltimore, Maryland 21202

Typeset by The Composing Room of Michigan, Inc.

Manufactured in the United States of America by Universal Lithographers, Inc., and The Maple Press Co.

Library of Congress Cataloging in Publication Data

Main entry under this title:

Neurophysiology II.

(International review of physiology; v. 10)
Previous ed. by C. C. Hunt.
Includes index.
1. Neurophysiology. I. Porter, Robert, 1932-
II. Hunt, Carlton C., 1918- Neurophysiology.
III. Series.
QP1.P62 vol. 10 [QP355.2] 599'.01'08s [599'.01'88]
ISBN 0-8391-1059-6 76-13039

Consultant Editor's Note

It is now two years since the first series of the *International Review of Physiology* appeared. This new review was launched in response to unfulfilled needs in the field of physiological science, most importantly the need for an in-depth review written especially for teachers and students of physiology throughout the world. It was not without trepidation that this publishing venture was begun, but its early success seems to assure its future. Therefore, we need to repeat here the philosophy, the goals, and the concept of the *International Review of Physiology*.

The *International Review of Physiology* has the same goals as all other reviews for accuracy, timeliness, and completeness, but it also has policies that we hope and believe engender still other important qualities often missing in reviews, the qualities of critical evaluation, integration, and instructiveness. To achieve these goals, the format allows publication of approximately 2,500 pages per series, divided into eight subspecialty volumes, each organized by experts in their respective fields. This extensiveness of coverage allows consideration of each subject in great depth. And, to make the review as timely as possible, a new series of all eight volumes is published approximately every two years, giving a cycle time that will keep the articles current.

Yet, perhaps the greatest hope that this new review will achieve its goals lies in its editorial policies. A simple but firm request is made to each author that he utilize his expertise and his judgment to sift from the mass of biennial publications those new facts and concepts that are important to the progress of physiology; that he make a conscious effort not to write a review consisting of an annotated list of references; and that the important material that he does choose be presented in thoughtful and logical exposition, complete enough to convey full understanding, as well as woven into context with previously established physiological principles. Hopefully, these processes will continue to bring to the reader each two years a treatise that he will use not merely as a reference in his own personal field but also as an exercise in refreshing and modernizing his whole store of physiological knowledge.

A. C. Guyton

Contents

Preface

Neurophysiology has come to embrace such a wide spectrum of scientific endeavor that the term neuroscience has been preferred by many people. Fundamental studies on the physics and chemistry of nerve cell membranes, experimental observations on synaptic transmission in the peripheral and central nervous systems of vertebrates and invertebrates, anatomical investigation of nerve cell connections, stimulation and recording experiments, behavioral observations and pharmacological manipulations of central nervous state have all been grouped under the single heading. Moreover, the involvement of nervous action in control of bodily functions such as respiration or peristaltic movements of the gut, the role of nervous detectors in temperature regulation or the control of blood pressure, and the interrelations between nervous control and humoral control of secretory or endocrine function have extended neurophysiology into all aspects of physiological study. Finally, model builders and those who would provide predictive theories of function have enquired of the significance of neurophysiology in communication. Here there are interrelations between neurophysiologists, biophysicists, mathematicians, and engineers.

A single volume review can scarcely attempt to address itself to representative parts of this whole spectrum of neurophysiological activity. Moreover, one must recognize that a major motivation in the study of neurophysiology is an attempt to understand the mechanisms by which the nervous system of an organism allows that organism to relate appropriately and react adequately to the environment in which it is situated. So it might be appropriate to select areas for comment which examine progress in understanding some of the most dominant nervous mechanisms which are involved in this totality.

The neurophysiology volume in the first series of this review did, in fact, approach nervous function in this way. The contributions of the study of identified neurons and their behavior in invertebrates were examined, the basic mechanisms of synaptic transmission and of muscle contraction were reviewed, the afferent system into which the vestibular organs feed was analyzed, and an overview of somatic sensation and movement performance in vertebrates was provided. In addition, the functional relationships of an anatomical structure in the brain, the superior colliculus, were described.

The present volume is arranged to complement the first. It considers some of the consequences of the fundamental organization of nerve cells with branching dendrites and possibly strategically located synaptic contacts. These considerations will have an essential contribution to make to an understanding of nervous integration if a proportion of this integration occurs on the surface of individual neurons. It demonstrates the contributions of biochemical observations on nerve cell function and axonal transport to the study of nerve cell connections—a new era of histology and histochemistry which has had important consequences for neurophysiological thoughts as well as for neuroanatomy. It reviews the long-term modifications in synaptic transmission which occur at peripheral synapses with use and disuse. These modifications, if they have counterparts in the central nervous system of higher animals as they appear to do in invertebrates, could be of the greatest significance in understanding changes and adaptations in nervous system function.

From these fundamental considerations the volume proceeds to examine some classical problems for the physiology of sensation. What is known about the neurophysiological basis of pain? What is the significance of the receptive field of neurons in the visual pathway? What is known of the anatomical and physiological organization of central neurons concerned with hearing in vertebrates and what is the meaning of tonotopic organization?

Finally, the volume addresses itself to a major function of nervous system action. The function is locomotion–the coordinated activity of nervous and musculoskeletal apparatus in the production of appropriate progression at a controlled speed toward or away from environmental stimuli. Following the studies of Sherrington and Graham Brown, locomotion received relatively little neurophysiological attention until recent years. But some challenging studies now command that it be re-introduced as a topic for scientific debate and endeavor. The study reveals the coordinated interaction of central "programs," peripheral feedback, and control structures such as the cerebellum in producing an organized motor output.

This volume has been prepared with the needs of graduate students in mind. Authors have been asked to provide opinion and critical comment rather than straight documentations of fact. Of course, the chapters contain the factual evidence for the comments which are made. But authors have been encouraged to be selective, to draw attention to the significant advances in their historical context and to show what the study of these aspects of neurophysiology has demonstrated at the present time.

R. Porter

International Review of Physiology
Neurophysiology II, Volume 10
Edited by Robert Porter

University Park Press Baltimore

1
A Quantitative Approach to Integrative Function of Dendrites

S. J. REDMAN
Monash University, Clayton, Victoria, Australia

INTRODUCTION

Neuronal integration is a concept which is probably familiar to all neurophysiologists. Sherrington (1) introduced this term to describe the summation of converging excitatory and inhibitory influences in a neuron long before any postsynaptic potentials were recorded, and long before the electron microscope had been used to reveal the structural characteristics of synapses. Yet as Sherrington (1) pointed out, the use of the term integration "is hardly sufficient for a definition." The main purpose of this article is to attempt to provide a theoretical and experimental framework for a quantitative approach to neuronal integration. The aim has been to relate modulations in neuron discharge frequency to alterations in impulse activity on one or several projections converging onto the neuron. The intermediary in this relationship is synaptic charge.

In developing this approach it has been necessary to discuss the mechanisms by which charge crosses synaptic junctions in dendrites, and then spreads to the soma where it combines with charge spreading from other dendrites. Cable models of neurons have proven to be useful tools in developing understanding of the spread of charge in dendrites. Two reviews (2,3) on cable models of neurons have recently been published, and these articles provide an excellent exposure to the theoretical and experimental development of neuron models. In this review the emphasis has been placed on the use of neural models as aids in understanding the importance of dendritic geometry, dendritic membrane properties, and synaptic location on dendrites for the accumulation and spread of charge in dendrites. The models also provide a theoretical basis for formulating experimental techniques whereby much important data related to dendritic propagation can be measured.

Much of the experimental data drawn upon for this article was obtained from mammalian spinal motoneurons. This emphasis was unavoidable as it is only for this neuron that the quantitative data which are required in this approach to neuronal integration are available. Even then it is incomplete and applies only to the group la projection. Nevertheless, it is hoped that there is sufficient generality in the suggestions for a quantitative description of dendritic integration that will encourage others to plan experiments with similar objectives for other synaptic projections and for other neurons.

PASSIVE PROPAGATION OF SYNAPTIC POTENTIALS IN DENDRITES

Dendritic trees of neurons in the vertebrate central nervous system provide an extensive surface upon which tens of thousands of synaptic junctions are formed. The surface area of the dendritic tree is often very much larger than the surface area of the soma. (It is about 20 times greater for spinal motoneurons.) The density of synaptic contacts does not vary greatly over the entire neuron surface (4). The spread of synaptic current within the dendrites, and across the somatic membrane, is central to any understanding of the summation of synaptic potentials arising from impulse activity at these dendritic synapses. The spread of current away from a synapse depends upon the cable properties of the dendrites and the junctional mechanism occurring at the synapse. Two basically different modes of propagation in dendrites can be identified, and each has been observed in neurons of the CNS. These different modes are usually referred to as passive and active propagation. In the latter, the membrane conductance is voltage dependent.

Our understanding of the factors affecting passive propagation is much more advanced than for active membrane processes. For this reason, much more emphasis is given in this review to passive spread of current. Passive propagation is influenced by neuronal geometry, membrane properties, and the location of synapses on the dendritic tree. These in turn affect the time course, peak amplitude, and synaptic charge of synaptic potentials as they spread from the synapse of origin throughout the dendrites and to the soma.

Cable Properties of Dendrites

The geometrical structure of neurons is usually very complicated, and can be very diverse for different neuron types (5,6). The number of dendrites, their length, orientation, and branching patterns all need to be considered in any quantitative description (7). Much diversity may also exist in the electrical properties of the membrane of the dendritic trees of different neurons, and perhaps in different regions of the same dendrite. While geometrical details of dendrites may be reasonably well classified by staining and dye injection techniques, the electrical characteristics of dendritic membranes are extremely difficult to determine. The basic problem is that successful microelectrode penetrations are most likely to occur in the soma-proximal dendrite region. Electrical measurements made in this region are sensitive to local membrane properties, but become progressively less sensitive to membrane properties at more distal regions.

Further difficulties arise in formulating a mathematical description of the cable characteristics of branching structures, especially when they may have non-uniform cross section and non-uniform membrane properties. A mathematical model is only useful if it is experimentally testable. This usually requires that the number of independent parameters involved be small. For this to be so, many simplifying assumptions must be made, and a simple cable description of the dendritic geometry is required.

The Rall Model of the Motoneuron The only neuron model which has been extensively investigated both mathematically and experimentally is the Rall model of the motoneuron (8–20). Apart from these publications, this model has recently been reviewed by Rall (3) and by Jack et al. (2). In this section the model is briefly described, and its present strengths and weaknesses, as revealed by experiment, are discussed.

The important simplifying assumptions in this model are:

1. The unit of membrane area can be represented as a resistance in parallel with a capacitance. The resistance is voltage independent (passive). Uniform resistance and capacitance per unit area of membrane are assumed for the entire neuron surface.
2. The internal surface of the soma is isopotential. Hence the soma membrane is represented by a single resistance-capacitance parallel combination.
3. The external surface membrane of the neuron is equipotential.
4. Each dendritic tree reduces to a cylindrical cable structure of uniform cross section. These dendritic cylinders are in parallel with each other and can be lumped together to form a single cylinder which is electrically equivalent at its connection with the soma to the entire dendritic tree. This cylinder may be finite or infinite in electrical length and is connected to the R-C circuit representing the soma.

The parameters which are necessary to define this model are: (*a*) the membrane time constant (τ_m); (*b*) the electrical length of the equivalent dendritic cylinder (L); (*c*) the ratio of the steady state input conductance of the dendritic tree, and the steady state conductance of the soma (ρ).

The reduction of the dendritic tree to an equivalent uniform diameter cylindrical cable requires a particular branching structure. The dendritic diameter before branching, raised to the 3/2 power, must equal the sum of the diameters of the branches distal to the branch point, when each of these diameters is raised to the 3/2 power. (This geometry gives what is known as an "impedance matching" at the branch points, such that any electrical energy spreading distally is transmitted with maximum efficiency past the branch point.) That such a geometry does in fact occur in motoneuron dendrites was originally suggested from the Golgi material of Aitken and Bridger (21) and this was the basis of Rall's original derivation of the equivalent dendritic cylinder. Subsequently, intracellular injection of radioactive glycine has confirmed this geometry (22), at least out to branch diameters of 2 μm. However, Barrett and Crill (23) report considerable tapering of the equivalent dendritic cylinder, based on intracellular injection of procion dye. The marked difference between these two results has not been resolved. The voltage response to current impulses applied to the soma revealed no evidence for a significant tapering in the cross section of the equivalent dendritic cable (24).

Tapering in the equivalent dendritic cable can occur due to a reduction in dendritic branch diameter between branch points, a deviation from the 3/2 power law, and termination of different dendritic cylinders at different electrical lengths. Lux et al. (22) report a 20% variation in the electrical lengths of different dendrites from the same neuron. Lumping 10 or more such dendrites together to form an equivalent cylinder of a single electrical length is unlikely to cause significant error in the representation of the dendritic tree. However, the effects on the spread of synaptic potentials due to cross-sectional reduction in proximal dendritic trunks (23) when the working model assumes a uniform diameter cable, needs more careful assessment.

There is general agreement on the finite electrical length of the dendritic cylinder. This has important consequences for the spread of charge from dendrites to the soma, and for the summation of synaptic potentials within dendrites. The original calculations of Rall (12) suggested an electrical length of 1 to 2 space constants (λ) for the equivalent dendritic cylinder. This has subsequently been confirmed in a number of separate investigations (22–25). The electrical termination of the dendritic cable was originally assumed to be an open circuit or sealed end (12) due to final termination of dendritic branches. Lux (16) suggested that tonic synaptic activity on the fine terminal branches could cause a leaky dendritic termination. Analysis of the voltage response to brief intracellular current pulses (24) and to brief, localized, synaptic currents (26) indicates that the dendritic termination (in deeply anesthetized preparations) has more in common with a sealed termination than with a leaky termination.

The assumption of uniform membrane resistivity for the soma and dendrites has not been proven. The possibility that the specific membrane resistivity increases in distal regions of the dendrites was first raised by Fatt (27) and by Katz and Miledi (28). Subsequently Iansek and Redman (24) found tentative evidence for different resistivity of soma and dendritic membrane. However, it is difficult to separate the effects of a lower soma membrane resistivity and the effect of electrode damage at the soma. A higher membrane resistivity in distal dendrites would have important consequences for the attenuation of synaptic potentials, and for charge loss across dendritic membrane, during passive propagation.

Some preliminary analyses exist for the effect of different membrane resistivities on potential transients generated by brief current pulses at the soma membrane. Iansek and Redman (24) have studied a model in which the soma membrane resistivity differs from the dendritic cylinder resistivity (assumed uniform). The length of the dendritic cylinder is infinite. Jack (personal communication) has generalized this to cope with finite length dendritic cylinders. These analyses have confirmed that the time constant of final decay of potential transients generated at the soma, or propagating to the soma from distal dendritic regions, will show little difference in neurons with dendritic dominance and higher dendritic membrane resistivity. Injection of alternating currents and measuring phase-frequency relationships (16, 29) may provide a more sensitive method for detecting differences in soma and proximal dendritic resistivities.

The dominance of the dendrites in determining the passive cable properties of the motoneuron, as measured at the soma, is now clearly established. Several different attempts have been made to measure the dendritic to soma conductance ratio (ρ) (16, 24, 29) following theoretical suggestions by Rall (9, 14) and by Jack and Redman (17). There are difficulties in avoiding technical artifacts in these experiments, and there is some uncertainty as to how accurately (ρ) has been measured. However, independent calculations based on measurement of soma and dendritic surface areas (8) tend to confirm the experimental findings that (ρ) generally exceeds 10.

All of the geometrical and electrical parameters mentioned above have important effects on the spread of synaptically transferred charge to the soma-initial segment region. Before the cable properties of motoneuron dendrites were determined, it was generally believed that the electrical length of dendrites was many space constants, and therefore synapses on distal dendritic regions would be ineffective in altering somatic membrane potential (30). Alternatively, for these synapses to be effective, some active membrane response was necessary during dendritic propagation (31, 32). Such arguments are no longer necessary. Theoretical investigations on charge and voltage attenuation of synaptic potentials during dendritic propagation support the contention (18, 20, 33) that synaptic potentials generated by very brief synaptic currents (duration $\ll \tau_m$) and located at distances of λ to 2λ from the soma can still make significant contributions to the soma membrane potential. (This is discussed in detail in the sections on the amplitude and charge transfer accompanying synaptic potentials.)

Cortical Pyramidal Cells Little detailed analysis of the passive cable properties of other vertebrate central neurons has been achieved. One exception is the cortical pyramidal cell for which estimates of the steady state attenuation in the apical dendritic shaft, and of the total dendritic dominance, have been made (34, 35). These calculations suggest that the electrical distance to the first major branch point on the shaft of the apical dendrite is about one space constant. The steady state shunt that the apical dendrite and all basilar dendrites present to the soma membrane varies between 3 and 6. The relative contributions that the apical dendrite and basilar dendrites make to this dendritic shunt vary considerably, but they are of the same order.

Further progress towards establishing cable properties of central neurons with complicated dendritic trees awaits simultaneous injection of dye (or labeled amino acids) and electrophysiological measurements of steady state input resistance and response to brief current or voltage clamps.

Time Course of Postsynaptic Potentials Evoked in Dendrites

The summation of postsynaptic potentials in dendrites, and their propagation to the soma, depends upon the synaptic processes which generate them, and their time course and amplitude. As microelectrode recording from dendritic branches is not normally feasible, the voltage time course and amplitude of a dendritic potential at its point of generation have to be inferred from the time course and amplitude of this potential after propagation to the recording site.

Synaptic Potential Time Course at Synaptic Site The time course at the subsynaptic region depends upon: 1) the time course of synaptic conductance change, 2) the difference between the reversal potential for that synapse and the membrane potential, and 3) the *input impedance* of the dendrites at that point (18, 20).

The input impedance is determined by the dendritic geometry and membrane properties, and especially those geometrical features close to the synaptic site (e.g., proximity of synapses to branch points or dendrite terminations). Unfortunately, such detailed information is not available for any neuron. For this reason the time course of the postsynaptic potential at the recording site must be used as the basis for calculating the time course in the dendrites (18). This calculation also requires a description of the time course of synaptic current, and an electrotonic distance from the recording point to the synapse. The synapse is assumed to be at a point on a cylindrical cable which is electrically equivalent to those dendrites which receive synaptic connections. But the time course is independent of the fraction of the dendritic tree that this cylinder represents.

The cable properties of the neuron, and the time course of the postsynaptic potential at the recording point, may be used to calculate the electrotonic distance to the synapse, and the synaptic current time course (17, 24, 36). Such calculations have been made for unitary 1a e.p.s.p.s in spinal motoneurons (18, 36). The conclusion reached from these calculations was that e.p.s.p.s of dendritic origin have voltage time courses near their site of generation which are very brief compared to their time course at the soma, and in some cases they are very brief compared to the membrane time constant. Most e.p.s.p.s reached their peak and then decreased to less than half of peak amplitude in less than 1 ms (See Iansek and Redman, Figure 3 (36)).

The major qualification to these results is the need to assume a synaptic input on a portion of the equivalent dendritic cable, rather than on a single dendritic branch. Applying the synaptic current to the equivalent cable, at an electrotonic distance X from the soma, requires that all dendritic branches included in that cable at that electrotonic distance receive the synaptic input. For proximal synapses this may be true, but it would seem unlikely to regularly hold for synapses made distal to the first or second branch points. The effects of departures from this assumption on the calculation of dendritic time course (and amplitude) have not been carefully investigated. Intuitively, a deficit in synaptic projection to all branches of a dendrite at a particular electrotonic distance would aid rapid redistribution of charge away from synaptic sites, and so increase the rate of decay of the synaptic potential in these regions.

In other calculations (20, 33) a conductance change has been placed on a distal dendritic branch. A conductance time course compatible with the synaptic currents derived from experimental data on unitary 1a e.p.s.p.s has been assumed. The resulting voltage transients have time courses which are consistent with those derived from somatic time courses. In addition, the time course and amplitude of the postsynaptic potential have been calculated at other points in the dendritic tree (20). This will be valuable information when the additive effects of synaptic activity, dispersed over the entire dendritic tree, need to be calculated.

Implications of Brief Synaptic Potential Time Course for Repetitive Summation in Dendrites The brief time course of synaptic current which has been shown to apply for 1a synapses on motoneurons may not occur for other types of synapses on these or other types of neurons. Until this information can be obtained, it would be dangerous to generalize about the time course of dendritic postsynaptic potentials. However, there are implications for the summation of synaptic potentials. Repetitive activity in single group 1a fibers is unlikely to cause any significant summation of synaptic potential at the synaptic site, as the impulse rates in these fibers rarely exceed a few hundred per second. Consequently the driving potential for synaptic current will not alter significantly for successive impulses (*cf.* Ref. 37). Provided that the conductance change is the same for successive impulses, a fairly linear addition of these potentials can be expected. Similarly, if the synapses of two different 1a fibers are close together on the same dendrite, it will require nearly synchronous activation of these neighboring synapses for the resulting postsynaptic potentials to add in a nonlinear fashion by one e.p.s.p. reducing the driving potential for the second conductance change. These arguments apply to summation of potentials at the one synapse, or immediately adjacent synapses. Nonlinear summation of postsynaptic potentials does occur (see discussion under "Nonlinear Interactions between Postsynaptic Potentials") but it is more likely when stringent requirements on impulse timing are unnecessary. This will be so when one of the postsynaptic potentials involved has developed a more prolonged time course following electrotonic spread.

Amplitude of Postsynaptic Potentials in Dendrites

The peak amplitude of a postsynaptic potential at or near its point of generation is an important factor in determining the degree of interaction which could occur with other conductance changes in the summation of dendritic potentials, and the efficiency of charge transfer at its synapse of origin (33). It could also be helpful in establishing some bounds on the voltage insensitivity of various regions of dendritic membrane. Finally, this peak potential, together with dendritic cable properties, determines the peak potential after propagation to the soma.

Just as the time course of a dendritic potential has to be calculated from its time course at the soma, so must the peak amplitude in the dendrite be calculated from the recorded peak amplitude. Unfortunately, this calculation is very sensitive to the uncertainties in dendritic geometry and the location of synaptic projections to dendritic branches. In particular, the number of dendrites actually receiving the synaptic input is vital to this calculation (18). The number of dendrites which receive synapses from an afferent fiber cannot be determined from recordings taken from a central recording electrode at the soma. All dendrites appear as one to this electrode.

The recorded peak amplitudes of unitary 1a e.p.s.p.s in motoneurons were usually in the range 100 to 500 μV (26, 36, 38, 39). Surprisingly, no tendency was found for the peak amplitude of these e.p.s.p.s at the soma to become smaller as the distance to the synapses of origin increased (36, 40). (This observation is discussed in detail under "Charge Transfer and Redistribution at Dendritic Synapses".) One immediate implication is that, at least for 1a afferent fibers synapsing on motoneurons, there does not appear to be any spatial weighting attached to different fibers through the location of their synapses on the dendrites.

If the synaptic input is restricted to one of 10 equal diameter dendrites, then peak voltages in the distal half of the equivalent dendritic cable have been calculated to range from 3 to 20 mV (36). However, to convert these figures to the peak voltages which are generated in the component branches of this equivalent dendrite, it is necessary to know the number of branches. Between the first and second branch point, these voltages could be approximately doubled if only one branch receives a synapse, and both branches are roughly equal in diameter.

Another approach to calculating dendritic potentials (33) is to calculate the conductance change which must occur at the soma to generate a quantal somatic e.p.s.p. This same quantal conductance change can then be applied to various regions of a reconstructed dendritic tree. This calculation gives a peak potential in a distal dendrite of approximately 20 mV for a single quantum 1a e.p.s.p. Similarly Rinzel and Rall (20) have applied a current source to a single branch terminal and calculated the transient attenuation factor for transmission from this point to various other locations on the neuron surface. These voltage attenuation factors are generally very large when the synaptic input is so spatially restricted. For a realistic synaptic current time course, the peak voltage attenuation factor is 235 from branch terminal to soma. Thus a peak voltage of 100 μV at the soma becomes 23.5 mV at the synapse. (In section "Experimental Observations on Nonlinear Interactions in Dendrites," nonlinear summation of two unitary 1a e.p.s.p.s is analyzed to derive bounds for the peak voltage at each subsynaptic site. The results agree with those obtained using the procedures described above.)

These various approaches tend to suggest subsynaptic peak voltages in the range 20 to 40 mV in distal dendritic branches. Such potentials are sufficiently large to reduce the efficiency of charge transfer at these synapses (33). After electrotonic spread to adjacent branches, attenuated potentials of this initial magnitude will still provide a significant shift in driving potential at synapses in these regions (20) with a consequent change in their charge transfer efficiency. If such synaptic potentials contain quantal components, these will add nonlinearly (39).

Although the different approaches to calculating peak potentials in dendrites give consistent results, each method of calculation rests on at least one of several unproven assumptions. These include uniform membrane resistivity along the dendrites, similar junctional mechanisms at somatic and dendritic synapses, and synaptic projection to all sub-branches of a dendrite. Until firmer evidence can be obtained for these assumptions, these calculations must be regarded as first estimates only.

Charge Transfer and Redistribution at Dendritic Synapses

Calculation of Synaptic Charge There are many advantages in the use of charge associated with a synaptic potential, rather than voltage, when studying synaptic mechanisms at junctions remote from the recording site and on complicated dendritic structures. The calculation of either the potential in the subsynaptic region, or the net charge transferred at the synapse, requires the time course of the synaptic potential as basic data. However, the calculation of net charge transfer at the synapse requires fewer assumptions than the calculation of subsynaptic potential. No assumptions are necessary on the time course of the synaptic current and neither is the fraction of the dendritic tree that receives the synaptic input important. The assumptions involved in collapsing the dendritic tree to an equivalent uniform cylinder, with uniform membrane properties, are still required. (However, relaxing these assumptions would not lead to such mathematical difficulties for charge calculations as those that would arise for voltage calculations.) The charge that redistributes to the recording site is calculated from the time integral of the measured postsynaptic potential, divided by the input resistance at that point (2, 20, 33, 36). The net inward charge transferred at the synapse can be calculated from the somatic charge, provided that the synaptic location (X) and the electrotonic length (L) of the dendritic cable are known (2, 36).

The other important advantage of charge measurement is that the average amount of charge per unit time spreading to the soma from a synapse (or group of synapses) is the *current* that is provided by this pathway to the soma. Thus, a knowledge of net charge transfer, and impulse frequency, for a synaptic input can be directly converted to current supplied to the soma. This concept is used in section "Non-passive Propagation and Active Responses in Neuron Dendrites" in discussing the relationship between neuron discharge frequency and somatic current.

Experimental Observations on Synaptic Charge Transfer Just as the peak synaptic potential of unitary 1a e.p.s.p.s at the soma shows negligible dependence on synaptic location, so too is somatic charge almost independent of synaptic location. (The charge redistributing to the soma for 1a e.p.s.p.s in motoneurons may be obtained from Figure 8, Iansek and Redman (36), by multiplying each charge by the dendritic to soma conductance ratio ρ, which in almost all cases was 25.) If any dependence on synaptic location exists, it is that distal synapses provide more somatic charge than do proximal synapses. Although these are population results, it is unlikely that in any one neuron a systematic variation of charge with synaptic location occurs. This finding should dispel any residual generalizations that synapses on distal dendrites are ineffective in modulating current flowing into the soma region. The attenuation factors for charge spread in dendrites are the same as for d.c. potentials, and are very much less than those which apply to brief synaptic potentials (2, 20, 33). Applying these attenuation factors to somatic charge gives figures for net inward charge transfer at distal synapses which are 10 times larger than the net charge crossing somatic synapses. The accuracy of this calculation may be questioned, but some charge must be lost during passive dendritic propagation, and this loss will increase as the electrotonic distance to the synapse increases. Thus dendritic synapses are more effective sites for inward charge transfer than are somatic synapses (for the 1a pathway). It is tempting to develop a teleological argument to suggest that if synapses of the one kind are to be weighted equally in their effect on the discharge of an integrative neuron, then a compensatory mechanism has to be built into the more remote synaptic connections to provide the additional charge which will be lost during dendritic propagation.

The net (inward) charge crossing a synaptic junction can be increased by: 1) increasing the area of the junction; 2) increasing the depth and duration of the conductance modulation; and 3) shifting the equilibrium potential of the conductance modulation away from the membrane potential.

Some of these possibilities may be excluded. No evidence was found for longer duration synaptic currents with increasing electrotonic distance for the synapse (36). The size of 1a synapses is uniform over the motoneuron surface (41, 42), but synaptic area can be increased by increasing the number of terminals per afferent fiber. Information on the number and location of synaptic terminals on a single motoneuron, which arise from a single 1a afferent, is still sketchy. Iles (Ref. 43 and personal communication) reports an average of 1.8 terminals per afferent fiber for terminations on the proximal half of the dendrites. The total number of 1a terminals on a motoneuron is probably less than 100 (42). From the total number of 1a fibers projecting to a single motoneuron (38, 44, 45) an average number of boutons per 1a afferent per motoneuron of less than 2 seems likely. As the proximal regions have this average number, so too must the distal regions. For these reasons, total synaptic area cannot account for the extra charge injected. The remaining explanations are that the depth of conductance modulation increases and/or there is a shift in reversal potential to a more positive potential at dendritic synapses. This implies specialization of synaptic properties as a function of distance from the soma. It will be important to discover whether this occurs for other projections to motoneurons, or indeed for any neuron with passively propagating dendritic potentials.

Efficiency of Charge Transfer at Dendritic Synapses When two identical conductance changes, with equal reversal potentials for the ions involved, occur at geometrically different points on the neuron surface, different amounts of charge cross at these junctions. This is because the instantaneous synaptic current depends upon the instantaneous difference between the reversal potential and the membrane potential. The higher the input impedance, the smaller will be this difference for a given conductance change. Thus less charge will cross the synapse at the site of higher input impedance (20, 33).

The most efficient method for transferring charge at a dendritic synapse is to have numerous and dispersed synaptic contacts, rather than one or two large synapses covering the same junctional area. The larger currents which would flow for a highly localized junctional region would tend to reduce the driving potential during the generation of the postsynaptic potential much more than would numerous and less intense currents dispersed over a wider dendritic region. Barrett and Crill (33) calculate that a quantal conductance change of 40×10^{-10} mho ms, in the form of a rectangular pulse 0.2 to 0.5 ms in duration, and with a reversal potential of 0 mV, has an 81 to 86% charge transfer efficiency at a dendritic tip (input resistance 170 megohm). In contrast, the same conductance change is almost 100% efficient when located at the soma. There are several known instances where large e.p.s.p.s are produced by multiple and spatially dispersed synaptic contacts arising from a single afferent fiber. The synaptic connections of a climbing fiber with a cerebellar Purkinje cell are an obvious example. Single 1a fibers make multiple contacts with dorsal spinocerebellar tract neurons, and large e.p.s.p.s are evoked by impulses in these fibers (46, 47).

The reduction in driving potential during synaptic current flow also causes nonlinear addition of charge transfer during quantal fluctuations in conductance. This nonlinearity was originally recognized in the summation of postsynaptic potentials at the neuromuscular junction (48). For unitary 1a e.p.s.p.s evoked in motoneuron dendrites, a discrepancy was observed between the variability in peak amplitude calculated from the probability of failures (using Poisson statistics) and the measured variability (39). The compressing effect of nonlinear voltage addition was proposed as the explanation. The same nonlinearity affects charge transfer at distal dendritic synapses, and this nonlinearity can be very significant (33). For 4 quanta of conductance change, charge transfer efficiency ranges from 50 to 55%, compared with the figures given above (for one quantum) of 81 to 86%. Calculating charge transfer at dendritic synapses is a more secure procedure than calculating peak potentials at these synapses. Thus studies on fluctuations in synaptic potentials at such synapses may be more accurate if charge transfer is used as the measure of fluctuation (49).

LOCATION OF SYNAPSES ON DENDRITES

Any search for organization in the morphology of synaptic connections on dendrites should take into account the spatial location of synapses originating from a particular projection to that neuron. While the soma-initial segment region is an obvious reference zone, especially for a neuron with passive dendritic properties, it need not be the only one. The relative locations of the synapses originating from two different projections to the same neuron will also be important. If one of these projections makes inhibitory connections, then the effectiveness of this pathway depends on the relative location of the two types of synapses (see section "Theoretical Studies on Nonlinear Interactions in Dendrites"). Not all dendrites have passive membrane properties and, if any inward (membrane) current flow is to be initiated by a large dendritic depolarization, then the relative proximity of two converging excitatory pathways could also be important. (A discussion on these possibilities may be found in section "Control of Neuronal Discharge by Presynaptic Impulse Activity".)

Various analog and digital compartmental models of neurons with dendritic trees have been used to study the effects of spatial location (and timing) of different excitatory and inhibitory activity on the depolarization at the soma and the firing pattern of the neuron (12, 50–53). The results are largely hypothetical. These studies require accurate data on the spatial location and charge transfer capacities of different synaptic inputs. Then the effects of modulations in impulse activity (and timing effects) in these various pathways to the neuron can be determined.

In this section various approaches to the problems of determining the electrical distance from the soma to a synapse are described, together with some results obtained with these procedures.

Methods for Determining Synaptic Location

Histological identification of the synapses which a single fiber makes with a single neuron is obviously a tedious and difficult task. However, techniques which allow the axons and neurons to be filled with opaque dyes or various labeled substances are now commonly used. It is to be hoped that the combination of histology and the various electrophysiological methods described below will provide detailed information on the location of synapses arising from different projections to the neuron. This information must be obtained if a complete picture of synaptic interactions in dendrites is to be achieved.

Analysis of Time Course of Synaptic Potential The time course of synaptic potential at the soma depends upon the electrical distance (X) over which it has propagated, the time course of synaptic current, and the cable parameters of the neuron. If the cable parameters (ρ, L, and τ_m) are separately measured (24), then two independent measures taken from the time course of the synaptic potential will allow X, and some measure of the synaptic current, to be calculated. Usually, the two time course measures are 10 to 90% rise time and halfwidth. These must be normalized by the membrane time constant to enable the use of published data.

The basic assumptions of this method are: 1) that the synaptic current is brief (in comparison with the membrane time constant), and 2) that the various synaptic terminals (if more than one) are at the same electrotonic distance from the soma.

This is the most developed method for calculating synaptic location, and full details and a discussion on the accuracy of the calculation are given in Jack, Noble, and Tsien, Chapter 7 (2). A critical evaluation of this procedure requires joint histological localization of the synapses arising from the single fiber on the same cell to which the time course analysis was applied.

Comparison of Time Courses of Evoked Potentials within Same Neuron If two different postsynaptic potentials are evoked in the same neuron by stimulation of fibers in different pathways, then it may be thought that the potential with the briefer time course originates from a more proximal location (54–59). As the neuronal geometry and membrane properties (assumed uniform) are in common, then the time course of a synaptic potential is determined by the time course of synaptic current and the synaptic location. Thus it is only appropriate to use this comparison if the postsynaptic potentials are generated in a localized region, and the time course of synaptic current is the same for both synaptic inputs. This latter condition will not matter if the synaptic currents are both brief, and both synaptic regions are well removed from the soma. Under these conditions the exact time course of synaptic current is unimportant (60). However, evidence exists that different synaptic current durations can occur at different group 1a synapses on the same motoneuron (61). Thus it may not be correct to assume a similar time course of synaptic current for different projections to the same cell. In general this will leave a distance comparison based on time courses in the same cell open to error. Such a comparison should be reinforced with a further test, such as sensitivity to membrane potential changes (59).

Sensitivity of Postsynaptic Potentials to Changes in Membrane Potential Synaptic potentials which are generated by a conductance change to one or several ions can, in principle, be reduced to zero when the membrane potential at the synaptic site is displaced to the reversal potential (E_{rev}) for the ionic currents. The further a synapse is from the soma, the greater must be the membrane potential shift at the soma to alter the subsynaptic membrane potential to E_{rev} (62). Thus, if E_{rev} is known, if the somatic membrane potential at which the recorded e.p.s.p. becomes zero is obtained, and if the electrical length of the dendritic cable is determined, then it is theoretically possible to calculate the electrotonic distance to the synapse (2). In practice, this method presents many difficulties. In all neurons in which synapses are spatially dispersed, there appears to be no unequivocal determination of E_{rev}, especially for excitatory synapses (61, 63). A two electrode bridge is almost essential if membrane potential shifts need to be accurately known. Finally, for large changes in membrane potential the steady state current-voltage characteristic for neurons will almost certainly be nonlinear. This nonlinearity makes the potential-distance profile difficult to calculate (2). For excitatory synapses, it may not be possible to supply sufficient current to depolarize the membrane at the synapse to E_{rev}. Extrapolation of the reduction in peak potential may be attempted, but such extrapolation must not only take into account the potential-distance profile, but also the reduction in the passively propagated e.p.s.p. amplitude due to a non-uniform lowering of the dendritic membrane resistance. In other words, there are serious experimental difficulties in using this approach to obtain a quantitative distance measure.

When two different fibers which synapse on the same neuron are separately stimulated, it may be reasonable to detect the relative proximity of the two terminations by differential sensitivity to intracellularly applied currents. Provided both synapses have the same E_{rev}, then regardless of membrane nonlinearities and the exact value of membrane potential shift, the postsynaptic potential which shows the greater change for a given current will be the more proximal. While it has not been established that synaptic terminations of the same type (E or I) but of different origins have the same reversal potential, this method is attractive for practical reasons. It has been used in several investigations, but applied to composite postsynaptic potentials. In primate motoneurons, little difference in sensitivity to depolarizing currents was observed in the e.p.s.p.s evoked (in the one cell) by stimulation of rubro-spinal, group 1a, and cortico-motoneuronal pathways, even though these composite potentials had clearly different time courses (57). In contrast, hyperpolarizing currents applied to red nucleus neurons increased the amplitude of the synaptic potential with the faster time course (evoked by stimulating the interpositus nucleus) and did not affect the e.p.s.p. with the slower time course (evoked by stimulating the sensorimotor cortex) (59).

Sensitivity of Postsynaptic Potential to Ion Injection Coombs, Eccles, and Fatt (64) found that the group 1a e.p.s.p. in motoneurons was almost insensitive to ion injection. In contrast, the group 1a i.p.s.p. was very sensitive to chloride ions. As the internal chloride ion concentration increased, the inhibitory potential changed from a hyperpolarization to a depolarization. This observation has been used to determine the relative proximity to the soma of the inhibitory synapses arising from two different inhibitory pathways: the group 1a reciprocal inhibitory projection and the recurrent pathway originating from Renshaw cells (65). The relative times at which the two potentials reverse following chloride ion injection and diffusion along the dendrite are an indication of their relative proximity to the soma. (This assumes that both types of synapse are equally sensitive to chloride ion concentration.) The current inhibitory potential was slower to reverse, and it was concluded that this synaptic input zone was the more distal.

In principle this technique could be made quantitative, as the diffusion times could be calculated from the neuron geometry using equations similar to the cable equations. However, it would be necessary to know the diffusion coefficient for chloride in the dendritic core, the intracellular chloride concentration before injection, and the concentration at which the i.p.s.p. reverses.

Detection of Synaptic Conductance Change A localized synaptic conductance change at any point on a neuron surface will appear as an impedance change at the soma of the neuron. This impedance change will, in general, have a resistive and a reactive component. The time course and amplitude of each of these components depend upon the time course and magnitude of the conductance change, the electrotonic distance from the soma to the conductance change, and the cable properties of the neuron.

Smith, Wuerker, and Frank (66) developed an a.c. impedance bridge to record the resistive and reactive parts of the impedance change, as well as the postsynaptic potential. This technique required a phase sensitive detector and time averaging of transients so that a low frequency carrier (100 Hz) could be used to increase the distance over which conductance transients could be detected, and at the same time to preserve the time course of the conductance change. Their results showed that conductance changes accompanying various forms of inhibition could readily be detected, while group 1a e.p.s.p.s were not always associated with a detectable impedance change. Their qualitative interpretation of this result was that the conductance changes accompanying the inhibitory potentials were close to the soma, while some conductance changes associated with excitatory potentials were on dendrites and beyond the range of detectability.

This technique could be made quantitative by analyzing the time course and amplitude of the resistive and reactive components of the impedance transient, assuming a time course for the conductance change, and knowing the cable parameters for the neuron. However, the range over which conductance transients can be detected is limited by the frequency of the carrier, and the maximum carrier current which can be applied to the neuron without altering membrane properties.

Experimental Observations on Synaptic Location

Spatial Distribution of Synapses on Motoneuron Dendrites A quantitative description of the spatial distribution of synapses has only been achieved for the projections of Group 1a fibers onto motoneurons (26, 40). The method is based on analysis of the time course of unitary e.p.s.p.s, coupled with independent measurements of the cable parameters of the motoneuron (24).

The group 1a projections to motoneuron dendrites were found to be denser on the proximal half than on the distal half (24). The greatest concentration was in the range 0.2 λ to 0.8 λ from the soma (26, 40). No difference was observed in the spatial distribution of synapses from homonymous and heteronymous group 1a fibers (40). Furthermore, the peak voltage at the soma and the charge redistributed for these e.p.s.p.s was virtually independent of synaptic location (see sections "Amplitude of Postsynaptic Potentials in Dendrites" and "Charge

Transfer and Redistribution at Dendritic Synapses"). For these reasons, electrotonic distance from the soma to 1a synapses does not seem to be an important variable in determining the contribution of impulses on any one fiber to the depolarization of the motoneuron soma. However, it would be unwise to generalize to other pathways, and to other neurons, on the basis of this result. What may be important for the group 1a pathway to motoneurons is the relative location of these synapses to inhibitory synapses, such as those mediating 1a inhibition, and the recurrent inhibitory pathway.

Less precise location data are available for other synaptic projections to motoneurons. On the basis of relative Cl^- ion sensitivity, the reciprocal group 1a inhibitory synapses are on the soma and proximal dendrites, and the recurrent inhibitory synapses are on proximal dendrites (65). The reticular inhibitory projection is removed from the soma (67) but the inhibitory potential is sensitive to injected Cl^- ions and to hyperpolarizing currents. There are monosynaptic projections to primate motoneurons from the red nucleus and pyramidal tract. On the basis of time course comparisons (within the one cell) of composite e.p.s.p.s (57) it appears that the rubro-spinal projections synapse towards the distal margins of the group 1a synaptic zone. This spatial relationship between group 1a synapses and cortico-motoneuronal synapses is reinforced by comparisons of time course of unitary (or minimal) e.p.s.p.s from both of these pathways (55).

To use this information on synaptic locations (in the way that is suggested in section "Control of Neuronal Discharge by Presynaptic Impulse Activity") it is necessary to have further information. This includes the numbers of fibers in each projection, the range of impulse frequencies which may occur in each projection, and the charge transfer properties of each type of synapse. Furthermore, to study synaptic interactions (as proposed in section "Nonlinear Interactions between Postsynaptic Potentials") it is necessary to know the reversal potential for each type of synapse, precise details of electrotonic distances, and whether any synaptic arrangements are dendrite specific.

Spatial Distribution of Synapses on Other Neurons It is not the purpose to review here all histological and electrophysiological data pertinent to synaptic location in the CNS. Rather it is simply to point out that there is now ample evidence for spatially distinct zones of synaptic termination for different projections to many CNS neurons (Scheibel and Scheibel (6) have reviewed these data for some neurons). In particular, pyramidal cells of the cerebrum and hippocampus, red nucleus neurons, and cerebellar Purkinje cells have distinct spatial organization in termination patterns. The importance of these patterns is just beginning to be assessed. One striking feature for each of these neuron types, and also for the motoneuron, is that the soma and proximal dendrites always receive powerful inhibitory projections. Theoretical approaches to find the optimum location for inhibitory synapses to produce specific or nonspecific inhibitory actions are reviewed in section "Theoretical Studies on Nonlinear Interactions in Dendrites." For many neuron geometries, this location will be the soma and proximal dendrites. However, if the inhibitory action is to be specific to a particular excitatory pathway, with a termination pattern which is dendrite specific, the inhibitory projections will also need to be dendrite specific.

NONLINEAR INTERACTIONS BETWEEN POSTSYNAPTIC POTENTIALS

If the current flowing to the soma of a neuron were simply a linear sum of the currents generated by all convergent excitatory and inhibitory activity, a quantitative description of this integration process would appear to be within reach. Predicting the neuronal firing rate would then be a matter of knowing the charge transfer at synapses on the different pathways, the afferent frequencies and the number of synapses involved, together with the frequency-current relationship (see section "Control of Neuronal Discharge by Presynaptic Impulse Activity"). Neuronal integration, even when only passive dendritic propagation is involved, is undoubtedly more complex than this. Some of the complexities reside in the nonlinear mechanisms of charge transfer at synapses, and interactions between postsynaptic potentials.

There is now evidence for a great variety of junctional mechanisms (30, 68), and the effects that neurotransmitters can have on the postsynaptic membrane (69–72). The most common junctional mechanisms can be detected as an increase in postsynaptic conductance, and have a reversal potential for the charge crossing the synapse. When two regions of membrane, in close electrical proximity to each other, have simultaneous conductance changes, the resulting charge transfer (or postsynaptic potential) is not the linear sum of the charge transfer (or potential) generated by each synapse alone. There are two reasons for this (10, 12):

1. The input impedance detected at the first synaptic location is reduced when the second conductance change is occurring, resulting in a smaller postsynaptic potential for the first synapse (and conversely).
2. When both conductances increase, the reduction in driving potential is greater than when each operates in isolation. As a result, less synaptic current and charge cross the synapse.

The reason for attaching considerable importance to synaptic interactions in the integrative operation of a neuron is that the ability of a synapse to transfer charge to or from the soma not only depends upon the properties of that synapse and its location, but on the presence of other synaptic activity. Conductance changes mediating both excitation and inhibition which are near an excitatory synapse between the synapse and the soma, on the soma, or distal to the synapse on the same dendrite, can reduce the capacity of that synapse to supply charge to the soma, or to depolarize the soma membrane.

Theoretical Studies on Nonlinear Interactions in Dendrites

A theoretical study of nonlinear synaptic effects in dendrites is much more complicated than when an isopotential patch of membrane is considered. The degree of interaction between synapses depends not only upon the magnitude of the conductance increase and the input impedance of the dendritic tree at each synapse, but also upon the spatial separation between the two synapses and the general geometry of the neuron. The mathematical difficulties in obtaining analytical expressions for transients generated by conductance changes in cable structures with finite geometries are considerable. The compartmental model used by Rall (12, 13) is one numerical approach to this problem. This model breaks the dendritic cable into a series connection of isopotential regions, with resistive coupling between each compartment. Conductance changes, each in series with a reversal potential, can be attached to any compartment and the voltage response can be measured in any compartment.

Optimum Location of Inhibitory Synapses Rall (12) has considered the effect of varying the location of an inhibitory conductance on the voltage at the soma compartment produced by an excitatory conductance at the center of the dendritic cable (Ref. 12, Figure 8). The conductance changes are brief (0.25 τ_m) rectangular pulses. Placing the inhibitory conductance at the distal end of the chain of compartments has no effect on the peak amplitude of the depolarization in the soma compartment. However, the decaying phase of this potential is more rapid. The current to the soma from the excitatory synapse has components which arise from reflections at the sealed end of the cable, after first spreading distally from the excitatory conductance (2). These reflection terms are reduced by the low resistance at the distal end. Consequently the decay of potential at the soma is more rapid. When the inhibitory conductance is located at the same point as the excitatory conductance, the soma response is reduced, except for the early rising phase. The greatest reduction occurs when the inhibitory conductance is placed at the soma.

These calculations raise the important question of the appropriate location for an inhibitory conductance to maximally reduce the peak voltage and charge at the soma following an excitatory conductance change. This location will depend upon the geometry of the neuron and the placement of the excitation. A steady state analysis of this problem is given in Jack et al. (2), Chapter 7, and the results are therefore directly relevant to charge interactions. If the postsynaptic inhibitory action is to be specific to a particular excitatory input, then the most effective arrangement is for the inhibitory synapses to be located alongside the excitatory synapse. The effectiveness of this arrangement increases as the synapses are placed on more distal dendritic branches. If the inhibition is to be nonspecific, but is to have maximum effect on excitatory inputs to the proximal regions of all the dendrites, then the optimal location for the inhibitory synapses will be the soma. However, when general excitatory inputs are restricted to distal branches, even more distal inhibitory terminations will be more effective. The distance that the excitatory synapses must be from the soma for the optimum location of inhibitory synapses to switch to the distal end of the dendritic cable depends on the dendritic to soma conductance ratio (ρ) and the electrotonic length of dendritic cable. For motoneurons with values of ρ in excess of 4, inhibition is best located at the soma unless the excitatory synapses are on the distal 0.25 λ of the dendritic cable.

The pattern becomes more complex if the excitatory input is restricted to a particular dendrite, and the inhibitory input has to be optimally placed to diminish the effects of this excitation. Then the optimum position can shift from the soma to the site of the excitatory synapses. The conditions underlying this change in inhibitory location are the relative location of the excitatory synapses, the fraction of dendritic cable involved, and the values of ρ and L (dendritic length). Take as an example a motoneuron for which $\rho \approx 5$, L= 1.5, and the inhibitory conductance is twice the excitatory conductance (Ref. 2, Figure 7.41). If the excitatory conductance is placed on 1/10 of the total dendritic cable and 0.5 λ from the soma, an inhibitory conductance at or near this point on the same fraction of dendrite is much more effective than when applied to the soma. If the fraction of dendritic cable is raised to 0.5, then little difference is observed in the somatic depolarization when the inhibition is located anywhere between the soma and along the receiving dendrite fraction out to the point of excitation.

These calculations apply for steady state, and hence charge transfer, in dendrites. Some other simplifications have been made. The inhibitory action is restricted to a conductance change, and the additional effects of a hyperpolarizing potential have not been included. Furthermore, the excitation is a constant current source, rather than a conductance change. As a result, this will overemphasize the effects of the inhibitory conductance in reducing the input impedance when it is located close to the excitatory current source.

This is the first methodical analysis of the optimum relative locations of excitatory and inhibitory conductance changes for either maximal inhibition of a specific synapse, or maximal inhibition of a general arrangement of synapses. It must be the forerunner to much more complicated analyses, in which transient conductance changes are used, and reversal potentials for both excitatory and inhibitory currents are included.

Distributed Structures Activated by Time Varying Conductances The analytical problems in analyzing the spread of potential in a cable structure when the cable is activated by a time varying conductance, in series with a battery, are more severe than when the cable is activated by a current source. One approach to this problem by Barrett and Crill (33), Rinzel and Rall (20), and Walmsley (61), has been to use numerical methods to solve the integral equation for the voltage and current transients in the cable. Walmsley (61) has developed this analysis to cope with two time varying conductances at variable electrotonic distances from each other, and from the recording point, in an infinite uniform cable. From these calculations, the degree of nonlinear addition in voltage and charge has been determined as a function of depth of conductance modulation, time course of conductance change, electrotonic separation between synapses, and asynchrony in synaptic activation times. This theoretical work has been applied to the study of nonlinear summation of unitary group 1a excitatory potentials in motoneurons (73). The degree of nonlinearity observed was used to attach bounds to the peak synaptic potentials at their point of generation in the dendrites, and the amount of dendrite common to both synaptic inputs (see section "Experimental Observations on Nonlinear Interactions in Dendrites").

The same analytical procedure but with a single conductance change (20, 33, 61) has been used to investigate the efficiency of charge transfer across a conductance change, as a function of input impedance. The nonlinear relationship between peak conductance and peak postsynaptic potential is also a function of input impedance. Solutions to this problem have been obtained analytically (2, 33) and using the compartmental model (13). The solutions to this problem are also directly applicable to nonlinear summation of quantal potentials and quantal charge at dendritic synapses (33, 39).

Effect of Changes in Specific Membrane Resistivity The effect of specific membrane resistivity on the efficiency of charge transfer across dendritic synapses, and on its subsequent redistribution to the soma, can be derived from the voltage transient following a transient conductance change (20, 33; Jack et al. (Ref. 2, Chapter 7)). A widespread synaptic projection to a dendrite, with repetitive activity on all pathways, would have the effect of lowering the average specific membrane resistance in these regions. Barrett and Crill (33) calculate that for a resting membrane resistivity of 2000 ohm cm^2, 15 quantal conductance changes/s at each bouton represents a drop in membrane resistivity to 600 ohm cm^2. The synaptic density is 20 boutons/100 μm^2. The quantal conductance change is 80×10^{-10} mho ms for 0.5 ms. Figure 6 of Barrett and Crill (33) shows an increase in charge attenuation of about 50% for a synapse at 500 μm from the soma, due to this decrease in R_m. Furthermore, if the tonic synaptic activity were excitatory, then the resulting steady state depolarization would considerably reduce the charge transferred at the synapse, and in effect further reduce the synaptic charge transfer efficiency.

Analog Cable Models Because the mathematical problems associated with conductance transients in neuron cable models are formidable, and because a large number of free parameters are often involved, many physiologists may find much more value and intuitive understanding to be gained from studies on analog cable models activated by conductance changes (53, 61). Electronic circuits which generate transient conductance changes proportional to a controlling voltage are not difficult to implement. A dendritic cable is made up of a series of isopotential compartments, each represented by a parallel RC network and connected with series resistance. Dendrites with non-uniform passive membrane properties, or non-uniform geometry, can also be easily constructed (24). Families of voltages and current transients in the cable structure can be conveniently displayed on a storage oscilloscope as various parameters are altered.

These theoretical approaches to dendritic interactions must be regarded as introductory to what will eventually become a large body of theoretical information. At the moment, the main impediment to further development of this theoretical work is the lack of electrophysiological and histological data to which it can be applied.

Experimental Observations of Nonlinear Interactions in Dendrites

Burke (74) reported nonlinear summation of some pairs of composite e.p.s.p.s evoked in motoneurons by multi-fiber stimulation of group 1a afferents. The summed potentials were always less than or equal to the linear sum of the separate e.p.s.p.s. The maximum reduction in peak amplitude over the linear sum was 16%. To account for this reduction, it was suggested that some of the synapses generating each e.p.s.p. must be on the same dendrites, and sufficiently close to each other to allow this nonlinear interaction. This was early electrophysiological evidence for the dendritic location of group 1a synapses as well as for the synaptic mechanism involving a conductance change. It was also strong evidence for passive membrane properties, as there were no examples of summed potentials exceeding the linear sum.

In an attempt to refine this experiment, and to make it more amenable to cable analysis, summation of unitary group 1a e.p.s.p.s was investigated in motoneurons (61, 73). The degree of nonlinearity was measured by the reduction in peak voltage and also in total charge. Immediately the experiment is restricted to two unitary e.p.s.p.s, the chance of finding two which have synaptic terminals sufficiently close together on the dendritic tree to make nonlinear effects possible becomes very small. Of 33 pairs of unitary e.p.s.p.s which were summed (in different motoneurons) only 4 pairs showed a nonlinear addition of charge which exceeded 5%. The maximum nonlinearity was 11.5%. Different latencies were used to allow for different conduction times in the afferent nerves and electrotonic delays in the dendrites, to find the temporal relationship which gave the maximum nonlinearity. A full cable analysis of these motoneurons was carried out (24, 61) to obtain the electrotonic distance to the point of origin of each e.p.s.p. This information, together with the cable parameters of the neuron, and the peak amplitudes (at the soma) of each e.p.s.p, could be used to attach bounds to the fraction of the dendritic cable receiving the synaptic input, and the peak voltage generated at each synapse. There is one major uncertainty; that is, if each afferent fiber has more than one synapse, are all synapses on a common fraction of the equivalent dendritic cable? Or is the nonlinearity due to interaction between some of the synapses arising from each fiber which share dendritic cable? This cannot be answered from the electrophysiological data. Proceeding on the basis that the interaction involves all synapses on the one fraction of cable, this leads to a maximum fraction of cable, and a minimum peak voltage at each synaptic site.

The largest nonlinearity arose when two e.p.s.p.s generated at 0.8 λ and 1.1 λ from the soma were added. These e.p.s.p.s had peak voltages of 293 μV and 92 μV at the soma. The reduction in peak amplitude over the linear sum was 11.4%, and the reduction in charge over the linear sum was 11.5%. Cable analysis of this situation indicated that the major nonlinear effect (10%) was the loss of driving potential at the more distal synapse, due to the large voltage at the proximal synapse, and the small attenuation in these distal regions. The peak potential at the proximal site was calculated as 40 mV, and 10 mV at the distal synapse. These potentials corresponded to all synapses being on 1/40 of the equivalent electrotonic cable for the dendritic tree. This would correspond approximately to a second order branch of one dendrite. If the assumption that each fiber synapses only on the common branch is incorrect, then the synapses which are on a common branch have to be placed on an even smaller dendritic branch, with larger subsynaptic potentials than those indicated above. (Complete details of this calculation, and others involving nonlinear effects, are in preparation.)

Again, the major qualification on this type of calculation is the lack of histological data on the precise neuronal geometry, and the synaptic termination patterns for the single fibers. A definitive experiment requires these data for the same neuron to which the cable calculations are applied.

As well as interactions between two excitatory synapses, significant nonlinearities can occur in the addition of excitatory and inhibitory postsynaptic potentials (65, 67, 75–77). Again, composite potentials have been evoked for each pathway. The variation in the amount of nonlinearity observed can best be explained by variation in relative locations of inhibitory and excitatory synapses. However, because of the large number of synapses involved in generating these potentials, a quantitative analysis of these data is not possible.

These results again emphasize the need for information on synaptic location; not just the electrical distance from soma to synapse, but the number of dendrites which receive synaptic terminals from an afferent fiber, and whether any particular organization is dendrite specific. Electrophysiological data cannot distinguish between potentials which are generated in different dendrites, although interaction experiments allow some bounds to be placed on the amount of common dendrite for two synaptic potentials.

NON-PASSIVE PROPAGATION AND ACTIVE RESPONSES IN NEURON DENDRITES

The emphasis in this paper has been to review the theoretical framework and supporting experimental data relevant to integrative processes in neuron dendrites with passive membrane properties. This emphasis is not to deny the importance of active responses and non-passive propagation in dendrites of many types of neurons. It is simply a reflection of the much more advanced state of theoretical and experimental understanding of passive dendritic properties. The major problem in developing a theoretical basis for active processes in dendrites is the lack of quantitative data on the current-voltage characteristics, and their time dependences, of various regions of neuronal membrane.

One obvious role of action potentials and active responses in dendrites is that they could amplify synaptic potentials, or provide less attenuation of these potentials than occurs with passive propagation. Rall (15) has suggested that once the electrotonic length of a dendrite exceeds 2 or 3 space constants, passive attenuation becomes severe, and there is a need for active membrane responses if remote synapses are to be effective (This teleologism is not to suggest that neurons with dendrites of shorter electrical length always have passive responses.) It is obviously necessary that before any such generalizations are made the electrical length of dendrites be determined in other neurons in which dendritic nonlinear responses can be observed.

Conditions for Generation and Propagation of Active Responses in Distributed Structures

Some discussion on the necessary conditions for action potential generation in a distributed structure is given in Chapters 7, 9, 10, and 12 of Jack, Noble, and Tsien (2). These conditions are quite different, and more complicated than for an isopotential patch of membrane. Threshold potential for isopotential membrane is that potential at which the net ionic current becomes inward. For a distributed structure, it is necessary that the net ionic current, integrated over the total surface of the structure, becomes inward. To meet this condition, it is necessary that the inward current at the point or region where the impulse is to be initiated be much greater than the current threshold for that particular patch of membrane. This additional current depends upon the electrical load that other regions of the membrane carrying outward current present to this region, and the current-voltage relationships for the membrane. As little is known about these I-V relationships and how they may vary over the neuron surface, it is necessary to restrict the discussion to geometric factors.

The input impedance of a dendrite will increase as the site of synaptic action moves from proximal dendrites to the fine branches. The peak postsynaptic potential generated by a conductance change depends upon the input impedance, and the time course of the conductance change (20, 61). The greater synaptic potentials generated in distal dendrites may be offset by higher voltage thresholds. The passive load that the branching structure presents to the possible impulse generation site depends on the detailed branching arrangements. If the membrane area per unit of electrical length remains the same, or reduces, distal to a branch point, then the further away the site is from a parent branch, the lower will be the electrical load presented by this parent branch and other sibling branches. On the other hand, in dendrites where the effective surface area per unit of electrical length increases towards the periphery, such as in Purkinje cell dendrites, then the loading increases in moving distally, until final terminating branches are reached. These arguments assume uniform membrane (passive) resistivity, and could be drastically altered if this were not the case.

Although a full action potential may not be generated because of the excessive electrical load, the total net inward current in the proximity of the synapse will be greater, and more prolonged, than the net inward current crossing the synaptic conductance change if the local voltage threshold is exceeded. Passive attenuation of potentials in dendrites is greatly reduced as the duration of injected current is prolonged (18). Such localized active responses, supplying more net inward charge than the synapse alone, and prolonging the inward current time course, could partly overcome severe attenuation in dendrites with excessive electrical length.

Another form of active response which does not require the generation of an action potential, but which causes less attenuation than passive propagation, was suggested by Lorente de Nó and Condouris (32). This form of propagation requires an I-V characteristic with quasi-instantaneous kinetics which show a negative slope resistance, but do not have a threshold potential. That is, over some range of depolarization, increase in depolarization leads to less outward current flow. Such an I-V characteristic (for an isopotential membrane) will occur if the inward current ionic channels are similar to those for non-myelinated nerve, but are so sparse that the instantaneous linear I-V characteristic for potassium dominates the instantaneous response. Depending upon the density of the inward channels, the current will always be outwards for depolarization beyond the resting potential, but less than that for a linear membrane resistivity. Again, such membrane characteristics, when applied to a distributed structure, cannot be considered in isolation of electrical loading.

Propagation of Action Potentials through Branch Points

The theoretical problem of propagation of action potentials through branch points, and along cylinders with non-uniform cross section, has recently been studied (78). When the 3/2 power law for branching (8) holds, an impulse will propagate centrifugally through the branch point without change in time course or conduction velocity. This is not the case if there is an effective diameter change in the direction of impulse propagation. Impulses originating in a single dendritic branch, and propagating centripetally, will normally strike a discontinuity in equivalent dendritic cable diameter at the first proximal branch point. The other sibling branches and the parent branch will combine in parallel to form a cable with a larger equivalent diameter than that of the sibling branch which is activated. Goldstein and Rall (78) calculate that when the ratio of the diameter of the equivalent cable representing the proximal electrical load, and the diameter of the active branch, exceeds about 2.5, forward conduction fails. This would correspond to an impulse originating in a dendrite of 2 μm diameter and propagating to a branch point where it and a second 2 μm diameter branch arise from a 4.5 μm (or larger) diameter branch. These calculations of propagation past branch points assume uniform membrane properties. If the voltage dependent conductances become more sensitive to membrane potential in the centripetal direction, or around the branch points, then this branching diameter condition could be further relaxed.

Non-Uniform Excitability of Dendritic Membrane

The question of the relative excitability of dendritic membrane in motoneurons has not been resolved. It is fairly clear that differences in excitability exist between the initial segment and the soma of the neuron (79–81). No direct experimental evidence is available concerning the excitability of dendritic membrane. However, calculations by Dodge and Cooley (82) have considered the relative density of sodium channels in the membrane of the initial segment, soma, and dendrites, to determine if such variations, together with the geometry of the neuron, can account for the observed time course of the antidromic action potential. The results of these calculations suggest that the density of sodium channels in dendritic membrane is negligible relative to the density in the soma membrane, which in turn is about 10 times less than the corresponding density in the initial segment.

The above theoretical considerations are largely tentative, because of our ignorance of the voltage dependent conductances of neuronal membrane, and their time dependences. It is to be hoped that eventually a reliable histological marker for sodium and potassium channels can be developed so that the relative densities of these channels at different points on dendritic membrane can be established. With this information the interpretation of I-V curves, measured at the soma, would become possible. For distributed systems in which a point voltage or current clamp is applied, the actual I-V curves for an isopotential patch of membrane can be calculated if the geometry is known, and the membrane has uniform electrical properties (2, 83). These techniques could be extended if the electrical properties were non-uniform, assuming that the distribution of these properties was established by histological methods.

Experimental Evidence for Nonlinear Processes in Neuron Dendrites

There exists convincing evidence that inward current can be activated in nerve cell dendrites. This current may be associated with a propagated action potential. Purpura (84) has reviewed the evidence for this, and subsequently Llinás (85) has discussed the recent evidence accumulated from intracellular recordings in Purkinje cell dendrites of alligator (86) and from chromatolysed motoneurons (87, 88). Evidence for active processes in dendrites from extracellular recording is much more difficult to interpret. An example of these difficulties is the recent controversy on the use of field potentials and current density analysis to show that action potentials were generated in reptilian Purkinje cell dendrites (62, 89–92). Important evidence for active dendritic responses was subsequently obtained by intradendritic recording (86).

Another subtle problem is to make use of the time course of a postsynaptic potential to deduce whether or not any active processes were involved in its generation and propagation. If the cable properties of the neuron have been determined, it is theoretically possible to deduce whether a recorded postsynaptic potential has a time course which is compatible with a brief synaptic current followed by purely passive propagation. In practice, this can be difficult, as the more remote the synaptic region in question becomes, the less sensitivity there is in the time course to the duration of current injection (60). In spinal motoneurons, most unitary group 1a e.p.s.p.s have time courses which are compatible with the passive membrane hypothesis (26, 36). A small number have time courses which cannot be explained on this basis. To account for these potentials, it is necessary to allow for a synaptic current two to three times the duration of the longest duration currents generating e.p.s.p.s which fit the passive model. Alternatively, this extra current could be generated by a voltage-dependent membrane response. That the membrane of dendritic branches of normal motoneurons is almost always passive, at least when tested with evoked potentials, is reinforced by the results of summation of group 1a e.p.s.p.s (39, 73) in which the summed response did not increase over the linear sum. These results are in sharp contrast with the change that takes place in the responsiveness of the dendritic membrane following chromatolysis (87). In chromatolysed motoneurons, unitary e.p.s.p.s of 0.5 mV peak amplitude at the soma can generate an orthodromic action potential in separate dendrites (85). The passive cable properties of chromatolysed motoneurons have recently been investigated (Llinás, personal communication) together with much supporting histological and electrophysiological data. The chromatolysed motoneuron may well become the working model for understanding the generation and propagation of active responses in other neurons of the central nervous system.

Role of Active Responses in Dendrites

The existence of dendritic impulses and active responses seems contrary to Sherrington's original concept of integration in neurons. If a particular region of a dendrite can generate an action potential as a result of conventional integrative activity in that region, and this impulse can then fully propagate to the soma and discharge the cell, why not build that integrative process into a separate neuron whose axon has the same projection? This would avoid interference with the other, more conventional, integrative processes in the same neuron, but it increases the number of neurons. Such an arrangement begins to conform to the classic picture of neuronal integration if it is possible to prevent the impulses propagating to the soma, and to other dendrites, by inhibitory synapses located between the region of impulse generation and the soma. This arrangement has been referred to as "functional amputation" of the dendritic segment by Llinás and Nicholson (90). There is support for this concept in reptilian cerebellar Purkinje cells.

Other authors (93) have commented on the logical possibilities that arise when an active response or impulse in a single dendrite, or in a branch of a dendrite, is reinforced with almost coincident timing by a similar active process in another dendrite, or in another branch of the same dendrite. By such reinforcement the cell may be discharged, while a single dendritic spike would be unsuccessful. This argument can be extended to introduce inhibitory synapses at points of mutual reinforcement of active responses, so that a further timing element is introduced. Such arrangements could make the cell a detector of coincidences or near coincidences in impulse timing on several afferent fibers, rather than sensing the general level of impulse activity on many afferent fibers.

Dendritic action potentials, either locally generated, or spreading from other parts of the dendritic tree, may play a role in the operation of dendro-dendritic synapses (94–96). This suggestion is entirely speculative, as the relationship between presynaptic potential and transmitter release is not known for these synapses. Ralston (96) has suggested that it is possible that presynaptic potentials generated by passive membrane responses could be sufficiently large to cause some transmitter release, but the effectiveness of the synapse could be increased by an active response.

CONTROL OF NEURONAL DISCHARGE BY PRESYNAPTIC IMPULSE ACTIVITY

The final objective in this approach to neuronal integration is to relate the parameters of the neuronal discharge to the impulse activity in converging presynaptic fibers. Considerable emphasis has been given to the transmission of charge from the synaptic sites in dendrites to the soma. The ways in which charge can be lost during passive spread and through synaptic interactions have been explored. The remaining link is to relate this transferred charge to the firing pattern of the neuron. Experimental results relating intracellularly applied current and frequency of impulse generation are extremely helpful in establishing this link between charge and frequency.

Parameters of Neuronal Discharge

The train of impulses generated by a neuron may be described by various statistical measures. It is a difficult problem to determine which of these measures are relevant (or decoded) in the next transduction process. This transduction process most commonly involves a summation of muscle responses, or a summation of postsynaptic potentials. The relevant parameters of the impulse train description not only depend upon the nature of the transduction process, but also upon which response variables are considered important. These problems are carefully reviewed and discussed by Perkel and Bullock (97). In this paper we restrict attention to the average discharge frequency as an obviously important parameter. However, because of the general nonlinear nature of the summation of muscle tension, or of postsynaptic potentials, other statistical parameters (such as variability) should ultimately be considered.

Relationship between Average Discharge Frequency and Applied Current

An important experimental approach to relationships between synaptic activity and discharge of the neuron is to study the relationship between intracellularly applied current and discharge frequency (98–106). Such experiments will be most meaningful in those neurons in which the lowest threshold for nerve impulse generation is in the soma-initial segment region, and the cell is not normally discharged by action potentials propagation from dendrites.

Kernell's experiments (100–102) have made an important contribution to the measurement and understanding of the relationship between somatic current and firing frequency of the motoneuron. At low discharge rates the motoneuron is less sensitive to somatic current than at high discharge rates. When synaptic excitation is superimposed upon the applied current, there is often a change in discharge frequency (107–109). The frequency may increase or decrease, but the slope of the frequency-current (f-I) relationship remains constant. Inhibitory synaptic activity reduces the discharge frequency, again without affecting the slope of the f-I curve.

An important application of this type of experiment is that it provides a means of determining the average somatic current supplied by stimulation of a particular pathway to the neuron. The shift in frequency corresponds to a shift in current on the f-I curve, and this current shift will be the "effective somatic current" generated by synaptic activity. The real current reaching the soma as a result of synaptic activity may be larger than this effective current, as any lowering of membrane resistance in the soma-proximal dendrite region will cause less depolarization due to the electrode current. This decrease will be offset by some or all of the synaptically generated current.

Another method of measuring somatic current supplied by impulse activity on an afferent pathway would be to voltage clamp the soma membrane at a potential below threshold. The clamp current would then be the maximum current that the synaptic pathway could supply to the soma. No current loss would occur in proximal dendrites due to a spread of high conductance into these regions following the generation of each impulse. Dendritic synapses would operate with maximum charge transfer efficiency (see section "Passive Propagation of Synaptic Potentials in Dendrites"), as a voltage clamp close to the resting membrane potential would partly restrain large membrane potential changes in proximal dendrites.

Relationship between Somatic Current and Synaptic Charge

An important and practical aspect of these measurements of synaptically evoked somatic current is that they provide a fairly direct connection with charge transfer at the active synapses. Current is simply total charge per unit time. Charge transfer to the soma from different types of synapses can be calculated from the time course of unitary postsynaptic potentials recorded at the soma. Initially, linear summation of charge can be assumed. If the number of afferent fibers in a pathway is known and the frequency of impulses can be estimated, then this allows a calculation of the maximum current which would spread to the soma from this input alone. This could then be compared against results obtained from a frequency shift on an f-I curve, or from a voltage clamp, when the afferent pathway was activated in identical fashion. Differences between the clamp current and the calculated current would reflect nonlinear interactions in dendrites between conductance changes. Larger differences would be expected between calculated current and that measured from the shift on the f-I curve, due to the shorting effect of membrane conductance in proximal dendritic regions during each action potential, as well as the effect on the applied current of a lowered input resistance due to soma and proximal dendritic conductance increases.

Illustrative Calculation

Consider the activity on the group 1a projection to a soleus motoneuron during a large maintained stretch of the triceps surae. The number of monosynaptic group 1a afferents projecting to this motoneuron will be approximately 110, made up as follows (44, 45): 1) from a total of 56 soleus group 1a fibers, 56×0.9 or 50 fibers synapse on a soleus motoneuron; and 2) from a total of 100 lateral and medial gastrocnemius group 1a fibers, 100×0.6 or 60 fibers synapse on a soleus motoneuron.

The charge transferred to the soma from group 1a synapses (without consideration of muscle of origin) ranges from 0.25 to 2.5 pC, with an average of approximately 1.1 pC (36). In deriving these figures, allowance has been made for the hyperpolarizing potential which follows the depolarization (the "undershoot"). Charge transferred to the soma is approximately independent of synaptic location so that each of the 110 fibers is considered equipotent in supplying charge. Thus, if the impulse frequency on each of these 110 fibers averages 50 impulses/s, the charge propagating to the soma is (assuming linear summation) $110 \times 50 \times 1.1 \times 10^{-12}$ coulombs/s, i.e., average somatic current = 6.1 nA.

Assuming the neuron is in the primary range of firing, the effects of this current may be obtained from the f-I curve in this range. Different neurons have different f-I curves, and clearly measurements of charge, and f-I characteristics, should be obtained for the same neuron. However, for the purposes of illustration, consider the f-I curves for ankle extensors in Figure 1 and 3 of Granit et al. (108). A 10 mm maintained stretch of gastrocnemius-soleus caused an increase in frequency equivalent to a current increase of between 5 and 10 nA. These results are in basic agreement with the calculation assuming that the average of 50 impulses/s for each fiber is realistic.

Of course there are many qualifications which can be directed at a calculation such as this. For reasons mentioned earlier, the current calculated is the maximum which could reach the soma. The actual current will have a noisy time course, due to asynchronous synaptic activity.

Harvey (110) has calculated the discharge rate of a model neuron when excited either by direct current, or by 100 asynchronous synaptic inputs which generate the same mean membrane current as the direct current. The model has a threshold equal to the sum of 100 unit responses. When the mean of the 100 asynchronous postsynaptic responses equals the rheobasic current, the model discharges at 10/s. This difference in sensitivity to a noisy current and a direct current becomes less marked as the intensity of excitation increases. Stretching the muscle by 10 mm causes other muscle afferents (groups 11 and 1b) to be active, and these will have reflex effects upon the extensor neuron. For this reason the comparison with experimental data on the basis of purely group 1a activation is questionable.

The main reason for including this calculation is that it illustrates the central objective of this review. Integrative neurons are caused to discharge by maintaining a current to the soma via a number of converging dendrites. The relationship between this current and the discharge frequency of the neuron can be measured. This relationship may then be used to predict the variations in neuron discharge rate which could be effected by modulation of impulse activity on one or several projections to the neuron. The basis of this prediction is an understanding of the transferral of charge from synapse to soma for each synaptic projection. This information can be obtained from measurement of postsynaptic potentials and the cable properties of the neuron. Departures from algebraic summation of charge can be explained on the basis of nonlinear interactions between conductance changes in dendrites. Finally, all synapses on the neuron surface should be considered in this calculation. The traditional belief that distal synapses provide only background modulation to somatic membrane potential has been disproved for the spinal motoneuron, and may not be true in general.

ACKNOWLEDGMENT

I am grateful to Dr. R. Harvey for his comments on this article.

REFERENCES

1. Sherrington, C. (1947). The Integrative Action of the Nervous System, 2nd Ed., Yale University Press, New Haven.
2. Jack, J.J.B., Noble, D., and Tsien, R.W. (1975). Electric Current Flow in Excitable Cells. Clarendon Press, Oxford.
3. Rall, W. (1976). Core conductor theory and cable properties of neurons. *In* E. Kandel (ed.), The Nervous System, Vol. 1, Cellular Biology of Neurons. American Physiological Society, Washington, D.C. In press.
4. Wyckoff, R.W.G., and Young, J.Z. (1956). The motoneuron surface. Proc. Roy. Soc. (Lond.) **B144**:440.
5. Ramon-Moliner, E. (1962). An attempt at classifying nerve cells on the basis of their dendritic patterns. J. Comp. Neurol. **119**:211.
6. Scheibel, M.E., and Scheibel, A.B. (1970). Of pattern and place in dendrites. Int. Rev. Neurobiol. **13**:1.
7. Gelfan, S., Kao, G., and Ruchkin, D.S. (1970). The dendritic tree of spinal neurons. J. Comp. Neurol. **139**:385.
8. Rall, W. (1959). Branching dendritic trees and motoneuron membrane resistivity. Exp. Neurol. **1**:491.
9. Rall, W. (1960). Membrane potential transients and membrane time constant of motoneurons. Exp. Neurol. **2**:503.
10. Rall, W. (1962). Theory of physiological properties of dendrites. Ann. N.Y. Acad. Sci. **96**:1071.
11. Rall, W. (1962). Electrophysiology of a dendritic neuron model. Biophys. J. **2**:145.
12. Rall, W. (1964). Theoretical significance of dendrites for neuronal input-output relations. *In* R. Reiss (ed.), Neuronal Theory and Modelling, pp. 73–97, Stanford University Press, Stanford.
13. Rall, W. (1967). Distinguishing theoretical synaptic potentials computed for different soma-dendritic distributions of synaptic input. J. Neurophysiol. *30:*1138.
14. Rall, W. (1969). Time constants and electrotonic length of membrane cylinders and neurons. Biophys. J. **9**:1483.
15. Rall, W. (1970). Cable properties of dendrites and effects of synaptic location. *In* P. Andersen and J.K.S. Jansen (eds.), Excitatory Synaptic Mechanisms, pp. 175–187, Universitetsforlaget, Oslo.
16. Lux, H.D. (1967). Eigenschaften eines Neuron modells mit Dendriten begrenzter lãnge. Pflügers Arch. **297**:238.
17. Jack, J.J.B., and Redman, S.J. (1971). An electrical description of the motoneuron, and its application to the analysis of synaptic potentials. J. Physiol. (Lond.) **215**:321.
18. Redman, S.J. (1973). The attenuation of passively propagating dendritic potentials in a motoneuron cable model. J. Physiol. (Lond.) **234**:637.
19. Rall, W., and Rinzel, J. (1973). Branch input resistance and steady attenuation for input to one branch of a dendritic neuron model. Biophys. J. **13**:648.
20. Rinzel, J., and Rall, W. (1974). Transient response in a dendritic neuron model for current injected at one branch. Biophys. J. **14**:759.
21. Aitken, J.T., and Bridger, J.E. (1961). Neuron size and neuron population density in the lumbosacral region of the cat's spinal cord. J. Anat. (Lond.) **95**:38.
22. Lux, H.D., Schubert, P., and Kreutzberg, G.W. (1970). Direct matching of morphological and electrophysiological data in cat spinal motoneurons. *In* P. Andersen and J.K.S. Jansen (eds.), Excitatory Synaptic Mechanisms, pp. 189–198. Universitetsforlaget, Oslo.
23. Barrett, J.N., and Crill, W.E. (1974). Specific membrane properties of motoneurons. J. Physiol. (Lond.) **239**:301.
24. Iansek, R., and Redman, S.J. (1973). An analysis of the cable properties of spinal motoneurons using a brief intracellular current pulse. J. Physiol. (Lond.) **234**:613.
25. Burke, R.E. and ten Bruggencate, G. (1971). Electrotonic characteristics of alpha motoneurons of varying size. J. Physiol. (Lond.) **212**:1.
26. Jack, J.J.B., Miller, S., Porter, R., and Redman, S.J. (1971). The time course of minimal excitatory postsynaptic potentials evoked in spinal motoneurons by group 1a afferent fibers. J. Physiol. (Lond.) **215**:353.
27. Fatt, P. (1957). Sequence of events in synaptic activation of a motoneuron. J. Neurophysiol. **20**:61.

28. Katz, B., and Miledi, R. (1963). A study of spontaneous miniature potentials in spinal motoneurons. J. Physiol. (Lond.) **168**:389.
29. Nelson, P.G., and Lux, H.D. (1970). Some electrical measurements of motoneuron parameters. Biophys. J. **10**:55.
30. Eccles, J.C. (1964). The Physiology of Synapses. Springer, Berlin.
31. Lorente de Nó, R. (1938). Synaptic stimulation as a local process. J. Neurophysiol. **1**:194.
32. Lorente de Nó, R., and Condouris, G.A. (1959). Decremental conduction in peripheral nerve; integration of stimuli in the neuron. Proc. Natl. Acad. Sci. USA **45**:592.
33. Barrett, J.N., and Crill, W.E. (1974). Influence of dendritic location and membrane properties on the effectiveness of synapses on cat motoneurons. J. Physiol. (Lond.) **239**:325.
34. Lux, H.D., and Pollen, D.A. (1966). Electrical constants of neurons in the motor cortex of the cat. J. Neurophysiol. **29**:207.
35. Jacobsen, S., and Pollen, D.A. (1968). Electrotonic spread of dendritic potentials in feline pyramidal cells. Science **161**:1351.
36. Iansek, R., and Redman, S.J. (1973). The amplitude, time course and charge of unitary excitatory postsynaptic potentials evoked in spinal motoneuron dendrites. J. Physiol. (Lond.) **234**:665.
37. MacGregor, R.J. (1968). A model for responses to activation by axodendritic synapses. Biophys. J. **8**:305.
38. Mendell, L.M., and Henneman, E. (1968). Terminals of single 1a fibers: distribution within a pool of 300 homonymous motor neurons. Science **160**:96.
39. Kuno, M., and Miyahara, J.T. (1969). Non-linear summation of unit synaptic potentials in spinal motoneurons of the cat. J. Physiol. (Lond.) **201**:465.
40. Jack, J.J.B., Miller, S., Porter, R., and Redman, S.J. (1970). The distribution of group 1a synapses on the lumbrosacral spinal motoneurons in the cat. *In* P. Andersen and J.K.S. Jansen (eds.), Excitatory Synaptic Mechanisms, pp. 199–205. Universitetsforlaget, Oslo.
41. Illis, L. (1967). The relative densities of the monosynaptic pathways to the cell bodies and dendrites of the cat ventral horn. J. Neurol. Sci. **4**:259.
42. Conradi, S. (1969). On motoneuron synaptology in adult cats. Acta Physiol. Scand. **Suppl. 332.**
43. Iles, J.F. (1973). Demonstration of afferent termination in the cat spinal cord. J. Physiol. (Lond.) **234**:22P.
44. Mendell, L.M., and Henneman, E. (1971). Terminals of single 1a fibers: location, density and distribution within a pool of 300 homonymous motoneurons. J. Neurophysiol. **34**:171.
45. Kuno, M., and Miyahara, J.T. (1969). Analysis of synaptic efficacy in spinal motoneurons from "quantum" aspects. J. Physiol. (Lond.) **201**:479.
46. Réthelyi, M. (1970). Ultrastructural study of Clarke's column. Exp. Brain Res. **11**:159.
47. Kuno, M., Muñoz-Martinez, E.J., and Randic, M. (1973). Synaptic action on Clarke's column neurons in relation to afferent terminal size. J. Physiol. (Lond.) **228**:343.
48. Martin, A.R. (1955). A further study of the statistical composition of the end-plate potential. J. Physiol. (Lond.) **130**:114.
49. Edwards, F.R., Redman, S.J., and Walmsley, B. (1974). Statistical fluctuations in unitary 1a e.p.s.p.s recorded in cat spinal motoneurons. Proc. Aust. Physiol. Pharmacol. Soc. **5 (1)**:66.
50. Fernald, R.D. (1971). A neuron model with spatially distributed synaptic input. Biophys. J. **11**:323.
51. Kernell, D. (1971). Effects of synapses of dendrites and soma on the repetitive impulse firing of a compartmental neuron model. Brain Res. **35**:551.
52. Barnwell, G.M., and Cerimele, B.J. (1972). A mathematical model of the effects of spatio-temporal patterns of dendritic input potentials on neuronal somatic potentials. Kybernetik **10**:144.
53. Pottala, E.W., Colburn, T.R., and Humphrey, D.R. (1973). A dendritic compartment model neuron. I.E.E.E. Trans. Biomed. Eng. **20**:132.
54. Fadiga, E., and Brookhart, J.M. (1960). Monosynaptic activation of different portions of the motor neuron membrane. Amer. J. Physiol. **198**:693.

55. Porter, R., and Hore, J. (1969). Time course of minimal corticomotoneuronal excitatory postsynaptic potentials in lumbar motoneurons of the monkey. J. Neurophysiol. **32**:443.
56. Mendell, L., and Weiner, R. (1975). Analysis of pairs of individual 1a e.p.s.p.s in single motoneurons. J. Physiol. (Lond.) In press.
57. Shapovalov, I.A., and Kurchavyi, G.C. (1974). Effects of transmembrane polarization and TEA injection on monosynaptic actions from motor cortex, red nucleus and group 1a afferents on lumbar motoneurons in the monkey. Brain Res. **82**:49.
58. Tsukahara, N., Toyama, K., and Kosaka, K. (1967). Electrical activity of red nucleus neurons investigated with intracellular microelectrodes. Exp. Brain Res. **4**:18.
59. Tsukahara, N., and Kosaka, K. (1968). The mode of cerebral excitation of red nucleus neurons. Exp. Brain Res. **5**:102.
60. Jack, J.J.B., and Redman, S.J. (1971). The propagation of transient potentials in some linear cable structures. J. Physiol. (Lond.) **215**:283.
61. Walmsley, B. (1975). An analysis of unitary 1a e.p.s.p.s evoked in cat spinal motoneurons. Ph.D. thesis, Monash University.
62. Calvin, W.H. (1969). Dendritic spikes revisited. Science **166**:637.
63. Edwards, F.R., Redman, S.J., and Walmsley, B. (1974). The effect of membrane potential on the amplitude of unitary 1a e.p.s.p.s evoked in cat spinal motor neurons. Proc. Aust. Physiol. Pharmacol. Soc. **5 (2)**:197.
64. Coombs, J.S., Eccles, J.C., and Fatt, P. (1955). Excitatory synaptic action in motoneurons. J. Physiol. (Lond.) **130**:374.
65. Burke, R.E., Fedina, L., and Lundberg, A. (1971). Spatial synaptic distribution of recurrent and group 1a inhibitory systems in cat spinal motoneurons. J. Physiol. (Lond.) **214**:305.
66. Smith, T.G., Wuerker, R.B., and Frank, K. (1967). Membrane impedance changes during synaptic transmission in cat spinal motoneurons. J. Neurophysiol. **30**:1072.
67. Terzuolo, C.A., and Llinás, R. (1966). Distribution of synaptic inputs in the spinal motoneuron and its functional significance. *In* R. Granit (ed.), Nobel Symposium 1: Muscular Afferents and Motor Control, pp. 373–384. Wiley, New York.
68. Bennett, M.V.L. (1972). A comparison of electrically and chemically mediated transmission. *In* G.D. Pappas and D.P. Purpura (eds.), Structure and Function of Synapses, pp. 221–256. Raven Press, New York.
69. Weight, F.F., and Votava, J. (1970). Slow synaptic excitation in sympathetic ganglion cells: evidence for synaptic inactivation of potassium conductance. Science **170**:755.
70. Siggins, G.R., Oliver, A.P., Hoffer, B.J., and Bloom, F.E. (1971). Cyclic adenosine monophosphate and norepinephrine: effects of transmembrane properties of cerebellar Purkinje cells. Science **171**:192.
71. Engberg, I., and Marshall, K.C. (1971). Mechanism of noradrenaline hyperpolarization in spinal cord interneurons of the cat. Acta Physiol. Scand. **83**:142.
72. Weight, F.F., and Padjen, A. (1973). Slow synaptic inhibition: evidence for synaptic inactivation of sodium conductance in sympathetic ganglion cells. Brain Res. **55**:219.
73. Edwards, F.R., Redman, S.J., and Walmsley, B. (1974). Nonlinear summation of post-synaptic potentials due to interacting conductance changes. Proc. Aust. Physiol. Pharmacol. Soc. **5 (1)**:65.
74. Burke, R.E. (1967). Composite nature of the monosynaptic excitatory postsynaptic potential. J. Neurophysiol. **30**:1114.
75. Curtis, D.R., and Eccles, J.C. (1959). The time course of excitatory and inhibitory synaptic actions. J. Physiol. (Lond.) **145**:529.
76. Rall, W., Burke, R.E., Smith, T.G., Nelson, P.G., and Frank, K. (1967). Dendritic location of synapses, and possible mechanisms for the monosynaptic e.p.s.p. in motoneurons. J. Neurophysiol. **30**:1169.
77. Tsukahara, N., and Fuller, D.R.G. (1969). Conductance changes during pyramidally induced postsynaptic potentials in red nucleus neurons. J. Neurophysiol. **32**:35.
78. Goldstein, S.S., and Rall, W. (1974). Changes of action potential shape and velocity for changing core conductor geometry. Biophys. J. **14**:731.
79. Fuortes, M.G.F., Frank, K., and Becker, M.C. (1957). Steps in the production of motoneuron spikes. J. Gen. Physiol. **40**:735.
80. Coombs, J.S., Curtis, D.R., and Eccles, J.C. (1957). The interpretation of spike potentials of motoneurons. J. Physiol. (Lond.) **139**:198.

81. Araki, T., and Terzuolo, C.A. (1962). Membrane currents in spinal motoneurons associated with the action potential and synaptic activity. J. Neurophysiol. **25**:772.
82. Dodge, F.A., and Cooley, J.W. (1973). Action potential of the motoneuron. IBM J. Res. Dev. **17**:219.
83. Kootsey, J.M., and Johnson, E.A. (1972). Voltage clamp of cardiac muscle. A theoretical analysis of early currents in the single sucrose gap. Biophys. J. **12**:1496.
84. Purpura, D.P. (1967). Comparative physiology of dendrites. *In* G.C. Quarton, T. Melnechuk, and F.O. Schmitt (eds.), The Neurosciences. A Study Program, pp. 372–393. Rockefeller University Press, New York.
85. Llinás, R. (1975). Electroresponsive properties of dendrites in central neurons. *In* G.W. Kreutzberg (ed.), Advances in Neurology, Vol. 12, pp. 1–13. Raven Press, New York.
86. Llinás, R., and Nicholson, C. (1971). Electrophysiological properties of dendrites and somata in alligator Purkinje cells. J. Neurophysiol. **34**:532.
87. Kuno, M., and Llinás, R. (1970). Enhancement of synaptic transmission by dendritic potentials in chromatolysed motoneurons of the cat. J. Physiol. (Lond.) **210**:807.
88. Kuno, M., and Llinás, R. (1970). Alterations of synaptic action in chromatolysed motoneurons of the cat. J. Physiol. (Lond.) **210**:823.
89. Llinás, R., Nicholson, C., Freeman, J., and Hillman, D.E. (1968). Dendritic spikes and their inhibition in alligator Purkinje cells. Science **160**:1132.
90. Llinás, R., and Nicholson, C. (1969). Electrophysiological analysis of alligator cerebellum: a study on dendritic spikes. *In* R. Llinás (ed.), Neurobiology of Cerebellar Evolution and Development, pp. 431–465. American Medical Association, Chicago.
91. Calvin, W.H., and Hellerstein, D. (1969). Dendritic spikes vs cable properties. Science **163**:96.
92. Zucker, R.S. (1969). Field potentials generated by dendritic spikes and synaptic potentials. Science **165**:409.
93. Arshavskii, Y.I., Berkinblit, M.B., Kovalev, S.A., Smolyaninov, V.V., and Chailakhyan, L.M. (1965). The role of dendrites in the functioning of nerve cells. Dokl. Akademii Nauk SSSR **163**:994. Translation in Doklady Biophysics, Consultants Bureau, New York.
94. Rall, W., Shepherd, G.M., Reese, T.S., and Brightman, M.W. (1966). Dendro-dendritic synaptic pathway for inhibition in the olfactory bulb. Exp. Neurol. **14**:44.
95. Ralston, H.J. (1968). The fine structure of neurons in the dorsal horn spinal cord. J. Comp. Physiol. **132**:275.
96. Ralston, H.J. (1971). Evidence for presynaptic dendrites and a proposal for their mechanism of action. Nature **230**:585.
97. Perkel, D.H., and Bullock, T.H. (1968). Neural coding. Neurosci. Res. Program Bull. **6**:3.
98. Fuortes, M.G.F., and Mantegazzini, F. (1962). Interpretation of the repetitive firing of nerve cells. J. Gen. Physiol. **45**:735.
99. Granit, R., Kernell, D., and Shortess, G.K. (1963). Quantitative aspects of repetitive firing of mammalian motoneurons caused by injected currents. J. Physiol. (Lond.) **168**:911.
100. Kernell, D. (1965). Adaptation and relation between discharge frequency and current strength of cat lumbrosacral motoneurons stimulated by long lasting injected currents. Acta Physiol. Scand. **65**:65.
101. Kernell, D. (1965). High frequency repetitive firing of cat lumbosacral motoneurons stimulated by long lasting injected currents. Acta Physiol. Scand. **65**:74.
102. Kernell, D. (1965). The limits of firing frequency in cat lumbosacral motoneurons possessing different time courses of afterhyperpolarization. Acta Physiol. Scand. **65**:87.
103. Koike, H., Mano, N., Okado, Y., and Oshima, T. (1970). Repetitive impulses generated in fast and slow pyramidal tract cells by intracellularly applied current steps. Exp. Brain Res. **11**:263.
104. Eide, E., Fedina, L., Jansen, J., Lundberg, A., and Vyklický, L. (1969). Properties of Clarke's column neurons. Acta Physiol. Scand. **77**:125.
105. Kuno, M., and Miyahara, J.T. (1968). Factors responsible for multiple discharge of neurons in Clarke's column. J. Neurophysiol. **31**:624.

106. Schwindt, P.C., and Calvin, W.H. (1972). Steps in the production of motoneuron spikes during rhythmic firing. J. Neurophysiol. **35**:311.
107. Granit, R., Kernell, D., and Lamarre, Y. (1966). Algebraic summation in synaptic activation of motoneurons firing with the "primary range" to injected currents. J. Physiol. (Lond.) **187**:379.
108. Granit, R., Kernell, D., and Lamarre, Y. (1966). Synaptic stimulation superimposed on motoneurons firing in the "secondary range" to injected currents. J. Physiol. (Lond.) **187**:401.
109. Kernell, D. (1969). Synaptic conductance changes and the repetitive impulse discharge of spinal motoneurons. Brain Res. *15*:291.
110. Harvey, R. (1976). Patterns of output firing generated by a many-input neuronal model for different model parameters and patterns of synaptic drive. Manuscript in preparation.

International Review of Physiology
Neurophysiology II, Volume 10
Edited by Robert Porter
 University Park Press Baltimore

2 Axonal Transport and Neuronal Dynamics: Contributions to the Study of Neuronal Connectivity

B. G. LIVETT
Monash University, Clayton, Victoria, Australia

INTRODUCTION

Correct functioning of the nervous system depends on the formation, during development, of correct connections between neurons, many of which may differ in their chemical and physiological characteristics. Ultimately, a knowledge of brain function demands a knowledge of the detailed neuronal circuitry, and physiologists and biochemists have traditionally looked to the neuroanatomists to provide this information from classic neurohistological staining techniques. Over the last 10 years, however, the disciplines of neurophysiology and neurochemistry have advanced rapidly to a stage where developments in these disciplines are now making their own contributions to neuronal mapping. By contrast with classic neurohistological methods, which rely on the selective staining of dead (i.e., fixed) neurons, much of the new information about neuronal circuitry and function has come through the study of neuronal dynamics and axonal flow in living neurons within intact and functioning nervous systems.

One has only to look at the process of neurite elongation during development and regeneration to recognize that neurons are very dynamic cells. It is now known that the developing neurite grows by addition of membrane and other components which are synthesized in the perikaryon and transported to the growing tip by a system of axonal transport (1). A supply of materials to the neurite from the cell body of the neuron is essential for neurite growth and, hence, for the final specification and selection of correct connections.

In this review, I present a view of the neuron as a dynamic cell involved in the continual synthesis and axonal transport of components essential for the maintenance of its processes, and will indicate where this new knowledge about neuronal dynamics has contributed in a practical way to neuronal tracing techniques and our knowledge of brain circuitry and function.

NEURONAL DYNAMICS

The study of neuronal dynamics was catalyzed by early phase contrast observations of neurons grown in tissue culture. These studies showed a rapid bidirectional saltatory movement of particles within axons. Since then a variety of experimental techniques including: 1) gross morphological observations of nerve fibers central to a constriction; 2) the accumulation of endogenous transmitter enzymes, and other axonal components, or of radioactively labeled components just proximal to a constriction; 3) the distribution of labeled axonal components among consecutive segments of nerves analyzed at various times after incorporation of labeled precursors into their cell bodies; and 4) ultrastructural and direct observations (by Nomarski optics) of axonal components in living nerve fibers, have all been used to provide a dynamic picture of many different kinds of neurons. Not surprisingly, the interest shown in axonal flow is reflected by the increasing number of investigators involved in this area and the number of recent review articles collating their achievements (2–7).

Synthesis of Neuronal Components

Most of the knowledge that we have about neuronal dynamics and axonal transport has come from studying the synthesis and elaboration of macromolecules, especially proteins. The reasons for this are that proteins are relatively easily labeled with radioactive amino acids, and are relatively stable molecules which biochemists can isolate and characterize. Since proteins are synthesized on ribosomes which are located in rough endoplasmic reticulum and in mitochondria, synthesis of proteins occurs in the perikaryon and in mitochondria within the axon (Figure 1). There is also the possibility of transference into the axon of macromolecules synthesized on ribosomes in either Schwann cells or in the postsynaptic cells. For reviews see Refs. 6, 8, and 9.

The fate of proteins in the axon has been studied after labeling with tritiated amino acids (to detect protein transport) or tritiated fucose (to detect glycoprotein transport) by: 1) injection into the Edinger-Westphal nucleus of the chick (10) whose axons project to the ciliary ganglion, 10 mm away, where they terminate as huge calyces, 17,000 μm^3 in volume, which are quite suitable for quantitative autoradiography (Figures 2 and 3); 2) injection into the posterior chamber of the eye of mice (11), rabbits (12), pigeons (13), chicks (14), and other species (6), so that the proteins and glycoproteins labeled in the retinal ganglion cells can be traced along the optic pathways to the terminals in the optic tectum; 3) injection into the spinal cord at the level of the motoneurons of sciatic nerve fibers in chicken (15, 16), cat (7, 17), and mice (18, 19); and 4) the placement of a pad of gel foam impregnated with the radioactive precursor into the nasal cavity of toads, mice (20), and the long-nosed garfish (21, 22).

These radioactive labeling studies have shown that the major site of synthesis of axonal and nerve terminal proteins is the perikaryon. The present view (9, 22) is that polypeptides are synthesized on the polyribosomes of the rough endoplasmic reticulum. The question of whether ribosomes exist in the axon is a subject of some controversy at the moment (23). Although ribosome-sized particles have been observed in the axon (in nodes of Ranvier) (24) there has been no positive identification of functional ribosomes in axonal membranes or axon terminals. Droz (8) has calculated the relative importance of synthesis of proteins in the three possible areas mentioned above, and shown that the

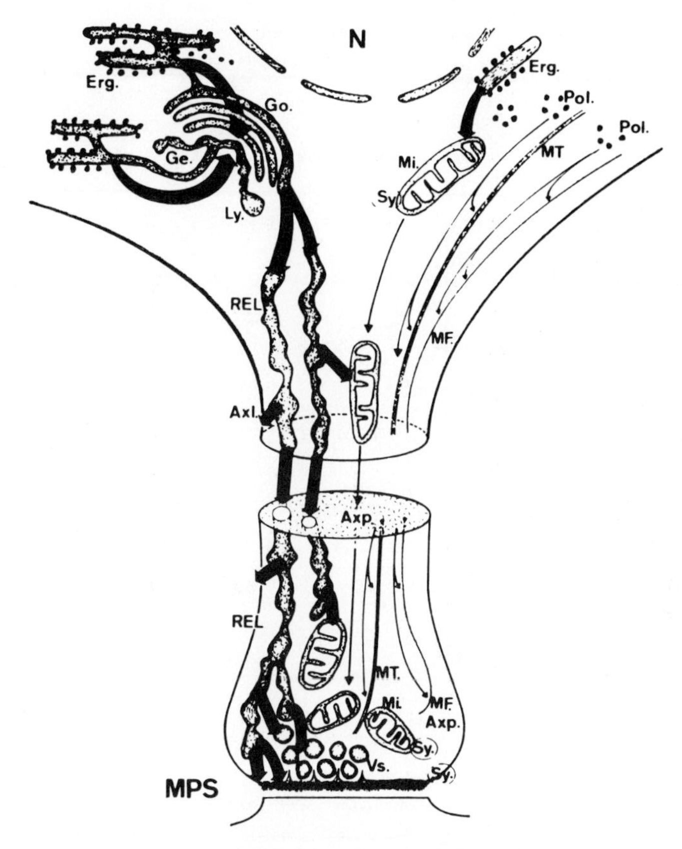

Figure 1. Schematic representation of the biosynthesis, axonal transport, and distribution of presynaptic proteins. The *thick arrows* on the left indicate rapid axonal transport; the *thin arrows* on the right, slow axonal transport. (From Droz et al. (9) by courtesy of Masson et Ce, Paris.) *Axl*, axolemma or axonal plasma membrane; *Axp*, axoplasm; *Erg*, systems of rough endoplasmic reticulum (Nissl bodies), composed of polysomes attached to membranes of the endoplasmic reticulum; *Go*, Golgi apparatus; *Ge*, GERL, or system "Golgi-Endoplasmic Reticulum-Lysosomes"; *Ly*, lysosome; *MF*, microfilament; *Mi*, mitochondria; *MPS*, presynaptic plasma membranes; *MT*, microtubule; *N*, nucleus; *Pol*, free polysomes; *REL*, smooth endoplasmic reticulum; *Sy*, local synthesis and recycling mechanisms; *Vs*, synaptic vesicle.

contribution of normal perikaryal protein synthesis to axonal protein is 98% and that from other sources only 2%. This implies that most if not all proteins found in axons and axon terminals are supplied by some form of axonal transport.

Within the perikaryon itself there are two routes for export of polypeptides synthesized on polyribosomes (see Figure 1).

Channels for Rapid Axonal Transport of Membrane Constituents and Vesicular Components Some of the proteins synthesized on polyribosomes pass into the Golgi apparatus and GERL where sugars such as glucosamine, galactose, and fucose are added to the proteins, thereby forming glycoproteins. Here they

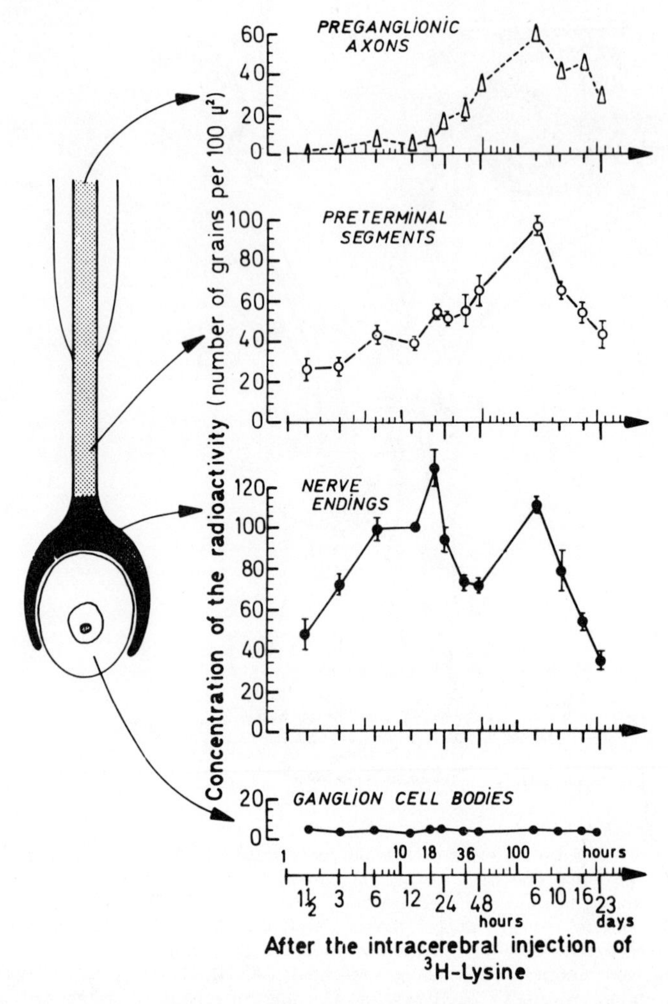

Figure 2. Quantitative radioautography of the chick ciliary ganglion. Time curves of the concentration of radioactivity in preganglionic axons, preterminal segments, calciform nerve endings, and ganglion cell bodies at various times after the intracerebral injection of [^{3}H]lysine. Note the sudden rise of label concentration in preganglionic axons after 24 hours. In nerve endings two peaks at 18 hours and 6 days are seen, whereas only the latter is observed in preganglionic axons and their preterminal segments. (From Droz et al. (108) by courtesy of Elsevier, Amsterdam.)

are either packaged into vesicular material (which could be synaptic vesicles, bags of smooth endoplasmic reticulum or lysosomes) and in this form are transported rapidly to the nerve terminals (10) (Figure 4), and the glycoproteins are incorporated directly into axonal membranes as integral proteins (Figure 5) and supplied to the synaptic terminals by lateral diffusion within the plane of preterminal axonal membranes (14, 22, 25). Excluding the microtubules and microfilaments, which do not appear to transport glycoproteins to any signifi-

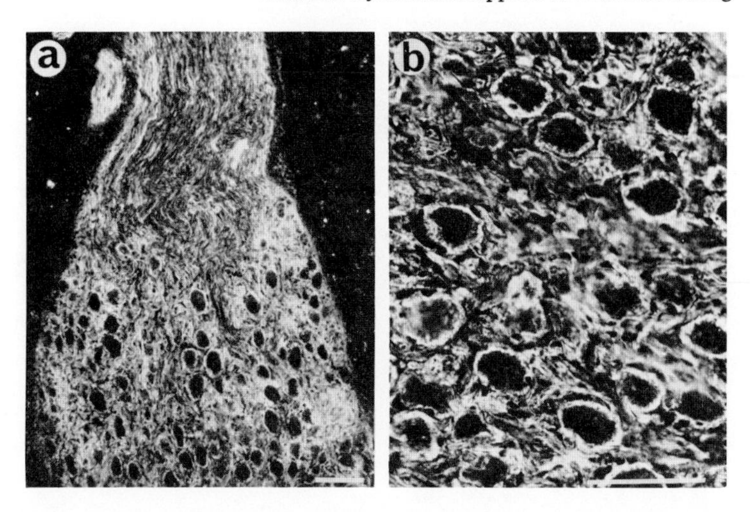

Figure 3. Immunofluorescence demonstration of presynaptic nerve-ending calices in the chick ciliary ganglion. Cryostat sections of day-old chick ciliary ganglion stained by the "sandwich" technique of immunofluorescence with an antiserum (25) against chick synaptic plasma membranes. *Bar* represents 100 μm. (By courtesy of John Rostas.)

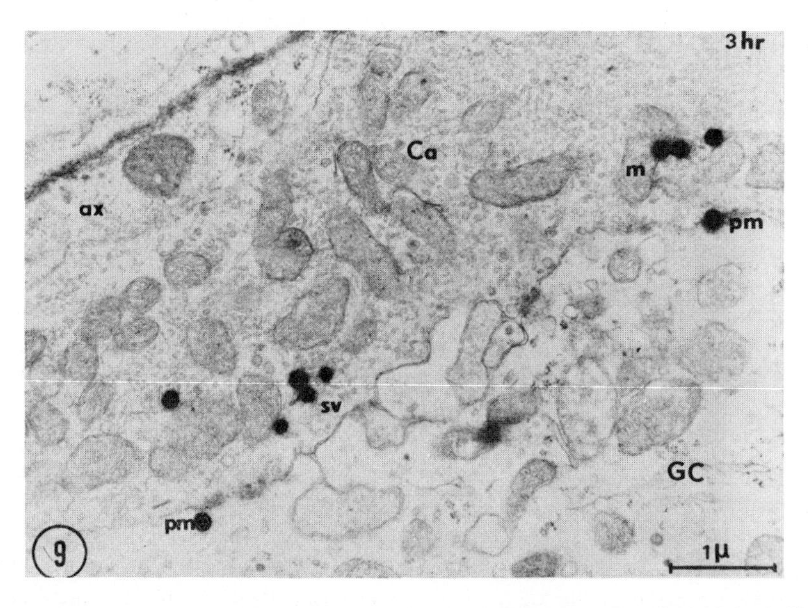

Figure 4. Electron microscope radioautograph of calciform nerve terminal (Ca) of ciliary ganglion 3 hours after intracerebral [^{3}H] fucose injection. Three hours after injection, most of the silver grains occur over the presynaptic plasma membrane (*pm*) and over regions of axoplasm rich in synaptic vesicles (*sv*) near the synaptic junction. Some are also near mitochondria (*m*), but none occur over regions of axoplasm devoid of synaptic vesicles (*ax*). The ganglion cell perikaryon (*GC*) exhibits almost no labeling. (From Bennett, et al. (10), by courtesy of Elsevier, Amsterdam.)

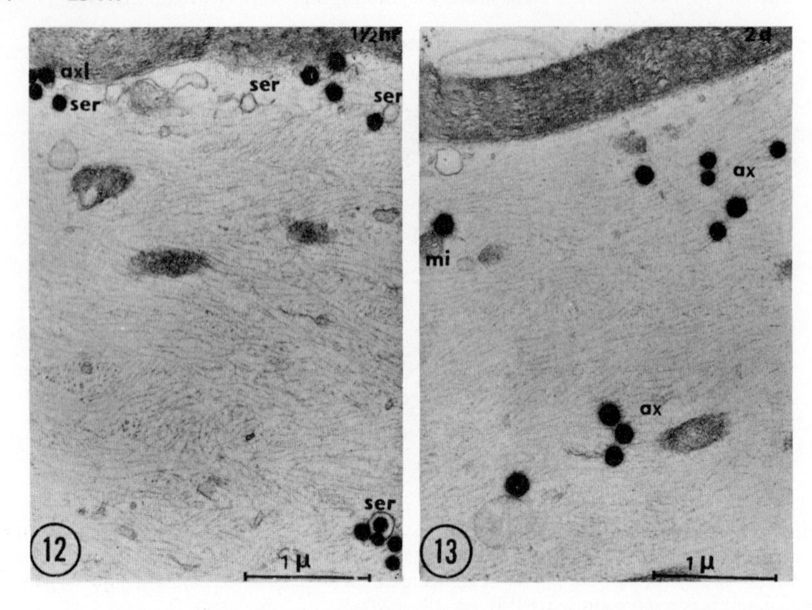

Figure 5. Electron microscope radioautographs of myelinated preganglionic axons 1.5 hours and 2 days after the intracerebral injection of [^{3}H]lysine. *Left,* at 1.5 hours, only occasional axons show some label. In this case silver grains are located over or near axolemma (*axl*) and profiles of smooth endoplasmic reticulum (*ser*). *Right,* by 2 days, the axons are heavily labeled. Numerous silver grains are distributed over the axoplasm (*ax*) containing microfilaments; some grains appear related to mitochondria (*mi*). (From Droz et al. (108) by courtesy of Elsevier, Amsterdam.)

cant extent (9), there are two continuous systems of axonal membrane linking the nerve cell body with its terminals in which this intra-membrane transport of glycoproteins could take place: (*a*) the axolemma, and (*b*) a continuous system of agranular endoplasmic reticulum.

Axolemmal Transport Glycoproteins (14) and other membrane proteins (25) destined for transport to the nerve terminals appear to be incorporated into the axolemmal membrane at sites distal to the axon hillock region (26). While it is not known just how this selective addition takes place, it is thought that transference occurs from channels of agranular endoplasmic reticulum present in the axon.

Transport within Agranular Endoplasmic Reticulum By use of high-voltage electron microscopy, Droz et al. (9) have been able to trace a continuous channel of agranular endoplasmic reticulum (in thick sections, 2 mm) from the perikaryon through the axon hillock of the initial segment and down the myelinated portion of the axon to the nerve terminals where it terminated just proximal to the store of synaptic vesicles. In the myelinated region of the axon the agranular reticulum was often observed to lie immediately beneath the surface of the axolemma; however, it is not known whether transference of membrane proteins occurs between the two membranes after direct contact of their surfaces and transient fusion of their membranes, or whether this transference takes place by diffusion across the axoplasm separating the two membrane systems.

The fate of axonally transported proteins within the axon has been investigated recently by studying the subcellular composition of different fractions of garfish olfactory nerve examined at various distances from the cell bodies after labeling with [^{3}H]leucine (22). By examining intact nerves and nerves cut from their cell bodies 6 hours after 1) a simple labeling with [^{3}H]leucine, or 2) a 1 hour pulse-chase label with [^{3}H]leucine, it was possible to differentiate between various factors affecting the distribution of labeled protein in the axon. The conclusions drawn from this study support those of Droz (8) with the chick ciliary ganglion system (Figure 6) in that both provided evidence that: 1) the bulk of rapidly transported proteins are incorporated into the axon as membranous axonal components; and 2) these membrane precursors are synthesized rapidly in the cell bodies, the majority being released into the axon immediately after synthesis and transported as particles rather than as subunits.

Whereas some 30% of the total protein transported along the axon reaches the nerve terminals, it had been calculated (22) that only 20% of the *membranous* material transported along the axon reaches the nerve terminals. Therefore,

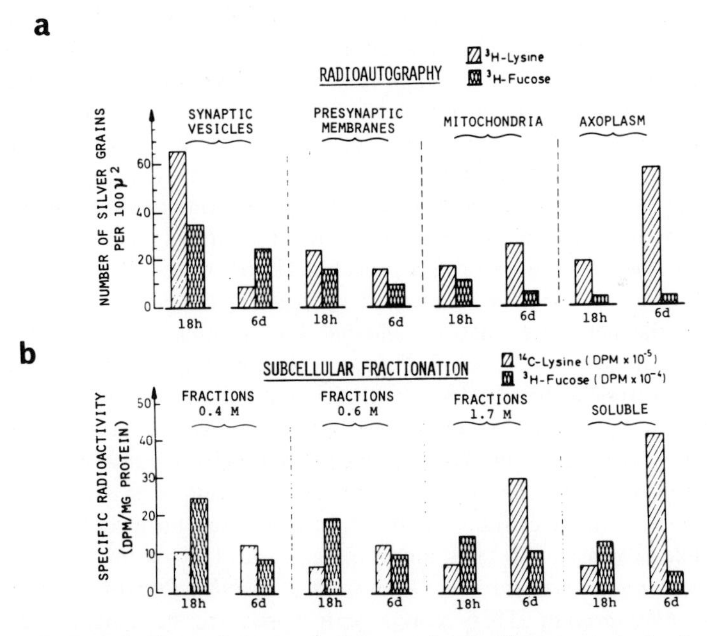

Figure 6. Contributions of fast and slow axoplasmic flow to renewal of proteins and glycoproteins in presynaptic nerve terminals of the chick ciliary ganglion. Comparison of (*a*) radioautography and (*b*) subcellular fractionation at 18 hours and 6 days after intracerebral injection of [^{3}H]lysine and [^{3}H]fucose into chicks. Rapid axonal transport delivers proteins and glycoproteins to the presynaptic membranes and synaptic vesicles, while slow axoplasmic transport contributes principally to the labeling of axoplasmic proteins and mitochondria. Some of the mitochondrial proteins are also supplied with the rapid phase. (From Droz et al. (9) by courtesy of Masson et Ce, Paris.)

it appears that the bulk of membranous components are deposited and remain more specifically within the axon.

As glycoproteins are integral components of neuronal plasma membranes and synaptic vesicle membranes (10, 13, 27) their supply is essential for maintaining the structure and function of neurons. For this reason glycoproteins may be especially important in developing neurons where they are thought to play a role in interneuronal recognition phenomena (28, 29). It has been proposed (26) that the axon hillock is a region of relatively non-fluid membrane that acts to partition specialized membrane components of the axolemma from those of the nerve cell body. Such a mechanism could be especially important in developing neurons in which partitioning of membrane components involved in intercellular recognition and orientation for axonal growth is thought to be a prerequisite to synapse formation.

It is now recognized that integral proteins within cell membranes enjoy a high degree of mobility (30), and it is possible that in nerve cells the mobility of glycoproteins within the membrane of the axolemma (31, 32) and agranular endoplasmic reticulum is high enough to permit a rapid delivery of these components to the nerve terminals (9, 14). In view of the abundance of agranular endoplasmic reticulum within axons it is probable that nerve cells represent a good example of the type of highly differentiated cell containing large amounts of internal membrane structures, for which it has been suggested (33) that the viscosity of the cytoplasm is so high that the movement by lateral translational diffusion within membranes is favored over movement through the cytoplasmic matrix. Although it seems likely that the channels of agranular endoplasmic reticulum provide the major pathway by which glycoprotein and other fast-moving proteins are transported to the nerve terminals, it is not known to what extent transport takes place within the intracisternal space and to what extent within the membranes of this reticulum. Furthermore, it is not known whether the synaptic vesicles are derived by pinching-off from the agranular reticulum in the nerve terminals or whether the agranular reticulum merely provides a channel for rapid renewal of synaptic vesicle components required for the assembly of vesicles formed by a local membrane recycling mechanism (see section "Release from Nerve Terminals"). This question is of interest in view of the observation (Figures 4 and 6) that much of the rapidly transported glycoprotein seen in nerve terminals is in close spatial relationship to synaptic vesicles and presynaptic membranes rather than evenly distributed throughout the axoplasm. Soluble axoplasmic proteins appear to be provided by another route.

An Axoplasmic Route for the Elaboration of Soluble Proteins, Neurofilaments, and Microtubules These components appear to be supplied to the terminals by a route that bypasses the Golgi apparatus (Figure 1). After release of the polypeptide chains from free polysomes in the perikaryon they migrate slowly into the axoplasm where some assemble as protein subunits, giving rise to the microfilaments and microtubules, and others are incorporated into mitochondria (Figure 7).

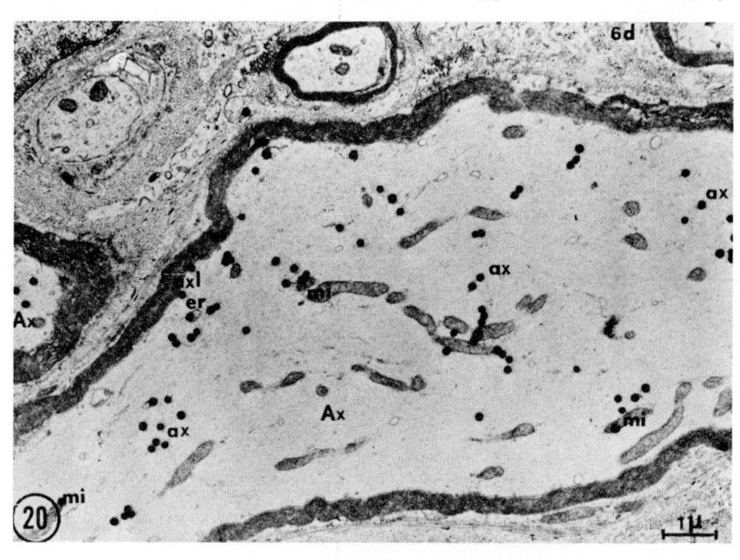

Figure 7. Electron microscope radioautographs of preganglionic axons 6 days after the intracerebral injection of [^{3}H]lysine. Two preganglionic axons (*Ax*) exhibit numerous silver grains distributed over the axoplasm (*ax*) and mitochondria (*mi*). A few silver grains are also seen over profiles of the endoplasmic reticulum (*er*), the axolemmal membrane (*axl*), and the myelin sheath. Three unlabeled postganglionic axons are visible in the upper part of the figure. (From Droz et al. (108) by courtesy of Elsevier, Amsterdam.)

Mitochondrial Proteins Mitochondrial proteins appear to be synthesized from two sources. They are formed by direct perikaryal synthesis in the nerve cell body and, in addition, some soluble proteins, glycoproteins, and amino acids transported by rapid axonal transport are supplied to the mitochondria during their migration along the axon (Figure 8). In addition, the mitochondria take up the amino acids from the axoplasm to synthesize membrane components (34, 35).

Rates of Transport and Turnover of Axonal Proteins

Axonal proteins, derived by the two routes outlined above, appear to be supplied to the axonal transport machinery and conveyed down the axon at two different rates. The microtubules and neurofilaments are transported at rates of about 1–6 mm/day, while the vesicular material moves at a much faster rate of between 40 and 600 mm/day. A wide range of different flow rates have been reported depending on the component and the system investigated (6).

Other evidence for this division in the rate of supply of proteins to nerve terminals has come from studies where the components transported at fast and slow rates have been analyzed after subcellular fractionation of the nerve (22, 36); e.g., glycoproteins in the synaptic plasma membrane are conveyed at a fast rate as are enzymes associated with membranes such as adenylate cyclase (37)

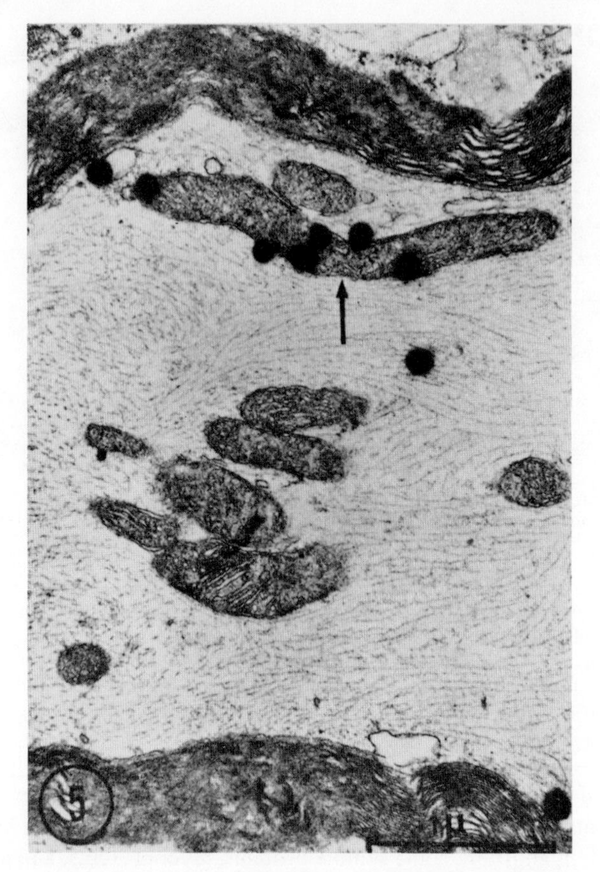

Figure 8. Electron microscope radioautograph of a preganglionic axon in a ciliary ganglion of a chicken at 6 days after intracerebral injection of [^{3}H] lysine. Heavily labeled mitochondria (↑) are frequently found among unlabeled or poorly labeled mitochondria. (From Di Giamberardino et al. (36) by courtesy of Elsevier, Amsterdam.)

and acetylcholinesterase (38). Grafstein (5, 39, 40) has shown that glycoproteins (5, 39) and phospholipids are transported exclusively at the fast rate of transport and that the lipids are transported in association with protein.

The general conclusion that may be drawn from this work is that rapid transport supplies glycoproteins and other proteins required for the maintenance of structures essential for impulse conduction (the axolemmal membrane) and transmission (mitochondria, synaptic vesicles, and synaptic plasma membranes), whereas the slowly transported material is required for the renewal of components essential for the maintenance of the structural integrity of cellular constituents such as mitochondria, axoplasm, microtubules, and soluble proteins.

Recent developments in cell fractionation techniques and in high resolution radioautography have permitted estimates of the rate of turnover of these neuronal components to be made (8). Depending on the method used to determine turnover times, rates ranging from hours to weeks have been obtained (8, 41). In general, rapidly transported proteins and glycoproteins are renewed within hours and some up to a few days, whereas synaptic proteins conveyed with the slow phase of axonal transport turn over much more slowly with half-lives of 1 to 10 weeks.

Factors Affecting the Rate of Axoplasmic Transport

Effect of Nerve Activity Early studies showed that transport of certain axonal components is enhanced under conditions of intense nervous activity. Electrical stimulation increases the transport of [^{14}C] glutamate in snail and frog neurons (42), and the transport of noradrenaline in rat sciatic nerves is accelerated by intense electrical stimulation (43) (see section "Contribution of the Cell Body and Terminals to Maintenance of Transmitter Levels in Peripheral Adrenergic Tissues") and during recovery from reserpine (44). Similarly, in the CNS the transport of [^{3}H] dopamine is enhanced after reserpine treatment (45). In neurosecretory neurons, osmotic stimuli, suckling and parturition stimulate the axonal transport of neurohypophysial hormones and their carrier proteins, the neurophysins (46, 47).

In considering quantitative aspects of axonal flow it is important to distinguish between the *rate* of transport of a substance and the *amount* transported. To this end experiments have been designed to examine the question of whether changes in neural activity affect the rate of incorporation of radioactive amino acids into protein in ganglion cell bodies and the subsequent rate of migration of protein out from the cell bodies. Increased or decreased nerve activity does not seem to affect the rate of migration out, though increased neural activity can increase the rate of incorporation of amino acids into protein in the perikaryon (48, 49). In the hypothalamo-neurohypophysial system, an increase in the rate of nerve impulse activity (induced physiologically by dehydration or salt-loading) results in an increased secretion of the neurohormones (vasopressin and oxytocin) and a compensatory increase in the rate of protein synthesis in the perikarya. This leads to an increase in the *amount* of vesicular protein transported but the rate at which it is transported remains constant (50). An interesting observation is that, following osmotic stress, there is a proliferation of microtubules in these axons (51). It has therefore been proposed (43) that increases in the amount of vesicular material transported depend on the formation of increased numbers of microtubules within the axon. Consistent with this hypothesis is the finding that injection of nerve growth factor into mice results in an increased incorporation of labeled amino acids into protein in the stellate ganglion (52) which is reflected by an increase in the amount of microtubular protein (53). This indicates that stimulation of the nerve cell body may result in an increased synthesis of the microtubular system responsible for axonal transport.

Temperature Metabolic studies have shown that the rapid phase of axonal transport requires a supply of energy in the form of ATP and that this is a temperature-sensitive process (6, 7, 54). In contrast, the slow phase is relatively temperature insensitive, indicating that the rapid and slow phases of axoplasmic flow are mediated by different mechanisms.

Chemical Lesions The slow phase of axoplasmic flow of protein can be blocked experimentally by acrylamide (55), a substance that induces a chemical neuropathy with axonal destruction similar in type to Wallerian degeneration; however, other chemicals, such as di-*iso*propylphosphorofluoridate (DFP) and tri*ortho*cresylphosphate (TOCP) that produce similar degenerative changes in axonal structure, have no effect on axonal transport (6).

Fast axonal transport of adrenergic vesicles is blocked by the chemical 6-hydroxydopamine, and this compound has been used extensively for mapping central adrenergic neurons (see section "Demonstration of Specific Transmitters") through the production of chemical lesions in selected regions of the brain (56, 57). Biochemical studies (58) have shown that 6-hydroxydopamine has a long-lasting degenerative effect on peripheral and central catecholamine nerve terminals. In the adult it causes selective damage to the terminals in the *peripheral* sympathetic system, but in the newborn the whole neuron is affected. In the adult *central* nervous system its effects seem also to include the perikarya of the noradrenergic and dopaminergic neurons (59). Long-lasting reductions in the levels of brain noradrenaline have also been brought about by the compound 6-hydroxydopa (60). It has been suggested that along with the depletion and preterminal accumulation of amine, this compound, like 6-hydroxydopamine, blocks axonal flow into the terminal plexus. Unlike 6-hydroxydopamine, 6-hydroxydopa is capable of passing through the blood-brain barrier and causes a selective and reversible depletion (possibly by destruction followed by regenerative sprouting) of noradrenergic terminals, without affecting dopaminergic terminals or the perikarya of noradrenergic and dopaminergic neurons. This compound has been used to provide a detailed mapping of central noradrenergic nerves (60) that is in essential agreement with observations from other chemical and surgical lesion studies of the central nervous system (57). In a similar manner, intraventricular administration of moderate doses (25–75 μg) of the compound, 5,7-dihydroxytryptamine (5,7-DHT) into rats causes pronounced damage to indoleamine axons in the central nervous system as reflected by the pile-up of large amounts of serotonin in the distorted and swollen axons (61). Since 5,7-DHT condenses with formaldehyde to form a light yellow compound visible by fluorescence microscopy, its uptake and accumulation into indoleamine terminals and axons have provided an important additional tool for mapping indoleamine neuron systems in the brain (62).

Development and Aging In general, the slow phase of axonal transport in young animals is faster (by 2- to 3-fold) than in adults (for review see Jeffrey and Austin (6) and Grafstein et al. (40)) and this probably reflects the processes of active outgrowth by immature axons, and the increased demands on axonal flow for the supply of materials required for neurite elongation and synapse formation. At about the time that synaptic structures are formed, there is a 2-fold increase in the rate of fast axonal transport and this may reflect the supply of transmitter vesicles and transmitter-synthesizing enzymes required for chemical neurotransmission.

Disease States One of the functions of axonal flow is thought to be to supply "trophic substances" from nerves to muscles (7, 338, 339). There is currently much interest in this question because of the possibility that certain degenerative diseases of muscle (e.g., muscular dystrophy) may have their origins in an altered axonal supply of these trophic materials. It has recently been demonstrated (63–65) that in the sciatic nerves of dystrophic mice there is an increase in the amounts of proteins, phospholipids, and cholesterol transported with the rapid phase. These alterations in axonal transport of membrane components are thought to contribute to the altered composition of neuronal and muscular cell membranes seen in this disease.

Neuronal Regeneration The axonal transport requirements of mature neurons are modulated not only in certain disease states, and during development and aging, but also under conditions of regeneration (1), where increased demands for membrane proteins have to be met. Increased *amounts* of protein are transported at both fast and slow rates in regenerating nerves (6) and this increase in transport of components is thought to reflect an increased perikaryal synthesis of proteins such as tubulin, which make up the axonal transport apparatus (52), rather than an increase in the rate at which axonal transport takes place.

Role of Microtubules in Axonal Transport of Vesicles

The axonal transport of proteins and vesicles is blocked by colchicine and vinblastine (66–69). These agents bind to microtubules suggesting that microtubules are involved in the transport process (6). Since microtubular proteins and noradrenergic vesicles move at slow (2 mm/day) (66, 67) and fast (120 mm/day) (70, 71) rates of transport, respectively, it is envisaged that if the vesicles are associated with the tubules they must move relative to them. Of a number of mechanisms that have been proposed to describe a role for microtubules in the rapid axonal transport of vesicles, two have received considerable attention.

In the first, Schmitt (72) proposed that microtubules contain bound ATP or GTP and that the vesicles (containing a binding site for the microtubule and an ATPase or GTPase) might move along the microtubules by energy-coupled hydrolysis of nucleotide, this process producing a motive force to the particle by conformational changes in the protein structure of the organelle membranes. By making and breaking bonds the vesicles would roll down the microtubules. If this were so, one might expect that changes in the intracellular Ca^{2+} concentration resulting from increased nerve activity might influence the interaction between vesicles and microtubules. However, marked changes in the Ca^{2+} concentration had little effect on the rate of accumulation of noradrenergic granular vesicles at a ligation of hypogastric nerves incubated in vitro (73) and there is no satisfactory explanation of this negative result to date. In addition, this model could not explain the fact that a variety of soluble proteins and polypeptides as well as vesicles are carried down nerve fibers by fast axonal flow.

To account for this phenomenon another hypothesis was required and a model for fast axoplasmic flow was advanced based on the sliding filament theory for muscle. In this model (7, 74) the sliding component is a protein "transporting filament" synthesized in the perikaryon and capable of carrying various components (mitochondria, soluble proteins, and vesicles) along the microtubules and/or neurofilaments. Cross-bridges between the transport filament and the microtubules and/or neurofilaments produce the movement in a similar fashion to the sliding filament theory of muscle. The energy required at the cross-bridges for such movement is supplied by ATP derived from oxidative phosphorylation. In early models (74, 75) the cross-bridges were assigned to the transport filaments; however, recent high resolution electron micrographs (76) have suggested a revision to the model as shown in Figure 9 in which the projections or side arms are located on the stationary microtubules and neurofilaments. The ATP appears to be hydrolyzed by an axonal Mg^{2+} or Ca^{2+} ATPase (77, 78) but it is not known whether this activity is associated with the microtubule cross-bridges in the same way that MgATPase dynein proteins comprise the contractile side arms on flagellae and microtubule systems in primitive motile systems.

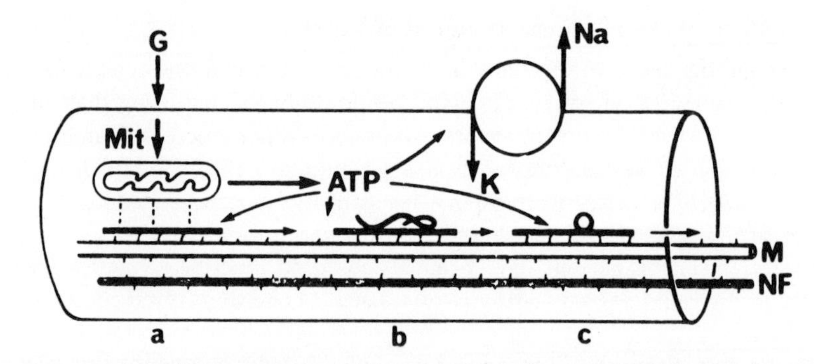

Figure 9. Transport filament hypothesis. Glucose (*G*) enters the fiber and after glycolysis and then oxidative phosphorylation in the mitochondrion (*Mit*), the resulting ATP supplies energy to the sodium pump shown controlling the level of Na^+ and K^+ in the fiber. ATP also supplies energy to the "transport filament," shown as *black bars,* to which various components are bound and so carried down the fiber. These components include the mitochondria (*a*) attaching temporarily as indicated by *dashed lines* to the transport filament and giving rise to a fast to-and-fro movement (though with a slow net forward movement as described in the text) and soluble protein (*b*) shown as a folded or globular configuration, as well as other polypeptides and small particulates (*c*). Simpler molecules are also bound to the transport filaments; thus, a wide range of components are carried at the same fast rate. Cross-bridges between the transport filament and microtubules and/or neurofilaments affect the movement in similar fashion to the sliding filament theory of muscle, the required energy supplied by ATP. The cross-bridges are shown arising from both microtubules as spurs or side arms seen in high resolution electron micrographs. (From Ochs (7), by courtesy of N.Y. Academy of Science.)

The idea that axoplasmic flow involves a contractile mechanism in which proteins with properties similar to actin and myosin interact with each other to hydroylze ATP to provide the energy for translocation (317) is a very attractive one. At present, progress in this area is made difficult by the ubiquitous distribution of Mg, Ca ATPases, the heterogeneity of neuronal proteins possessing this enzymic activity, and the difficulty of assigning physiological functions to a protein, or group of proteins, showing "actomyosin-like" properties in vitro (79).

Fate of Transported Components

A number of different cellular processes have been proposed to account for the turnover and disappearance of synaptic proteins (8). These include local hydrolytic breakdown, exocytosis of proteins contained in transmitter vesicles, and fast retrograde axonal transport of proteins from the axon endings to neuronal perikarya (Figure 10). The mechanisms involved in these processes are still uncertain as is the quantitative contribution that each process makes to the removal of specific proteins or even to the turnover of total protein in the terminals. As the processes of release and reuptake at nerve terminals have contributed to neuronal mapping, as, for example, in the release and trans-synaptic transfer of labeled amino acids, peptides, and nucleotides, and in the uptake of labeled neurotransmitters, amino acids, and exogenous proteins (such as horseradish peroxidase), these processes are reviewed here in more detail.

Release from Nerve Terminals Nerve stimulation results in depletion of synaptic vesicles in nerve terminals (332) (Figures 11 and 12). The synaptic vesicles appear to fuse with the plasma membrane releasing the neurotransmitter by exocytosis (83) (Figure 13). When sympathetic nerves are stimulated, vesicle proteins such as dopamine β-hydroxylase and chromogranins, which are in a soluble form within the vesicles, are released together with noradrenaline (80, 81). A similar process occurs in the release of neurophysins and their peptide hormones (oxytocin and vasopressin) from neurosecretory nerve terminals (82), and in the release of peroxidase from preloaded synaptic vesicles in cholinergic neurons at neuromuscular junctions (83, 84).

Studies of nerve terminals at the neuromuscular junction have shown that the presynaptic membrane contains specialized regions where the synaptic vesicles fuse with plasma membrane and release transmitter. These "vesicle release sites" (85), also termed "zones actives" (86) (Figure 14), are localized to regions immediately surrounding ridges, which can be observed by freeze fracture techniques as electron dense regions of the presynaptic membrane (87). These presynaptic ridges may correspond to areas of restricted mobility in the membrane which have been detected by fluorescence polarization spectroscopy (88).

After fusion of the vesicle membrane with the presynaptic plasma membrane at these sites, it has been proposed (on the basis of electron microscopic observation of stimulated synapses in the presence of exogenous horseradish peroxidase) that the vesicle membrane becomes incorporated into the presynaptic plasma membrane and diffuses away from the "vesicle release sites" to

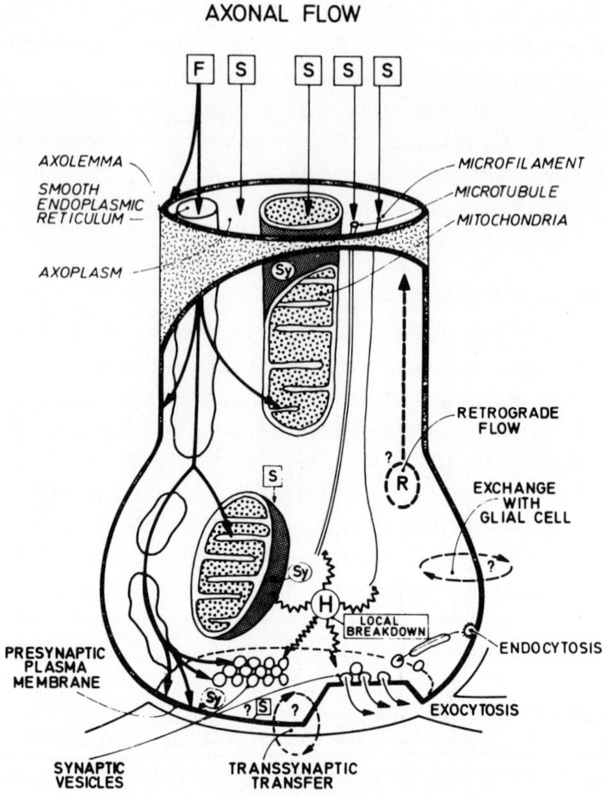

AXONAL FLOW

Figure 10. Diagrammatic representation of the dynamic condition of synaptic proteins in an axon terminal. *F,* fast phase of the axonal flow; *S,* slow phase of the axonal flow; *Sy,* sites of local protein synthesis or of local incorporation of labeled amino acids; *H,* hydrolytic enzymes such as proteinases and peptide hydrolases; *R,* retrograde flow. (From Droz (8), by courtesy of Elsevier, Amsterdam.)

"recovery sites" located at the periphery of the synapse (84) (Figure 15). At the "recovery sites," plasma membrane is returned to the inside of the terminal by endocytosis of coated vesicles. The coated vesicles coalesce (under conditions of rapid sustained stimulation) to form cisternae from which new vesicles are formed. In this way membrane recycling helps to maintain a supply of vesicles in the nerve terminal at rates proportional to transmitter release.

Other interpretations are possible: by analogy, one of the best studied secretory systems, from a biochemical point of view, is the exocrine pancreas (337). In this system, after stimulated exocytotic secretion of zymogen granules, it is found that redundant apical membrane is withdrawn by at least two routes:

1. During the *initial* rapid increase in the amount of apical cell membrane, withdrawal is accomplished by interiorization of luminal invaginations into smooth endocytotic vesicles, which in turn give rise to multivesicular bodies by infolding and subsequent fission of their limiting membrane.

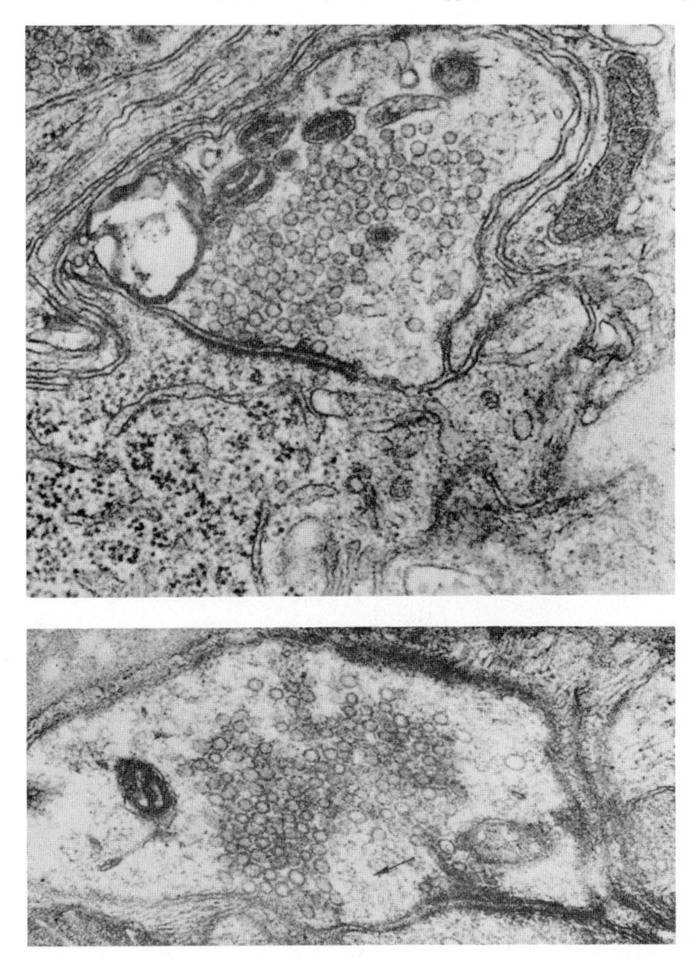

Figure 11. Effect of electrical stimulation on synaptic morphology. Cat superior cervical ganglion prepared for electron microscopy by perfusion fixation with glutaraldehyde in the presence of Mg^{2+}, to preserve the fine structure. *Top,* bouton in synaptic contact with a large dendrite from a ganglion stimulated at 1/s for 20 min. A diffuse myelin body occupies the area of bouton cytoplasm to the left of the vesicular region. Vesicles are lost from the periphery of the vesicular region except at the synapse and there is a loss of cytoplasmic matrix material. *Bottom,* bouton in synaptic contact with a dendrite at lower right of the field from a ganglion stimulated at 4/s for 20 min. Note loss of vesicles at the periphery of the vesicular region and small area devoid of vesicles *(arrow)* surrounded by vesicles except at the synaptic portion of the bouton membrane. (From Birks (332), by courtesy of Chapman and Hall Ltd., London.)

2. Once the bulk of stored secretions has been discharged, endocytotic coated vesicles predominate as carriers for redundant cell membrane which becomes incorporated in residual bodies and, presumably, digested by lysosomal degradation prior to re-use.

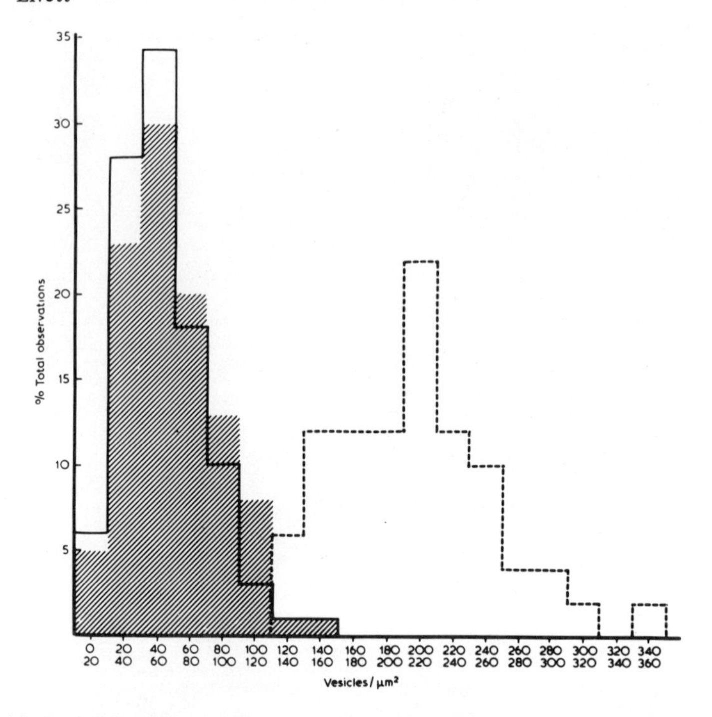

Figure 12. Reduction in synaptic vesicle numbers in terminals of the cat superior cervical ganglion following preganglionic nerve stimulation. Comparison of histograms of vesicle populations of boutons de passage in: unstimulated (*dotted line*), stimulated at 20/s for 20 min in the absence (*solid line*) and presence (*hatched*) of hemicholinium. Counts were sorted into successive cells of 20 vesicles/μm^2 of bouton area. Calculations based on a vesicle close-packing hypothesis gives a figure of 8000 vesicles per bouton. In ganglia stimulated for 20 min at 20/s the vesicle populations were reduced to 25%, and to 28.5% in ganglia in which acetylcholine synthesis was inhibited by hemicholinium. However, the fraction of ganglionic acetylcholine stores known to be released by this stimulation is substantially less than the fraction of vesicles lost, suggesting that some of the acetylcholine may also be released from a cytoplasmic pool during synaptic activation. (From Birks (332), by courtesy of Chapman and Hall, London.)

These morphological observations of membrane fusion at peripheral synapses and in the endocrine pancreas are supported by recent biochemical and immunological studies demonstrating fusion of vesicle and plasma membranes in intact synaptosomes isolated from chick forebrain, and in the adrenal medulla (89–91).

The postsynaptic membrane contains specialized functional regions too. In peripheral synapses this membrane is highly convoluted and contains regions of increased electron density ("thickenings") (92) (Figure 13). Recent studies (85) have shown that these regions, which are those on the crests of the postsynaptic folds closest to the presynaptic membrane, contain the highest density of acetylcholine receptors (Figure 14). By contrast, the troughs of the postsynaptic membrane folds contain acetylcholinesterase activity but few acetylcholine receptor sites. These acetylcholinesterase pits serve to remove acetylcholine after

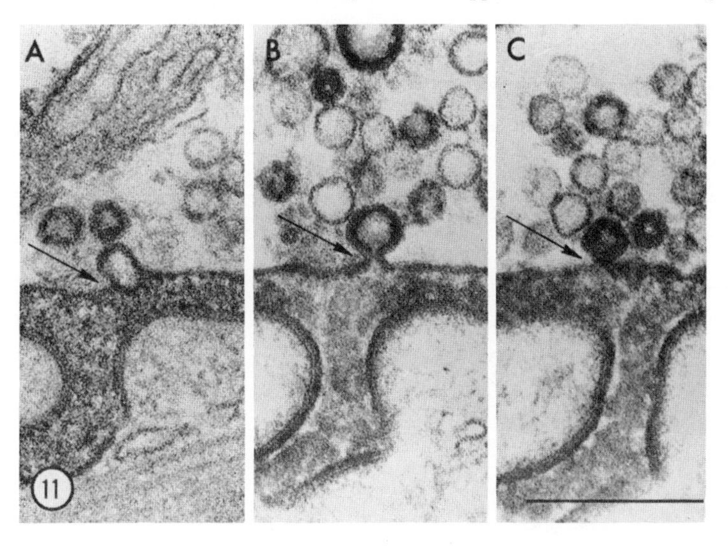

Figure 13. High power electron micrographs of portions of three different neuromuscular junctions at the level of the "active zone." Pectoris nerve-muscle preparations from frogs were stimulated for 2 hours at 2/s in curare (3×10^{-6} g/ml) plus horseradish peroxidase. Junctional clefts contain rich deposits of peroxidase reaction products. The figures show three degrees of association between peroxidase-labeled vesicles and the prejunctional membrane (*arrows*). In *A* the membrane of the vesicle is completely fused with the prejunctional membrane; in *B* the continuity of the two membranes is maintained through a short stalk; and in *C* the vesicle appears to be in the process of losing contact with the axolemma. *Bar* represents 0.25 μm; × 98,300. (From Ceccarelli et al. (83) by courtesy of the *Journal of Cell Biology,* Rockefeller Press, New York.)

release from receptor sites, and to remove acetylcholine that has not diffused away from the synaptic junction. While the main function of the pits is probably to remove acetylcholine and so avoid overstimulation of receptors, the possibility remains that such regions could be responsible for localized postsynaptic regenerative electrical phenomena (that is resistant to tetrodotoxin) which has been observed in endplates in rat diaphragm and frog sartorius muscle (93).

The organization of the postsynaptic membrane is clearly under presynaptic control since its structural and functional integrity is disturbed by denervation (94–98) or disuse (99). Indirect evidence for postsynaptic membrane fluidity is afforded by the demonstration that extrajunctional acetylcholine receptor sites (α-bungarotoxin binding sites) appear on the postsynaptic membrane following denervation, and that on reinnervation these are concentrated again in the junctional region (103); however, it is doubtful whether these are the same receptor molecules since on denervation there is a large increase in the total number of α-bungarotoxin binding sites (100, 101) and the turnover time of these extrajunctional receptor proteins is small (half-life, 19 hours) (102) by comparison with that of the junctional receptor (half-life > 7.5 days) or with the period (approximately 9 days) over which these redistribution phenomena are observed. By use of a fluorescent labeled antibody to rat muscle membranes, an

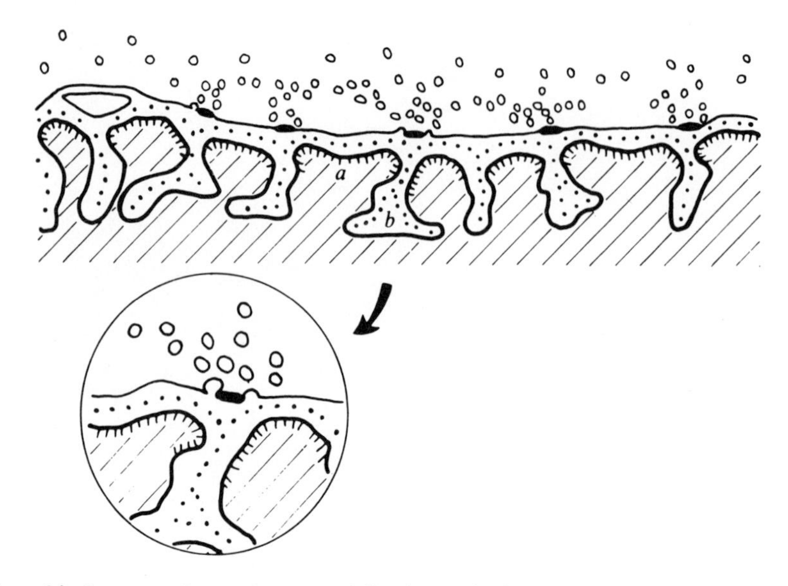

Figure 14. Postsynaptic membrane specialization at the frog motor endplate. Diagrammatic representation of part of the subneural apparatus to show the spatial relationships between sites of transmitter release and receptor concentrations. The striations along the postsynaptic membrane represent its "thickened" zone, rich in the receptors (i.e., the region referred to in the text as a "crest"). In the mammalian white fibers, the folds are usually more closely spaced than here, so that the crests are narrower and apical. *Dense bars* on the presynaptic membrane represent the ridges seen in freeze-etch preparations. *Dotted lines* show the extent of the stainable cleft substance, and, it is presumed, of the AChE. The *inset,* showing at higher magnification the release of a vesicle, is based on the presynaptic geometry of stimulated endplates, as determined by Couteaux and Pecot-Dechavassine (86) and Heuser et al. (87). A presynaptic ridge, seen here in cross-section as a dense amorphous zone, is associated with paired sites of vesicle release. Upon release from the receptors, that fraction of the ACh which does not immediately diffuse to the exterior, and which would otherwise tend to linger towards the base of each fold, is in a region which is poor in receptors but rich in acetylcholinesterase, so that it is hydrolyzed before it can give rise to excessive activation of receptors. (From Porter and Barnard (85) by courtesy of *J. Membrane Biol.*)

estimate of the fluidity of developing rat muscle myotube surface membranes in tissue culture has been obtained (103). The "diffusion constant" for Fab-antigen mobility in these surface membranes is 1.5×10^{-9} cm^2/s (which compares with the free diffusion of 100Å spheres in a medium slightly more viscous than olive oil, or of molten copper at 3000°C!). By contrast, in adult innervated muscle, the relatively rigid and specialized structure of endplate indicates that this region of the muscle membrane is very stable and probably less fluid.

This conclusion is supported by recent observations of the remarkable stability of the postsynaptic thickening (or "web") of synapses in the central nervous system following deafferentation (104). In this study it was shown that there existed a fixed number of postsynaptic sites in the septum which, when denervated, persisted, and could be recognized in the deafferented state by the presence of "vacated" synaptic thickenings on the postsynaptic cells. These

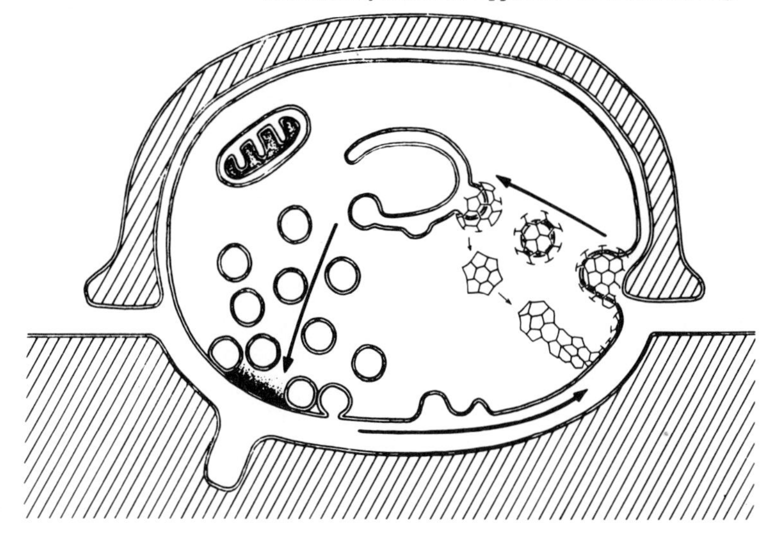

Figure 15. Diagrammatic summary of the proposed pathway for synaptic vesicle membrane recycling at nerve terminals. Synaptic vesicles discharge their content of transmitter by coalescing with the plasma membrane at specific regions adjacent to the muscle. Equal amounts of membrane are then retrieved by coated vesicles arising from regions of the plasma membrane adjacent to the Schwann sheath. The coated vesicles lose their coats and coalesce to form cisternae which accumulate in regions of vesicle depletion and slowly divide to form new synaptic vesicles. (From Heuser and Reese (84) by courtesy of *Journal of Cell Biology,* Rockefeller Press, New York.)

denervated sites are reinnervated by existing local axon terminals present in their immediate vicinity. This process, termed "collateral reinnervation," is very efficient and very rapid, reoccupation of nearly all the deafferented sites taking place almost immediately upon removal of the degenerating presynaptic terminals.

The postsynaptic membranes in central and peripheral neurons therefore appear to be more stable (less fluid) than the presynaptic membrane. The greater fluidity of the presynaptic membrane is no doubt of importance in the formation of exploratory growth cones in developing neurons (105), the establishment of functional connections, and in regeneration phenomena such as collateral sprouting. A full discussion of these events is beyond the scope of the present review, but readers interested in aspects of neuronal plasticity are referred to two recent review articles (1, 106). The study of membrane dynamics at the synapse is still at an early stage but should have important contributions to make to synaptic plasticity and related fields in the near future.

Trans-synaptic Migration The search for a transneuronal transfer of substances which can mediate the trophic influences of a neuron upon its effector cell (1) has led to the discovery that free amino acids (107) and small amounts of peptides (39, 108–113) and nucleotides (114) can be translocated from nerve terminals to other cells. This evidence rests on the autoradiographic demonstration of small amounts of labeled material in postsynaptic cells following presumed transference from axoplasmically labeled presynaptic nerve terminals in systems as diverse as motor nerves into muscle (109); from retina through the lateral geniculate to the striate cortex (39, 113) and in the reverse direction from the striate cortex to the lateral geniculate nucleus (114); and between neuroglia cells and neurons (107). However, it is not certain to what extent this represents specific transfer and to what extent a nonspecific transfer of labeled material or degradation products arising from turnover of labeled components in the presynaptic terminals.

Retrograde Transport of Proteins Since the demonstration by Kristensson et al. (115, 116) that exogenous proteins such as horseradish peroxidase (HRP) could be taken up by axon terminals and carried in a retrograde manner to the cell bodies, many investigators have used this technique to trace the origin of axons whose terminals end in a given region. When HRP is injected in the vicinity of axon terminals, the enzyme is taken up by pinocytosis into the axon terminals and into unmyelinated portions of axons in the region of the terminals. It is then compartmentalized in membrane-bound vesicles, tubules of agranular reticulum, multivesicular bodies and their precursors, and either these organelles or the HRP within them are transported back to the parent cell body at a rate of at least 72 mm/day (117, 118). It is not known whether the HRP found in the agranular reticulum is in a "continuous channel" running from the terminal to the smooth endoplasmic reticulum in the cell body, but it is thought likely that organelle transformations probably occur within the axons since no HRP-containing synaptic vesicles are found within axons and yet the tracer is transported back to the cell body. Two possibilities have been suggested: 1) that the HRP-vesicles coalesce in the terminals to form larger vesicles or fuse with multivesicular bodies, which are then transported back to the cell body; or 2) that the vesicular membrane which is recycled locally in the terminals (83, 84, 119–122) may form a different pool of membranes to those which are transported back to the cell body. The finding that retrograde transport of HRP is blocked by application of colchicine and vinblastine in both central (123, 124) and peripheral (125) neurons suggests that microtubules may be involved in the retrograde transport of HRP. However, these drugs could be affecting the system in less specific ways by limiting the metabolism of the cells, or by affecting uptake and membrane transport phenomena (126, 127). Similarly, the frequent physical association seen between microtubules and HRP-vesicles in chick ganglion cell axons (118) may be a consequence of the small diameter of these axons.

Retrograde intra-axonal transport may also be involved in the recovery of membrane components for re-utilization (128), and in limb regeneration as in the newt (129). Irrespective of its biological roles, as an experimental device, retrograde transport of exogenous proteins such as Evans blue albumen and HRP has provided much useful information about neuronal connectivity in the relatively short time of 5 years since it was first introduced. Some of these contributions are discussed in the next section.

TRACING OF NEURONAL PATHWAYS

Many different methods have been developed over the years for the mapping of central and peripheral neuronal pathways. While the cellular basis for these staining procedures is not always known with certainty, their selectivity for neuronal elements has been most useful in constructing a morphological picture of neuronal circuitry. In applying the more recent "dynamic" methods of tracing to the study of neuronal connectivity, much reliance has been placed on the earlier histological staining techniques, in particular for choosing suitable experimental systems with which to work. As results obtained with the more recent techniques are invariably compared with those obtained with conventional histochemical stains, a brief account of some of the more commonly used histological methods is given here. Emphasis will be given in this discussion to the histological basis of the staining method, the identity of the cellular components with which the stain reacts, and the relative advantages and disadvantages of the various methods. These aspects are summarized in Table 1.

Traditional Histological Staining Techniques

General Methods (130) There are many methods available for demonstrating nerve cell components. Among the general staining methods that give good results on histological sections of nervous tissue, Mallory's phosphotungstic acid hematoxylin (131) and Heidenhain's Azan (132) or osmic gallate are widely used. Osmic gallate is particularly good for displaying the nervous system in relation to its surroundings.

Selective Methods Even more specific for the nervous system are some of the silver impregnation methods (133, 134), although these do not stain the nervous systems of all animal species and may stain collagen, fibrin, and other connective tissue elements. For lower invertebrates special methods have been devised (148). By use of the methods described below, only a small proportion of the cells present are stained. Those that do stain, however, stain strongly and completely so that their entire course can be traced. In those that do not stain, the myelinated axon is refractory or the stain fails to penetrate the fine processes to their ends. Use of these stains therefore requires faith that impregnation is complete although no proof of this is available.

The Golgi-Cox method displays nerve cell bodies to best advantage, while the basic dyes demonstrate Nissl substance within the cells. Sulfur-rich neurosecretory substance can be shown after oxidation by Gomori's chromalum hematoxylin method and neurofibrils by Bielschowsky's and Von Braunmuhl's techniques.

Neurons The *Golgi silver methods* (135) demonstrate the external form of neurons and makes possible the tracing of individual axons and dendrites. The Golgi-Kopsch method involves treatment of the tissue block with an aldehyde-dichromate solution, or with a mixture of chromate and dichromate solutions, followed by impregnation with silver or mercury. It is selective, in that only a small proportion of nerve cells in any preparation are stained, but capricious in its results since in some circumstances it will also stain neuroglia. It was developed originally

Table 1. Some neurohistological methods

Stain	Components stained	Applications	Limitations
Neurons in general			
General: Mallory, Azan, Osmic gallate	Tissue elements in general: collagen; neuroglia cells and fibers; myofibrils and cross striations in muscle	Good discrimination of peripheral nervous system from surrounding tissue, such as bone, cartilage, etc.	Poor discrimination of individual neurons and glial elements within central nervous system
Silver impregnation			
Golgi (Golgi-Cox, Golgi-Kopsch, and rapid)	Selective impregnation of nerve fibers by soluble silver salts (particularly $AgNO_3$)	Displays external form of selected neurons. Can trace axons and dendrites	Capricious. Only small proportion of neurons stained. Neuroglia may also stain. Myelinated axons refractory but can be counterstained with Luxol fast blue.
Silver stains (Bodian, Cajal, Rasmussen, Davenport, Ungewitter).	Reacts with neurofibrils	Nerve cells and processes stain yellow, brown or black	Myelin not stained, but can be counterstained with Luxol fast blue.
Degeneration stains (Nauta and Gygax FinkHeimer, Eager, Deselin, and Glees)	Proliferation of neurofibrils in degenerating axon terminals (Glees method) or dense axoplasm (see Guillery, 158)	Study of degeneration in central nervous system of mammals. Degenerating terminals stain intensely	Not strictly selective for degenerating axons
Reduced methylene blue	Within the tissues, the methylene blue is reduced to a leuco form and is then reoxidized by atmospheric oxygen	Peripheral nerves and their endings, particularly visceral and sensory nerves	Not all the axons and endplates present may be demonstrated in the same preparation. Results depend on efficiency of perfusion and good oxygenation

Stain	Components stained	Applications	Limitations
Basic dyes Toluidine blue-O, Thionin, Cresyl violet, Nissl's methylene blue, Acridine orange (see below)	Stains basophilic compounds such as nucleoprotein and Nissl substance (RNA and DNA)	Best stains for Nissl bodies and nuclei. Used to detect chromatolysis and for *indirect* tracing after lesions	Do not stain axons or nerve endings. Cannot be used for *direct* observation of pathways. Chromatolysis does not always occur after axotomy.
Fluorescent dyes Acridine orange	RNA and DNA and other basophilic compounds	Differentiates DNA from RNA. Very sensitive fluorescence microscopy	Does not stain axons and terminals
Fast green F.C.F.	Cytoplasmic components	Localizing site of microelectrode placement	Relatively non-selective. Many tissue elements stain
Procion dyes		Localizing site of microelectrode placement Direct tracing of extent of neuronal processes	Limited to neuron in which it was injected
Staining myelin			
Normal (Weigert, Pal-Weigert, and Weil's hematoxylin) Luxol fast blue	Myelin (when fixed in Mueller's potassium bichromate solution) Distinguishes normal from degenerated myelin	Studying development and extent of myelination, and for tracing normal myelinated fiber tracts in brain and spinal cord. Can be used also for large peripheral nerves and ganglia	Structures other than myelin remain unstained unless counterstain has been used
Degenerated Weigert	Stains only normal tracts. (Determine degenerated tracts by difference)	Study of degenerated fiber tracts in which some normal undegenerated tracts remain	Does not display degenerated tracts directly
Marchi	Droplets of oleic acid in degenerated tracts stain in tissue fixed in Müller's fluid	Detection of myelin sheaths in degenerating axons	Applicability decreases with increasing age of lesion

for use with the light microscope (136–138), but can also be used at the ultrastructural level (139). The Golgi method has been widely used to identify cortical neurons, but is limited by the fact that it is only possible to trace the axons of these cells for a short distance.

Nissl Substance Basic dyes such as toluidine blue O, thionin, cresyl violet, and gallocyanine are used to monitor the condition of nerve cells since they stain Nissl bodies and nuclear chromatin–both sensitive indicators of the chromatolytic response of injured or dying neurons. However, not all injured neurons give a classic chromatolytic response and to date these stains have found most use with injured motoneurons, injured sensory neurons in the hypoglossal nucleus and those in the red nucleus although the magnitude of the response is species dependent. The original procedures of Nissl (140, 141), employing alkaline methylene blue, have fallen into disuse because newer methods (see Sheehan and Hrapchak (142)) are easier to handle.

Nerve Fibers and Nerve Terminals Reduced methylene blue (143) may be used after perfusion in vivo or in vitro, to demonstrate peripheral nerves and their endings (Figure 16), particularly visceral and sensory nerves. Nerve fibers may also be demonstrated by a number of staining techniques, the best known and most widely used being those of Ungewitter (130), Bielschowsky (149), Holmes (134), Bodian (144), Romanes (145), Nauta and Gygax (146), and De Myer (147). For technical details and further references see Ralis et al. (148).

Neurofibrils Neurofibrils may be demonstrated in the perikaryon and within the axon and dendrites of normal neurons by Bielschowsky's (149) silver impregnation method and also by Cajal's methods (286). Von Braunmuhl's (150) method is used extensively to demonstrate senile plaques and neurofibrillary tangles in pathological tissues.

Myelin: the Weigert or Pal-Weigert Method (130) Brain or spinal cord that has been treated for several weeks with a solution containing potassium bichromate (Müller's fluid) acquires a special affinity for hematoxylin such that the myelin sheaths become deep blue in color when stained according to Pal's (151) modification of Weigert's procedure (152). Similar results may be obtained more rapidly with Weil's method (153, 154) in which tissues are fixed in formalin and stained with iron-hematoxylin.

Axons, nerve cells, and all other tissue elements remain colorless unless the preparation has been counterstained. The method has been widely used to study the development and extent of myelination and for tracing myelinated fiber tracts.

Degenerating Nerve Fibers These fibers may be differentiated from normal nerve fibers by their greater avidity for ammoniacal silver solutions. The various methods (Eager's (155); Hjorth-Simonsen's (156) modification of Fink-Heimer's (157) method; Guillery, Shirra, and Webster's (158) are all based on the silver impregnation methods of Nauta and Gygax (146, 160). The Glees' silver method (159) stains the neurofibrillar hypertrophy that occurs after nerve lesion, and the accumulation of neurofibrils that occurs within terminals of axons of the type which exhibits this phenomenon. This hypertrophy has been postulated to be a

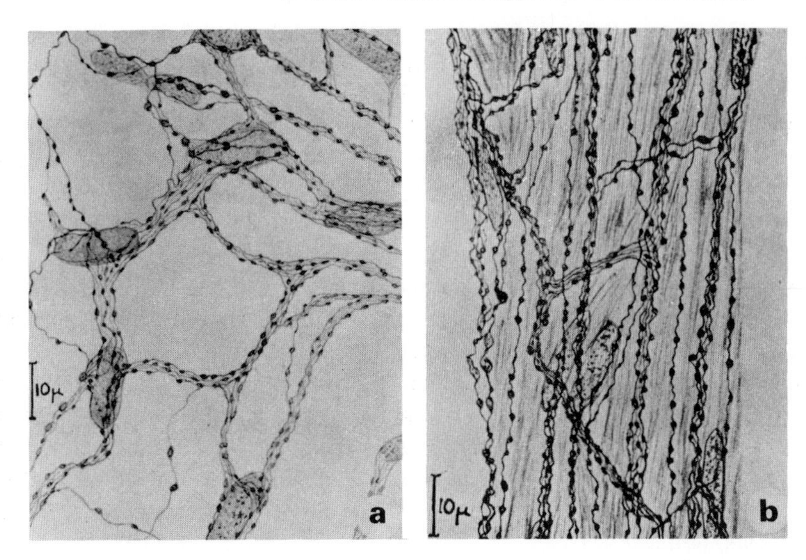

Figure 16. Methylene blue staining of peripheral autonomic nerve terminals. *a,* organization of sympathetic nerve terminals in the autonomic ground plexus of the rat iris with stained Schwann cell nuclei. (From Hillarp (333), by courtesy of *Acta Physiologica Scandinavica,* Stockholm.) *b,* nerve ground plexus in smooth muscle of a small trabecula in the urinary bladder of the frog. (From Hillarp (334), by courtesy of S. Karger, S.A. Basel.)

type of regenerative phenomenon, similar to the hypertrophy in the developing nervous system (161).

The process of demyelination involves the breakdown of compound lipids to simple lipids which are eventually phagocytosed by microglia. This forms the basis for a differential stain for degenerating fibers (the Marchi method) in which the tissue is fixed in a solution containing potassium bichromate (Müller's fluid). Degenerating myelin contains oleic acid, which is not oxidized by chromium salts and will therefore be blackened by osmium tetroxide after treatment with chromium salts, whereas normal myelin will not reduce osmium tetroxide with chromium salts. The result is that normal myelinated fibers appear light yellow and degenerated fibers appear as rows of black dots that can be traced in serial sections. Degenerated myelin can be demonstrated by the Marchi method between 7 and 10 days and 2, 4, or 12 months after damage to the nerve.

Degenerated fiber tracts may also be studied by the Pal-Weigert (151, 152) technique. In preparations in which the normal fiber tracts are well stained, the degenerated tracts remain colorless. Although the Marchi method is still the best for demonstrating degenerated myelin (within a certain time period), frequent artifacts produced by this method present difficulties in interpretation. While axons and axon terminals of cortical neurons have been analyzed successfully with anterograde silver methods or the Marchi technique, a limitation of both techniques is that the cell bodies from which these axons and terminals are derived cannot be demonstrated.

Degeneration following Axotomy in Young Animals (Gudden Technique) (162) When peripheral vertebrate axons are injured by sectioning or crushing, characteristic morphological changes occur in and around their cell bodies. Within a few hours or days of injury a complex series of changes (commonly referred to as "the retrograde response," or perhaps more appropriately as the "axon reaction") occurs in the neuronal cell bodies. This results in a swelling of the cell bodies, in nuclear eccentricity, in nucleolar enlargement, and in chromatolysis (163). Chromatolysis refers to the accompanying fragmentation and redistribution of the granular endoplasmic reticulum, observed histologically as a dissolution of basophilic Nissl bodies, and to the increase in cell volume, as reflected by an *apparent* reduction in cytoplasmic basophilia. Chromatolysis is therefore a stress reaction and is characteristic of more advanced stages of neuronal development. Lavelle (164) has suggested that the increasing ability of neurons with age to survive axotomy is related to the cytological maturation of the nucleolar apparatus and rough endoplasmic reticulum, the cellular organelles most responsible for protein synthesis and therefore with the neuron's capacity for regeneration.

It is now recognized that, in peripheral neurons at least, the perikaryal reaction is associated with a *regenerative* response of the cell to the axonal injury, directed towards the reconstitution of lost axoplasm and re-establishment of interrupted peripheral connections. In biochemical terms, this response involves an increased metabolism of RNA and proteins (165–168) and these regenerative responses to axotomy appear to be more vigorous in young animals than in old. Related to this is the suggestion (169) that there are two microsomal systems in developing brain: one, synthesizing *structural* proteins for cell growth, makes use of the unattached ribosomes that predominate in immature neurons (170); the other, which eventually takes over from the first, is thought to utilize bound ribosomes (that increase in number with age) to synthesize proteins for specific functions concerned with cell maintenance.

Neurons that are injured close to their cell bodies display a more rapid onset of perikaryal responses, and this probably reflects the shorter distance to be covered, and hence the earlier arrival in the cell body of the signal generated by the injury. Cragg (171) has calculated, from Watson's (167) microchemical data, that the signal ascends at a rate of 4–5 mm/day which is very much slower than the rate at which acetylcholinesterase (38), or exogeneous proteins such as horseradish peroxidase (115) taken up by the axon terminals, is transported to the perikaryon.

In contrast to the *regenerative* chromatolytic response seen in more mature neurons, axotomy in fetal, new-born, and very young animals causes a massive *degeneration* resulting in nucleolar loss and cell death. This nucleolar loss is greatest when the injury is inflicted during the early phases of nucleolar formation and segregation. Not surprisingly, the use of young animals for research into cell localization has become an important neuro-anatomical technique (162). In a series of now classic experiments involving selected cortical ablations in young animals, Gudden (162) mapped the projection of a number of thalamic subsystems upon the cortex. Since the Gudden technique takes advantage of the sensitivity of young neurons to axotomy, its applicability to neuronal tracing is consequently limited by the fact that this sensitivity gradually decreases with age; however, this technique has been particularly useful for localizing neurons whose perikaryal responses to axotomy in the adult are difficult to detect histologically. The brains of experimental animals are usually examined within a week or two of placing the lesion, since by this time most of the affected neurons have already disappeared (172). In this way, Grant (173) has recently mapped a variety of nuclei in the young kitten brain and has applied Gudden's method to the study of the terminal distribution of the central axons of primary *sensory* neurons. Unfortunately, the general applicability of this method to trace other pathways in the CNS is limited by the fact that central neurons do not react uniformly to lesions of their axons. While some show a series of histological changes very like those seen in peripheral neurons, others react by a very rapid degeneration, or respond with a progressive atrophy, while a few show minimal or no detectable change. Lieberman (163) has recently reviewed this work and discussed some of the cellular and molecular mechanisms that have been proposed to account for the observed histological changes following axotomy. A few systems which have been traced successfully by this approach are summarized in Table 2.

Dye Injection Methods These methods have become popular in recent years for identifying single neurons that have been impaled by a microelectrode. Dyes such as fast green F.C.F. (174) and large molecular weight, fluorescent, procion dyes (175) may be injected directly into cells either by pressure from a microsyringe or iontophoretically by passing a current through a dye-filled glass microelectrode.

Procion dyes are usually confined to the cell injected: if 15–96 hours are left after the injection, the macromolecular dye is found to have been carried to the terminals by axonal transport, enabling the entire neuron to be visualized by fluorescence microscopy.

Demonstration of Specific Transmitters

Histochemical Visualization of Catecholamine-containing Neurons The mapping of adrenergic pathways in the peripheral and central nervous system was made possible by the introduction of a sensitive fluorescence histochemical method. This technique, introduced some 15 years ago by Falck and Hillarp, permitted direct visualization of catecholamine-containing neurons (176).

Formaldehyde Fluorescence Technique: Histochemical Basis and Some Recent Modifications The Falck-Hillarp technique is based on the transformation of the endogenous catecholamines (β-phenylethylamines) into intensely fluorescent isoquinolines through combined cyclization-dehydrogenation reactions with gaseous formaldehyde (Figure 17). The chemical mechanisms underlying the histochemical method are now well understood and have recently been reviewed (177, 178). The *primary* catecholamines, noradrenaline and dopamine, can be differentiated from one another after treatment with HCl which reacts

Table 2. Central neuronal systems mapped by neuronal degeneration following axotomy (Gudden's method)

System	Age of animal	Days after lesion	% cells degenerated	References
Mice and rabbits:	Adult	8–16	30–35	172
cerebellar lesions	8–11 days	8	100	
Rabbits:	3 weeks	8	>90	326
neurons of the inferior olivary nucleus		11–12		
Mouse:	>1 week	1–133	few	327
spinal motor neurons to hind limb musculature				
Axonal lesions	Newborn	1–133	90–100	
Guinea pig:	Prenatal	–	100	328
spinal cord	Birth		90	
neurons	Postnatal		50–90	
spinal cord	Adult		30–50	
transection	Senile		few	
Feline:	1–11	7	50	173
hypoglossal	>20	7	none	
nucleus	Adult	7	none	
hypoglossal nerve resection				
Hamster:	Adult	4–30	none	164
facial nucleus	>20	1	none	
facial nerve	In utero	4–30	100	
section	Postnatal		100	
	7–15		80	
Hamster:	Adult	4–30	extensive	329
facial nerve	24 days		granular ER disruption	
as above	15		slight	
Rabbits:	Newborn		breakdown of	330
avulsion of facial nerve	(6–9)		granular ER aggregates, etc.	
transection facial nerve	Adult	16–30	progressive atrophy gradual depletion granular ER	331

with a labile hydroxyl group in position 4 of noradrenaline and so reduces the intensity of its fluorophore. *Secondary* catecholamines, such as adrenaline, form fluorescent isoquinolines at a much slower rate than the primary catecholamines, and can be distinguished from them on the basis of the longer incubation time required to obtain equivalent fluorescence. Indolamines, such as 5-hydroxytryptamine and other β-(3-indolyl) ethylamines, also produce fluorescent derivatives

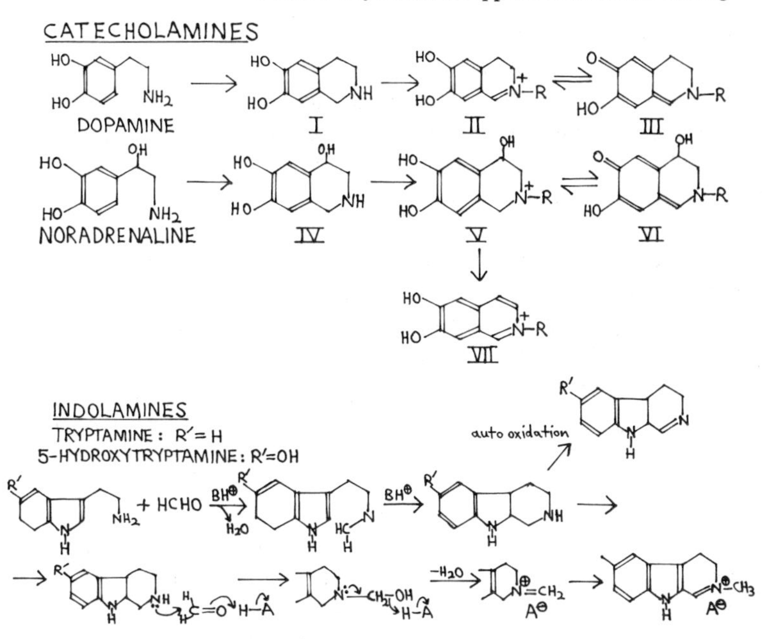

Figure 17. Sequence of reactions between catecholamines and indolamines with formaldehyde. The first step is a so-called Pictet-Spengler cyclization reaction. This results in the formation of 1,2,3,4-tetrahydroisoquinolines and 1,2,3,4-tetrahydro-β-carboline molecules which have in general either no or only very weak visible fluorescence. The fluorophore formation occurs in a second step where the tetrahydro derivatives are dehydrogenated to 3,4-dihydro compounds. Two alternate forms of the dehydrated isoquinolines are possible: either the 6,7-dihydroxy-3,4-dihydro derivatives (R=H), or the 6,7-dihydroxy-3,4-dihydro-2-methyl derivatives (R=CH_3). The second reaction pathway accounts for the observations that the fluorophore formation from the initially formed tetrahydroisoquinolines and tetrahydro-β-carbolines requires formaldehyde and is catalyzed by proteins or amino acids, and suggests that the 2-methyl-3,4-dihydro molecules might be the predominant fluorescent species in the Falck-Hillarp reaction. The figure also illustrates that these products are in a pH-dependent equilibrium through the 6-hydroxy group with their highly fluorescent tautomeric quinone structures (III and VI) and are capable of forming fully aromatic 6,7-dihydroisoquinolines (VII) through splitting-off of the 4-hydroxy group. (Modified from Björklund and Falck (178.)

(β-carbolines), but these are distinguished from the catecholamines by their greater emission maximum (525 nm; yellow) when compared to that of catecholamines (emission maximum 480 nm) which appear blue when short-wave-pass-interference filters are used or green to yellow-green when the conventional colored glass filters are employed.

The method is sensitive and reproducible and involves minimal expense; however, it is limited by the time taken to process freeze-dried paraffin-embedded tissues, by the diffusion of formaldehyde-catecholamine products away from the reaction site (especially when applied to cryostat sections), and by the low amounts of endogenous catecholamines present in preterminal axons. Two modifications of the Falck-Hillarp technique which go some way towards overcoming these problems have recently received attention from neurohistologists.

First has been the use of aqueous solutions of formaldehyde as an alternative to the more complex formaldehyde condensation technique. Following incubation (179, 180) or perfusion (181) with buffered formaldehyde, tissues are frozen, sectioned on a cryostat, and the pattern of adrenergic innervation observed by fluorescence microscopy after heating the specimens to 80°C. By eliminating the need for freeze-drying, studies of the effects of drugs or other procedures can be carried out in a single day on large tissue blocks that are free from the cracking artifacts inherent in the earlier procedures. To date, these techniques have not been successfully applied to tracing adrenergic neurons in the central nervous system.

Another modification of the Falck-Hillarp technique that avoids freeze-drying but still permits excellent visualization of both peripheral and central catecholamines makes use of a vibrating microtome knife (Vibratome) to section frozen tissue blocks at 0–5°C so that air-dried sections of unfixed or formalin-fixed tissue can be observed after reaction with formaldehyde vapors at 80°C (182). This technique is simple and has the added advantage that other staining techniques (e.g., silver impregnation, autoradiography, immunofluorescence, and retrograde horseradish peroxidase tracing) can be carried out on the same or on serial sections (183).

Further developments of this method have resulted in a combined formaldehyde ozone reaction (184, 185) and an acid-catalyzed formaldehyde reaction (186, 187) for the histochemical demonstration of small amounts of tryptamine and 3-methoxylated β-phenylethylamines in tissues.

In addition to the formaldehyde methods, fluorescence histochemical techniques have been developed for the demonstration of adrenaline by oxidation with gaseous iodine to form fluorescent trihydroxyindole derivatives (188), and of histamine by reaction with *o*-phtalaldehyde (189). The first technique has the advantage over the Falck-Hillarp formaldehyde fluorescence method in that it permits an improved differentiation of noradrenaline (emission in the blue spectrum) from adrenaline (emission in the green spectrum) and gives no visible fluorescent products with dopamine.

However, both the trihydroxyindole and *o*-phthalaldehyde techniques have a lower sensitivity than the formaldehyde techniques and have not been particularly successful for the demonstration of the lower concentrations of adrenaline and histamine present in axons and cell bodies of mammalian nervous systems.

Although the small amounts of transmitter found in preterminal axons have made the direct tracing of adrenergic fiber tracts difficult, the mapping of these pathways has been aided by the observation that catecholamine fluorescence accumulates central to a nerve ligation (190, 191). This observation has also stimulated interest in neuronal dynamics in adrenergic neurons since it implies that catecholamines are transported from the perikarya to the axon terminals by axonal flow.

Axonal Flow and Transmitter Histochemistry: Tools for Mapping Peripheral Noradrenergic Pathways Noradrenaline-containing vesicles found in adrenergic nerve terminals appear to be synthesized in the perikarya of sympathetic neurons, where they are assembled in the region of the Golgi apparatus, and then supplied to the axon and terminals by rapid axonal flow (4, 192) (Figure 18). The evidence for rapid axonal transport of noradrenergic vesicles rests on the observation that dense-core vesicles accumulate central to a ligature

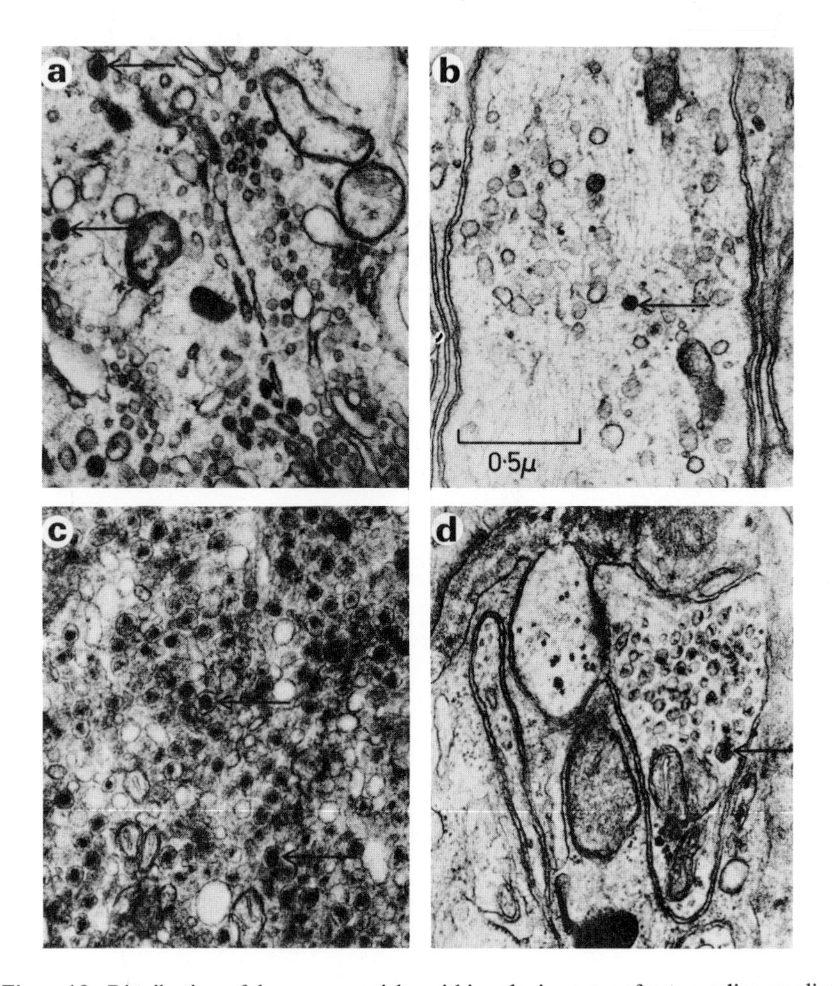

Figure 18. Distribution of dense-core vesicles within splenic nerves of cat. *a,* celiac ganglion, region of Golgi apparatus of perikaryon; *b,* splenic nerve trunk, unmyelinated axon; *c,* ligated splenic nerve, portion of swollen axon proximal to ligature; *d,* axon varicosity in spleen smooth muscle. Note that the small dense-core vesicles are confined to axon terminals and that the large dense-core vesicles (*arrows*) are distributed throughout the neuron and accumulate above a ligature as does norepinephrine. × 34,300. (Courtesy of Anna Ostberg; from Geffen and Livett (192) by courtesy of the American Physiological Society.)

placed on sympathetic nerves (193, 194), that noradrenaline levels central to a ligature increase linearly over a period of 48 hours (70), and that dopamine β-hydroxylase, a constituent of the vesicles, accumulates at a rate similar to noradrenaline (195–198).

Peripheral Adrenergic Pathways Axonal flow in adrenergic neurons was first studied in *peripheral* neurons of which two kinds–"long" and "short"– could be demonstrated by the formaldehyde fluorescence technique (Figure 19). The "long" adrenergic neurons have their cell bodies located in paravertebral and prevertebral sympathetic ganglia and fibers from these ganglia innervate smooth muscles and glands in many peripheral tissues. For example, ganglion cells in the celiac and lower thoracic ganglia give rise to the "long" adrenergic axons that comprise the splenic, and sciatic and splanchnic, nerves, respectively. These peripheral adrenergic systems have been studied extensively because of their accessibility and suitability for studies on axonal flow (Figure 20).

Another system, that consists of "short" adrenergic neurons with cell bodies located more peripherally in ganglia close to or within the pelvic organs of the urogenital tract, provides a dense innervation to the accessory male genital organs, the urethra, parts of the female reproductive tract, and to the trigonum area of the urinary bladder. This system is not so amenable to axonal flow studies although some evidence for axonal flow in "short" adrenergic neurons innervating the vas deferens has been obtained by fluorescence histochemical techniques following vasectomy (199).

Within certain experimental limits the fluorescence intensity observed in the microscope or recorded by microspectrofluorimetry is proportional to the amine content of the neurons, and, by combining the results from the Falck-Hillarp histochemical method with independent chemical estimation of tissue noradrenaline levels, much quantitative data have been obtained relating to the morphology and transmitter distribution in "long" and "short" adrenergic neurons.

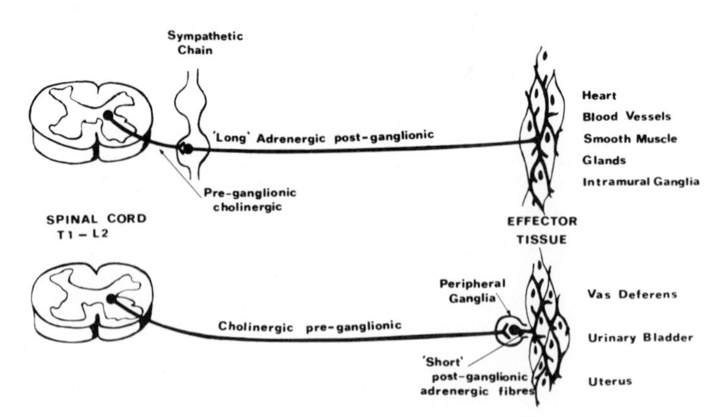

Figure 19. Organization of the peripheral sympathetic innervation showing the two systems of adrenergic neurons: "long" and "short." (From Livett (203), by courtesy of the British Council.)

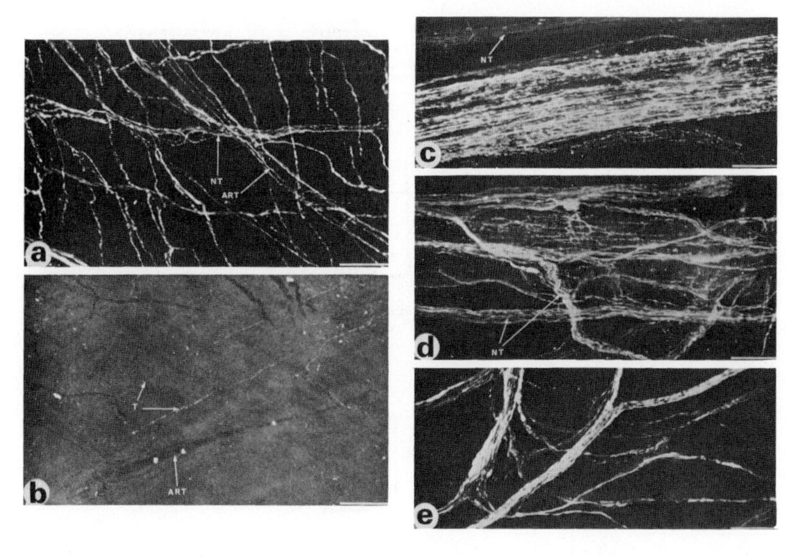

Figure 20. Formaldehyde fluorescence histochemical demonstration of noradrenergic nerves: effects of vinblastine. Fluorescence histochemistry of noradrenergic nerves in chicks (*a*) before and (*b*) after intravenous injections of vinblastine. This drug causes an irreversible blockade of axoplasmic transport of catecholamines in non-terminal axons, and also appears to have a direct toxic action on noradrenergic nerve terminals. *a,* oblique septum from an animal injected with saline. Note the loose-meshed plexus of non-terminal axon bundles (*NT*), and the occasional, varicose terminal fibers. Running across the picture is the perivascular plexus of an arteriole (*ART*). Stretch preparation. Calibration 100 μm. *b,* oblique septum from an animal injected 48 hours previously with vinblastine. A few, scattered terminal fibers (*T*) are seen, but are only faintly fluorescent. The nonspecific fluorescence of the background is more obvious than in the control tissues, and the blood vessels (*ART*) appear darker, due to the presence of erythrocytes within them. Note that the perivascular plexuses are not detectable. Stretch preparation. Calibration 100 μm. *c,* posterior mesenteric artery from an animal injected with saline. The bright fluorescent terminal fibers are aligned with the longitudinal axis of the vessel, and tend to obscure the loose perivascular plexus of non-terminal axon bundles. Some faintly fluorescent non-terminal axon bundles (*NT*) are seen accompanying the vessel. Stretch preparation. Calibration 100 μm. *d,* posterior mesenteric artery from an animal injected with vinblastine 48 hours previously. Note that the majority of terminal fibers are no longer detectable, although, in localized regions, some remain. The non-terminal axon bundles (*NT*) are now readily seen, due to the loss of terminal fibers and to the marked increase in fluorescence intensity of the axon bundles. Stretch preparation. Calibration 100 μm. *e,* posterior mesenteric artery from an animal injected 24 hours previously with 6-hydroxydopamine (100 mg/kg). Note the loss of terminal fibers and the increase in fluorescence intensity of the non-terminal axon bundles; compare with *d.* Stretch preparation. Calibration 100 μm. (From Bennett et al. (335), by courtesy of Springer-Verlag, Berlin.)

Several features of noradrenaline storage in "long" adrenergic neurons are depicted in Figure 21. First is the large separation between the cell body and its terminals, which may be up to 10^5 cell body diameters away. Note also the unequal distribution of transmitter—the terminals containing up to 300 times the amount of noradrenaline present in the cell body, in up to 1,000 times greater concentration. This pattern of transmitter distribution is also found in *central* noradrenergic and dopaminergic neurons (200, 201). The preterminal

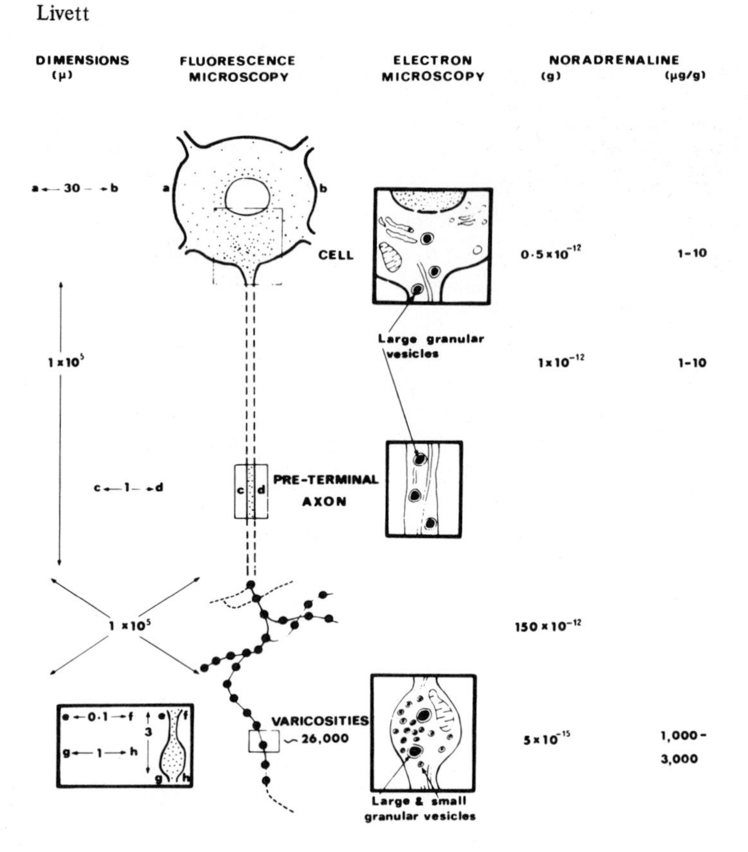

Figure 21. Schematic representation of a "long" adrenergic neuron, showing morphological and quantitative features of noradrenaline (NA) storage. (From Livett (203), by courtesy of the British Council.)

axons contain only low concentrations (1–10 μg/g) of noradrenaline. However, upon sectioning the preterminal axons or interfering with axonal flow by local compression of the axon, by injury or by local application of colchicine or vinblastine (Figure 20), the amounts of noradrenaline in the axons immediately central to the site of injury increase rapidly and linearly for the first 24 hours and reach concentrations high enough to be observed by the Falck-Hillarp fluorescence histochemical technique.

By observing the time course of accumulation of both endogenous (70) and radioactive (4, 71) noradrenaline in segments of ligated and unligated nerves at different time intervals after local application or injection of labeled noradrenaline and amino acids into sympathetic ganglia and by observing the accumulation of noradrenergic vesicles central to a nerve ligation by electron microscopy (202) (Figure 18) and the rate of recovery of noradrenaline stores after depletion with drugs such as reserpine and tetrabenazine that interfere with vesicle storage, it has been calculated that noradrenaline-containing vesicles are transported from the perikaryon to the terminals at a rate of about 5–10 mm/hour. (For reviews see Refs. 4, 192, and 203.)

Contributions of the Cell Body and Terminals to Maintenance of Transmitter Levels in Peripheral Adrenergic Tissues Synthesis of noradrenaline in ganglion cell bodies and its rapid axonal transport to the terminals make, however, little quantitative contribution to noradrenaline stores in the terminals. Whereas, noradrenaline in axon terminals is synthesized at rates of 3–30% of the terminal stores *per hour* (see Geffen and Livett (192)), the amounts of noradrenaline accumulating above a ligature constitute less than 3% of the stores *per day*. Thus, when the nerves to half the cat spleen were ligated for 24 hours, it was found that its noradrenaline content was unchanged compared with the intact portion, and that the amounts of noradrenaline which accumulated above the ligature in a day constituted less than 1% of that present in the nerve terminals (204). The major role of anterograde transport of noradrenergic vesicles appears to be the supply of vesicle proteins (such as the enzyme dopamine β-hydroxylase) required for the local synthesis and storage of noradrenaline in the terminals.

Nervous activity does not appear to be necessary for axonal transport of noradrenaline since decentralization does not affect its rate of transport in cat splenic nerves (204) and transport continues at a normal rate in hypogastric nerves isolated in vitro (69). Although nervous activity is not necessary for axonal transport, recent studies (43) indicate that intense nervous activity may increase the amount of noradrenergic vesicles supplied to adrenergic axons. The rate of spontaneous activity in adrenergic nerves rarely exceeds 10 Hz (205, 206) but if the lumbar sympathetic outflow from the spinal cord of the rat is stimulated continuously at 2 Hz or intermittently at 10 Hz, the rate at which noradrenaline and dopamine β-hydroxylase accumulate central to a ligature on the sciatic nerve is increased by approximately 40%. Adrenergic vesicles within sympathetic axons are not fully loaded with noradrenaline (207), so one explanation of the increased accumulation of noradrenaline proximal to the ligature could be that the intense nerve stimulation increased their loading with noradrenaline. However, the finding that stimulation increased the rates of accumulation of dopamine β-hydroxylase to an extent similar to that of noradrenaline suggests that the mechanism by which this was achieved involved the transport of an increased number of normally filled vesicles.

It is not known whether this is due to the vesicles being transported at a faster rate or to an increased number of vesicles per unit cross section of axon being transported at the same speed. Evidence from studies on axonal flow in other systems (see section "Effect of Nerve Activity") supports the latter interpretation.

A further question is whether *retrograde* transport of noradrenaline and other catecholamines occurs. The evidence for and against retrograde transport of noradrenaline has been reviewed by Dahlström (4) and by Geffen (208), the present view being that intravesicular retrograde transport of noradrenaline does not occur to any significant extent in unligated axons, and that observations of retrograde accumulation of noradrenaline at nerve ligatures reflect local injury phenomena in the axons rather than the interruption of a normal retrograde transport of transmitter.

As discussed above, decentralization has little or no effect on *anterograde* transport of noradrenaline but preganglionic nerve stimulation increases the quantity of noradrenergic vesicles transported to the terminals. In this way the transmitter economy of adrenergic nerve terminals can be modulated to keep pace with changes in sympathetic nerve activity arising from environmental and physiological factors such as exposure to the cold, exercise, and the circadian rhythm. The rate of noradrenaline synthesis may increase up to 5-fold following nerve stimulation and after exposure to drugs such as reserpine, phenoxybenzamine, and 6-hydroxydopamine that affect the economy of noradrenaline storage at axon terminals.

Both short- and long-term regulatory processes are involved in catecholamine synthesis in the peripheral and central adrenergic nervous system. Immediate control of synthesis at the terminals is mediated by rapid end-product inhibition of tyrosine hydroxylase, probably acting through a small regulatory pool of extravesicular stores (209). Long-term adaptation to prolonged increased sympathetic discharge is achieved by enzyme induction which results in increased amounts of tyrosine hydroxylase, dopamine β-hydroxylase (and possibly other enzymes and associated cofactors) being synthesized in the perikaryon (209–211). The synthesis of tubulin is also increased by increased nerve activity (51) so that an increased number of microtubules is probably made available for the rapid transport of transmitter-synthesizing enzymes to the terminals.

Central Adrenergic Pathways By placement of selective lesions in the brains of animals and observation of the accumulation of fluorescence attributable to noradrenaline or dopamine, a chemical mapping of central adrenergic pathways has been achieved. In many cases the observations made with the Falck-Hillarp technique have been supplemented with biochemical analyses of catecholamines in discrete regions of the brain (201). The results obtained with stereotaxic electrocoagulation and microblade lesions have been compared with other techniques in which stereotaxic injection of drugs such as 6-hydroxydopamine (212) has been used to produce selective lesions in the central noradrenaline and dopamine pathways without damaging other neuron systems (213).

Axonal flow has been used to trace the movement of protein (56, 214–216) and neurotransmitters (4, 45, 217) within nigrostriatal and other central nervous system tracts (218). The specificity of axonal transport for a given transmitter type has not been extensively investigated but, at least in the nigrostriatal tract, axonal transport of small molecules seems reasonably specific, being confined to the true transmitter (dopamine) and to other closely related compounds (e.g., noradrenaline and octopamine) (215).

The results of these different approaches are in accord with one another and are responsible for much of the detailed information we have about mapping of the central adrenergic pathways. In recent years, several improved techniques have contributed further to our knowledge of the detailed circuitry of central adrenergic pathways. A brief description of these new techniques is given here to assist the reader in interpreting the results from these methods illustrated in the accompanying figures.

The Glyoxylic Acid Method of Lindvall et al. (219) Because of the need for improved histochemical methods to study biogenic amines (such as tyramine, octopamine, and histamine) known to be present in mammalian tissues (220), but which could not be visualized with sufficient sensitivity by the existing formaldehyde fluorescence methods, Axelsson and colleagues (221) investigated the usefulness of a number of reactive carbomyl compounds (aldehydes, ketones, carboxylic acids, and γ-keto acids) as fluorescence histochemical detection reagents, and found that condensation with gaseous glyoxylic acid provided the most sensitive method for demonstration of the well known biogenic catecholamines and indolamines. Compared with the Falck-Hillarp formaldehyde fluorescence method, the glyoxylic acid method displays a considerably higher capacity for forming fluorophores with biogenic amines and catecholamines, and has the further advantage that it enables the histochemical detection of small amounts of *N*-acetylated indolamines (e.g., melatonin), methoxylated catecholamines (e.g., 3-methoxytyramine), and peptides with tryptophan in the NH_2-terminal or COOH-terminal position that cannot be demonstrated by the formaldehyde fluorescence technique.

Because of the increased sensitivity which allows fluorescence observations of the entire axon including the non-terminal portions to be made, the glyoxylic acid method has been particularly useful for demonstrating central noradrenaline and dopamine neuron systems (222). As glyoxylic acid does not readily penetrate tissue blocks the method gives best results when applied to Vibratome sections of fresh or glyoxylic acid perfused brain tissue (219). With this combination of techniques the terminal systems in certain brain regions have been shown to be more extensive than those demonstrated previously with the formaldehyde technique, and, due to the absence of diffusion of the glyoxylic reaction products, several delicate systems of central dopamine-containing pathways have now been traced throughout their entire course.

Immunohistochemical Procedures Additional information about the distribution and extent of noradrenaline-, dopamine-, and adrenaline-containing pathways in the peripheral and central nervous systems has come through the application of sensitive and specific immunohistochemical techniques with antisera raised against purified transmitter synthesizing enzymes. In this technique a specific antiserum to the enzyme is incubated with cryostat tissue sections and its site of reaction made visible either by direct coupling to a chemical tracer, such as horseradish peroxidase (341), or fluorescein isothiocyanate (FITC), or by a further incubation, this time with peroxidase or FITC-labeled antisera to γ-globulin from the species in which the first antiserum was raised. As with all immunological assays the most important component is the antiserum as this determines the sensitivity and specificity of the reaction. A recent review (208) discusses the applications and limitations of the immunohistochemical approach.

An immunofluorescence histochemical approach was first used on peripheral adrenergic neurons to study the localization and axonal transport of the noradrenaline-synthesizing enzyme, dopamine β-hydroxylase, in splenic nerves of the sheep (223). Monospecific antisera have now been raised to all six enzymes involved in the synthesis of the catecholamines (tyrosine hydroxylase, aromatic amino acid decarboxylase, dopamine β-hydroxylase, *p*-*N*-methyltransferase, monoamine oxidase, and catechol-*o*-methyltransferase), and the immunohistochemical technique has been used to determine the distribution of dopamine β-hydroxylase in peripheral and central adrenergic neurons (81, 224–228) and of tyrosine hydroxylase (229), aromatic L-amino acid decarboxylase (230, 231), and phenylethanolamine *N*-methyltransferase (227, 231, 232) in central adrenergic neurons.

Central Noradrenergic Pathways For the main part the central catecholamine pathways follow the course of non-adrenergic fiber tracts whose pathways are well known from classic neuroanatomy. It has been found that a relatively small number of cell body groups in the pons and medulla give rise to two main descending and two main ascending fiber tracts (Figure 22).

Descending Noradrenaline Pathways A descending bulbo-spinal pathway arises from the most caudal group of cell bodies (A1) and sends fibers into the ventral and dorsal horns of the spinal cord, while another more rostral pathway arises from the largest single group of noradrenergic cell bodies, the locus coeruleus (A6), and innervates the lower brainstem.

Ascending Pathways Rostrally projecting noradrenergic fibers from seven groups of cell bodies in the mesencephalic, pontine, and medullary regions give

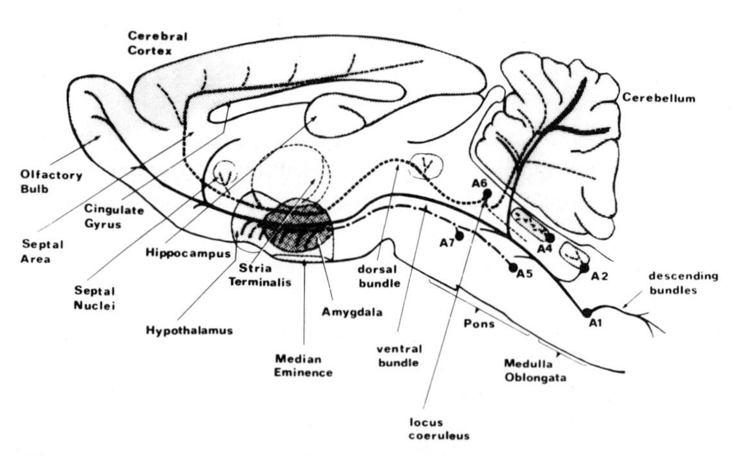

Figure 22. Sagittal representation of the rat brain, showing the principal ascending and descending NA pathways as displayed by the formaldehyde-fluorescence and immunohistochemical methods. Cell bodies in the locus coeruleus (A6) give rise to pathways (– – –) innervating all cortical areas of the brain. Cell group A1 (—) gives rise to a descending bulbospinal pathway. This ventral bundle, which comprises ascending fibers from cell groups A5 and A7 (·–·) as well as from A1 and A2, gives off branches to the lateral mammillary nuclei, the lateral and ventral hypothalamus and to large parts of the limbic forebrain. Another pathway arises in the A4 cell group and sends axons towards the A6 group. *Shaded areas* indicate regions of NA terminals. (From Livett (203), by courtesy of the British Council.)

rise to four major conduction pathways that innervate many regions of the brain.

1. *The central tegmental tract* consists of a ventral bundle of ascending axons that originate in cell groups A1, A2, A5, and A7 in the pons and medulla. This fiber tract increases in size as it traverses rostrally and collects adrenergic axons from the A5 group, from the subcoeruleus cell group and from some neurons in the locus coeruleus cell group. This ventral bundle gives off branches to the lateral mammillary nuclei, the lateral and ventral hypothalamus (the latter providing terminals to the median eminence and infundibulum), and to large parts of the limbic forebrain (including the anterior medial amygdaloid complex, the ventral medial septum, and the cingulate gyrus). Other pontine and medullary fibers in the central tegmental tract contribute to the ascending aminergic fiber system in the medial forebrain bundle. In addition, some of the fibers comprising the central tegmental tract pass rostrally into the supraoptic decussations (Figure 23), through the zona incerta and internal capsule, to innervate areas of the neostriatum and overlying cortex.

2. *The dorsal tegmental tract,* also termed the dorsal bundle of Ungerstedt (57), forms a separate tract of ascending noradrenergic fibers and is the major projection from the locus coeruleus (Figure 22 and 24*a*). This fiber tract ascends with the medial forebrain bundle from its cell bodies (A6) located in the floor of the fourth ventricle and innervates areas of the thalamus, hippocampus, cerebral

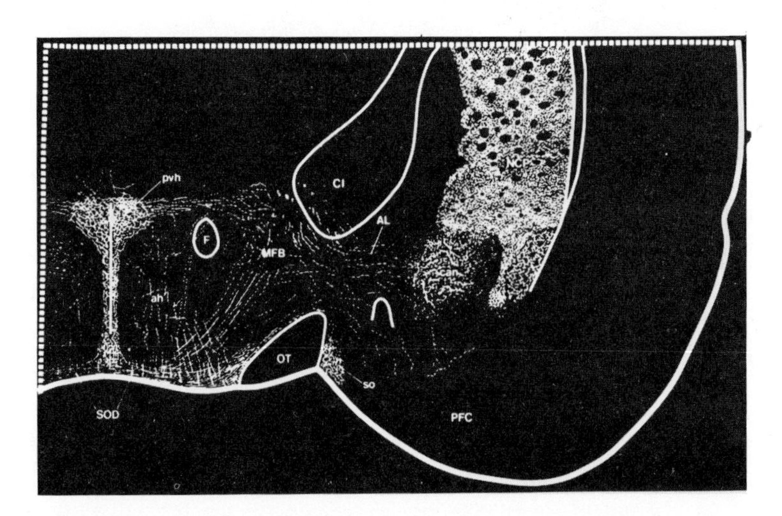

Figure 23. Semidiagrammatic representation of central catecholamine fiber systems displayed by the glyoxylic acid fluorescence technique. Catecholamine fiber systems running in the supraoptic decussations (*SOD*) and the ansa lenticularis (*AL*) as observed in a frontal section through the retrochiasmatic region. Composite drawing of slightly different frontal planes: *ah,* anterior hypothalamic area; *can,* central amygdaloid nucleus; *CI,* internal capsule; *F,* fornix; *MFB,* medial forebrain bundle; *NCP,* nucleus caudatus-putamen; *OT,* optic tract; *PFC,* piriform cortex; *pvh,* paraventricular hypothalamic nucleus; *so,* nucleus supraopticus. (From Lindvall and Björklund (222), by courtesy of *Acta Physiol. Scand.,* A.B. Grahns Boktryckeri, Stockholm.)

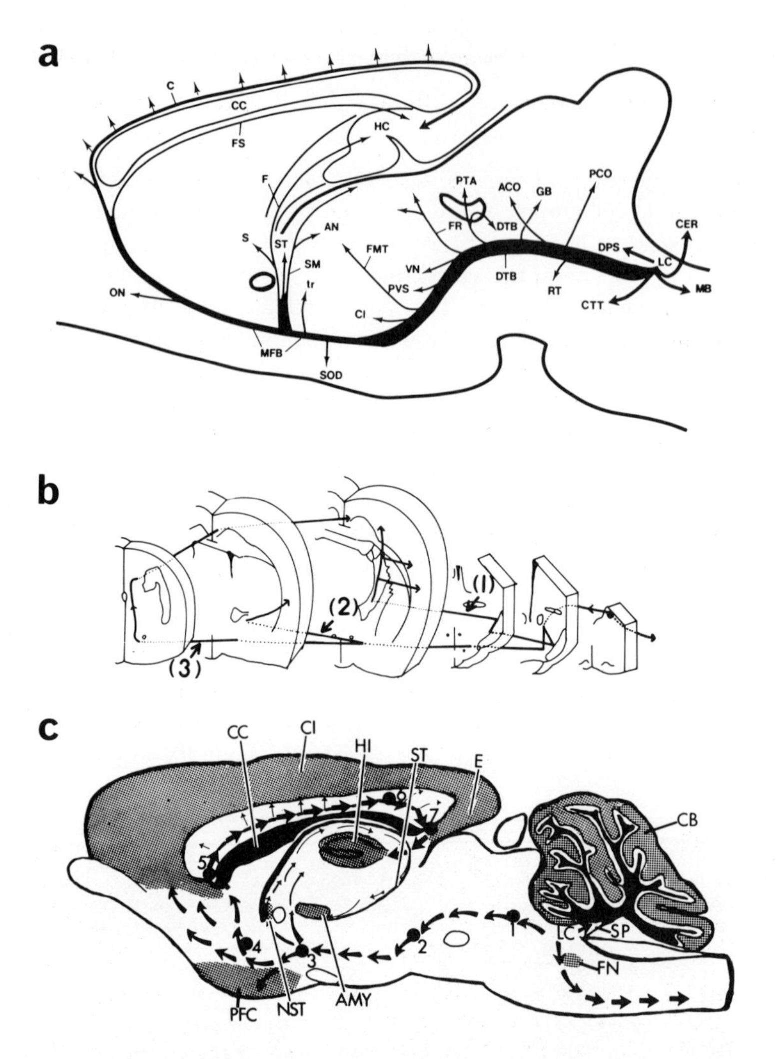

Figure 24. Systems of ascending catecholamine-containing pathways. *a,* schematic representation of the course and primary branchings of the dorsal tegmental bundle, originating in the catecholamine cell bodies of the principal locus coeruleus (*LC*), as demonstrated by the glyoxylic acid fluorescence method. For abbreviations, see Lindvall and Bjorklund (222). (From Lindvall and Bjorklund (222) by courtesy of *Acta Physiol. Scand.,* A.B. Grahns, Boktryckeri, Stockholm.) *b,* three ascending noradrenergic pathways from the locus coeruleus to the cerebral cortex. Schematic drawing of axons of the coerulocortical noradrenergic neurons of the rat as displayed by the formaldehyde-fluorescence method after administration of a large dose (300 mg/kg) of 6-hydroxydopa. The axons pass through 3 tracts to enter the cerebral cortex: (*1*) pathway through the internal capsule; (*2*) pathway through the external capsule; and (*3*) pathway through the anterior septal region. These axons give off the collaterals caudalward in the region just ventral to the locus coeruleus. (From Tohyama et al. (234), by courtesy of Elsevier, Amsterdam.) *c,* efferent pathways involved in the axoplasmic transport of [*3*H] proline from the locus coeruleus. Schematic representation on a sagittal plane of the main efferent pathways detected by radioautography following stereotaxic injections of [^{3}H] proline into the right locus coeruleus of rats. Note that the stria terminalis, amygdala, and pyriform cortex are lateral to the plane of the figure. *Arrows* indicate labeled fiber bundles running longitudinally. *Shaded regions* represent labeled terminal areas. For abbreviations, and radioautographs at positions 1–7, see Pickel et al. (218). (From Pickel et al. (218), by courtesy of the Wistar Institute Press.)

and cerebellar cortex. The glyoxylic acid method (Figure 24*a*) has shown that the medial forebrain bundle system is highly heterogeneous with respect to the chemical nature of its catecholamine fibers, and consists of both noradrenaline- and dopamine-containing fiber systems (222) (Figure 25). It has become evident as a result of this work that lesions of the so-called ventral bundle of Ungerstedt (57) in the region of the rostral mesencephalon, or the medial forebrain bundle, will affect several ascending catecholamine systems and will result in quite complex denervation patterns. More detailed information is required about the origin and fates of individual nerve fiber tracts.

One of the best studied groups of cell bodies with respect to the suborganization of its noradrenergic pathways is the locus coeruleus. By combination of a number of different techniques (fluorescence histochemistry (57, 233) (Figure 24*a*), controlled chemical lesions with 6-hydroxydopa (234) (Figure 24*b*), and radioautography following stereotaxic injections of [^{3}H]proline (218) (Figure 24*c*)), it has been possible to arrive at a detailed picture of the suborganization of efferent catecholamine pathways from the locus coeruleus. The results of fluorescence histochemical studies indicate that ascending noradrenaline axons originating from the locus coeruleus innervate the entire cerebral cortex. Cell bodies in the locus coeruleus (A6) also provide the major adrenergic innervation to the cerebellar cortex (57, 233). The dorsal ascending adrenergic bundle reaches the cerebral cortex by three separate routes: 1) by a pathway through the internal capsule, 2) by a pathway through the external capsule, and 3) by a pathway through the anterior septal region (Fig. 24*b*). This detailed mapping (234) was made possible by the observation that a single injection of a large dose of 6-hydroxydopa into rats induces a very early (3–6 hours) accumulation of noradrenaline throughout the entire course of the dorsal coerulo-cortical bundle, which precedes the pile-up of fluorescence resulting from retrograde axonal degeneration. These results extend the observations of others who employed methods involving electrocoagulation or longer exposure to 6-hydroxydopa and who had also noted pathways entering the cortex via the septum and cingulum (57, 235). Further confirmation that these pathways are taken by groups of axons entering the cerebral cortex comes from a recent autoradiographic study by Pickel et al. (236) who showed by axonal transport of radioactively labeled proline that, rostral to the anterior commissure, several labeled fiber bundles contributed to the innervation of the cerebral cortex, hippocampus, and amygdala (Figure 24*c*). These studies also clarified a controversy over the route by which peduncle and locus coeruleus fibers entered the cerebellum by confirming earlier fluorescence studies performed in conjunction with lesions of the cerebellar peduncles, in showing that these noradrenergic fibers entered the cerebellum via the *superior* peduncle. This study is an excellent example of the use of axoplasmic flow for tracing neuronal connections that have been difficult to visualize in their entirety by other histological or cytochemical methods, and the use of radioisotopes will no doubt find increasing applicability in mapping other areas of the central nervous system.

Recent improvements in immunohistochemical methodology (237) have permitted the tracing of very fine preterminal adrenergic axons containing

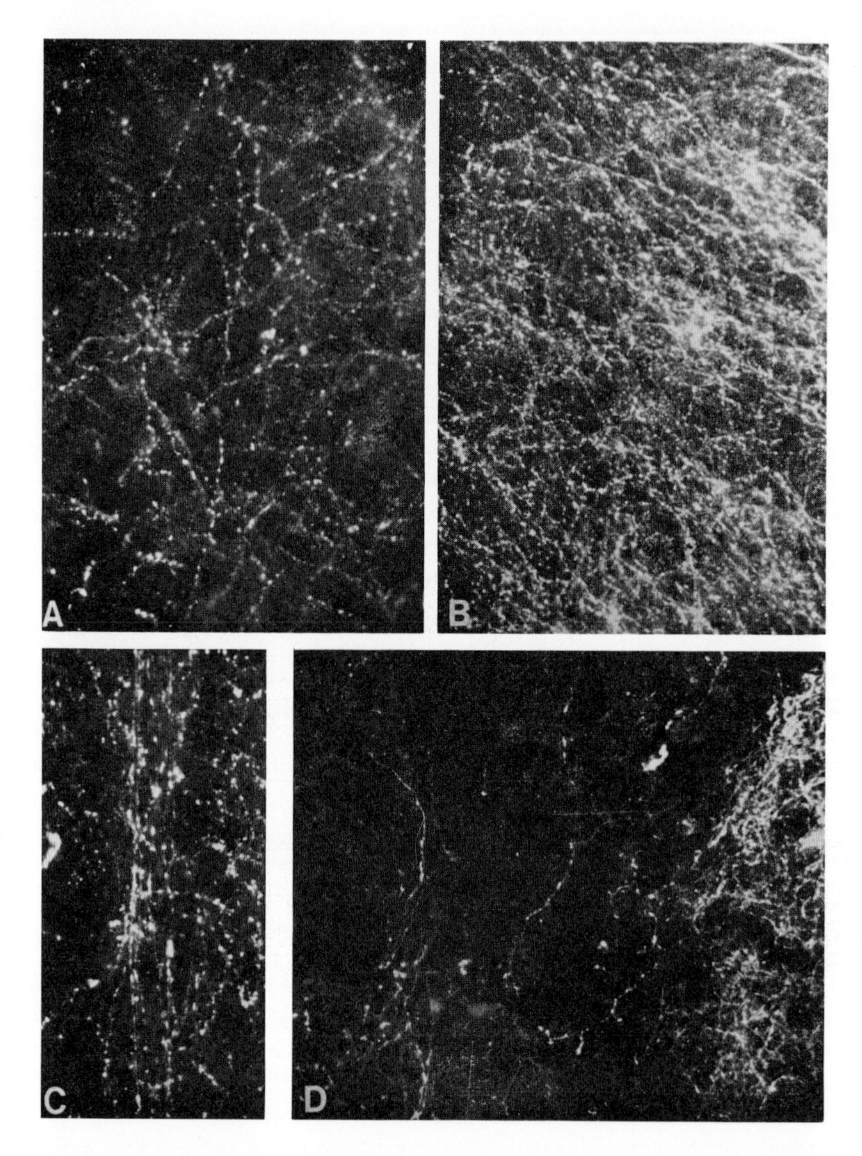

Figure 25. Glyoxylic acid-fluorescence demonstration of details of the two principal axon types innervating the neocortical areas. *A*, varicose terminals of the locus coeruleus, presumed noradrenaline-containing type in the molecular layer of the sensory-motor cortex (× 320). *B*, presumed dopamine-containing axon terminals in the deep layers of the frontal cortex (× 210). *C*, Non-terminal axons of the locus coeruleus type in the rostral part of the septum running towards the cingulum (× 280). *D*, Non-terminal presumed dopamine-containing axons running on the rostral side of the neostriatum (whose dense fluorescent terminal network is seen to the right in the picture) towards the frontal cortex (× 200). (From Lindvall and Björklund (222), by courtesy of *Acta Physiol. Scand.*, A.B. Grahns, Boktryckeri, Stockholm.)

dopamine β-hydroxylase in many cortical regions of the brain. The finding that many of these fibers terminate around small blood vessels within the brain parenchyma suggests that central noradrenergic neurons normally make both vascular and neuronal connections, and raises the possibility that central noradrenergic neurons may function to control cerebral microcirculation. Some highly vascular neuronal areas in the brain such as the supraoptic and paraventricular hypothalamic regions receive an especially dense adrenergic innervation (222, 228) (Figure 23), but it is not known whether the terminal processes belonging to a single neuron (with cell bodies in the locus coeruleus) make connections with both vascular and neuronal effector cells in these areas, or whether the neuronal and vascular effector cells are innervated by separate neurons. Some larger arteries in the brain are supplied by adrenergic fibers originating from the superior cervical ganglion, but as complete removal of this ganglion has little effect on cerebral blood flow and causes a complete disappearance only of the adrenergic terminals associated with pial blood vessels—dopamine β-hydroxylase-containing fibers associated with cerebral microcirculation remain unaffected (208)—it may be concluded that noradrenergic cell bodies in the locus coeruleus provide the major adrenergic innervation to vascular smooth muscle in the CNS.

Central noradrenergic neurons originating from the locus coeruleus also display regenerative sprouting and can be made to innervate smooth muscle explants introduced into the CNS (238). By analogy with the peripheral adrenergic nervous system in which a relatively small number of ganglia innervate blood vessels in a wide variety of tissues, the noradrenergic cell bodies in the locus coeruleus are found to innervate blood vessels in many different regions of the brain. This suggests that the locus coeruleus acts functionally as a central extension of the sympathetic chain (228).

3. *A periventricular system* of noradrenergic fibers has been traced by the glyoxylic method from its origins in cell bodies in the medulla oblongata, and from others located more rostrally along the periventricular and periaqueductal gray up to the rostral diencephalon and septal region (Figures 26 and 27). This fiber pathway is further subdivided into a dorsal and ventral system.

The *ventral periventricular system* can be traced from the rostral mesencephalon through the supramammillary region and posterior hypothalamic area into the medial part of dorsomedial hypothalamic nucleus where it receives fibers from the dorsal periventricular bundle and possibly also from the medial forebrain bundle to form a broad ascending system that innervates the medial and periventricular hypothalamic areas, and the caudal region of the septum.

The *dorsal periventricular system* is characterized by the fact that many of its cell bodies of origin are distributed diffusely along its extent through pons, mesencephalon and the posterior thalamus. These shorter adrenergic neurons project to tectal, pretectal, thalamic, epithalamic, hypothalamic, and possibly also septal regions. In addition, the dorsal periventricular system together with the dorsal part of the mesencephalic A10 cell group gives rise to a *paramedian fiber system* that is thought to form a major component of the mammillary

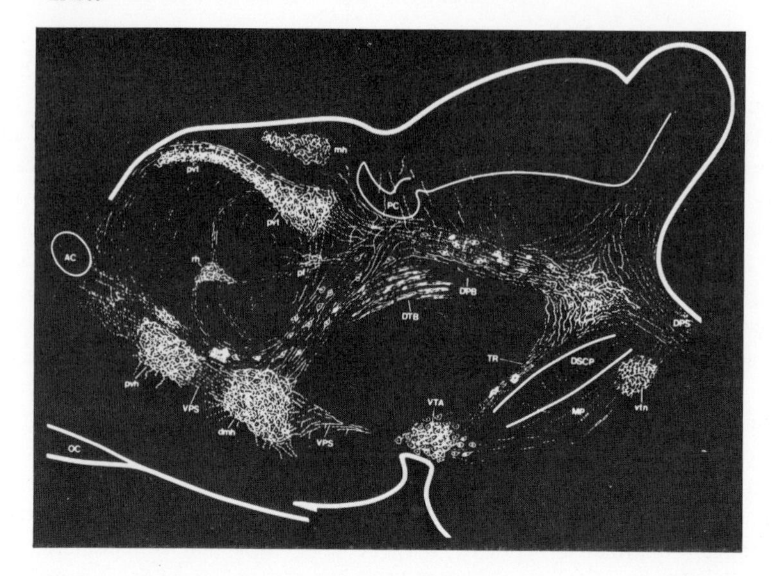

Figure 26. The ascending periventricular catecholamine fiber system. Semidiagrammatic representation of the periventricular catecholamine fiber system (rostral to the locus coeruleus), the medial fiber flow of the tegmental radiations (*TR*), and the fibers of the mammillary peduncle (*MP*). Composite drawing of somewhat different paramedian sagittal planes. For abbreviations, see Lindvall and Björklund (222); cf. Figure 27). (From Lindvall and Björklund (222), by courtesy of *Acta Physiol. Scand.*, A.B. Grahns, Boktryckeri, Stockholm.)

peduncle which has been shown with classic anterograde degeneration techniques to carry ascending fibers from the region of the dorsal and ventral tegmental nuclei to the mammillary nuclei.

The discovery that ascending adrenergic periventricular fiber systems together with the mammillary peduncle constitute a source of major connections from Nauta's "limbic midbrain area" to the hypothalamus is of great interest in view of the presumed roles of these systems in hypothalamic endocrine functions and of catecholamines in such central functions as food intake, motor activity, and rage reactions (239).

While the functional role of central noradrenergic neurons remains an area of intense investigation, the concept that central noradrenergic neurons innervate cerebral blood vessels and thereby control microcirculation in specific areas of the brain suggests that a re-evaluation of the roles of neuronal and vascular processes in neuroendocrine and behavioral phenomena mediated by catecholamines in the central nervous system may be valuable. Local changes in cerebral microcirculation mediated by catecholamines may have profound functional implications for the maintenance of homeostasis in the central nervous system and may also act as a modulatory mechanism by mediating longer term effects than those produced by point to point neuronal connections.

Immunohistochemical and glyoxylic acid procedures have contributed also to our knowledge of the distribution of dopamine and adrenaline pathways in the brain.

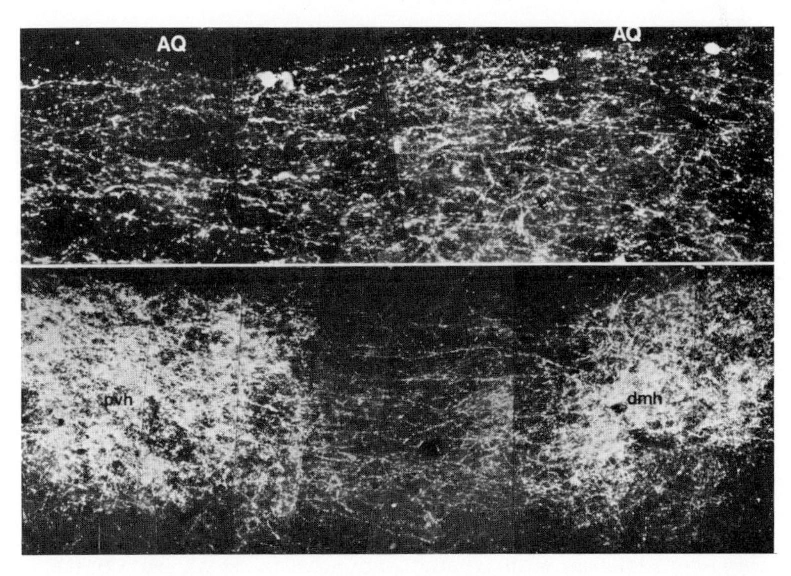

Figure 27. Photomontages of the dorsal and ventral periventricular CA fiber systems displayed by the glyoxylic acid fluorescence technique. The dorsal system (*above*) is shown in the horizontal plane, and the ventral system (*below*) in a paramedian sagittal section. Compare with Figure 26. The dorsal periventricular bundle is seen in the periaqueductal gray of the mesencephalon (*AQ,* aqueduct). Note the fluorescent cell bodies with a somewhat diffuse outline that project fibers rostrally (to the left) along the bundle (× 97). The ventral periventricular system is seen ascending broadly along the lateral aspect of the periventricular hypothalamic nucleus, between the dorsomedial (*dmh*) and the paraventricular (*pvh*) nuclei. These latter nuclei are heavily supplied with fluorescent CA terminal networks; cf. Figure 26 (× 51). (From Lindvall and Björklund (222), by courtesy of *Acta Physiol. Scand.,* A.B. Grahns, Boktryckeri, Stockholm.)

Central Dopamine Pathways Central dopamine pathways are less extensive than the noradrenergic pathways and both "long" and "short" dopamine-containing neurons are present in the brain (Figure 28). The nigrostriatal, mesolimbic, and mesocortical pathways represent systems of "long" dopamine neurons, while the "short" dopamine neurons are represented by the tubero-infundibular system. The application of formaldehyde (57), glyoxylic (222), and immunofluorescence (230, 231) techniques in conjunction with specific lesions has together contributed to the description and detailed mapping of these pathways (Figures 28 and 29). It was not possible to obtain this information earlier by the conventional staining and degeneration studies.

Nigrostriatal Pathway This pathway originates principally from two groups of neuronal cell bodies (A8 and A9) in the substantia nigra which send ascending fibers through the lateral and mid-hypothalamus to innervate the corpus neo-striatum.

Mesolimbic Pathway This pathway arises from a group of mesencephalic cell bodies (A10) surrounding the interpeduncular nucleus. Fibers from this group ascend through the medial forebrain bundle to innervate among others the nucleus accumbens, the olfactory tubercle, and the interstitial nucleus of the stria terminalis, all within the limbic forebrain.

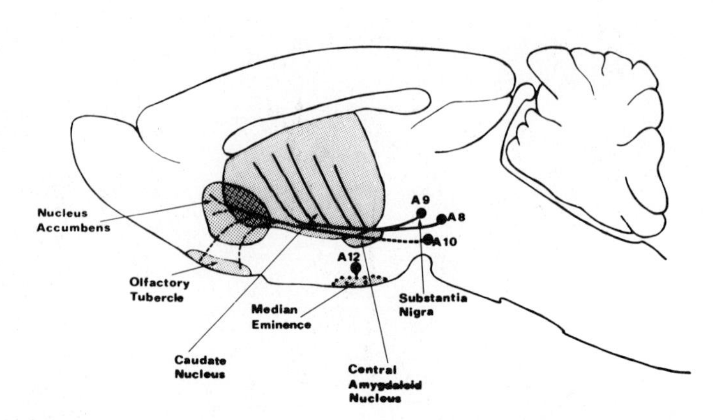

Figure 28. Sagittal representation of the principal dopamine (DA) pathways in the rat brain as displayed by the formaldehyde fluorescence technique. Shown are the "long" nigrostriatal fibers (—) originating in the substantia nigra (cell bodies A8 and A9), the "long" mesolimbic pathways (– – –) originating from DA cell bodies surrounding the interpeduncular nucleus (A10), and the "short" tubero-infundibular DA neurons (· · · ·) having their cell bodies in the arcuate nucleus (A12) in the hypothalamus. *Shaded areas* indicate regions supplied by DA terminals. (From Livett (203), by courtesy of the British Council.)

These two systems have recently been studied in detail in the hope that a better understanding of the neuronal mapping and interconnections may provide some clues to the clinical management of parkinsonism (in which the nigrostriatal system undergoes degeneration) (240) and might also help to explain the actions of a number of psychoactive drugs that have been shown to modify dopamine transmission in the central nervous system (241).

Mesocortical Pathway A third group of "long" dopamine-containing neurons innervate the cerebral cortex (242) and probably form part of the cortical system of presumed dopamine-containing fibers that fluoresce with the glyoxylic acid technique (222). According to lesion experiments, this newly discovered pathway to the frontal cortex originates in the A10 mesencephalic cell group (222). Unlike the noradrenergic neocortical innervation from the locus coeruleus, these mesocortical dopaminergic pathways are more restricted in their areas of projection in the neocortex, and their terminal density is correspondingly higher. This difference in the density of innervation may explain the more intense staining given by terminal regions of dopaminergic neurons than noradrenergic neurons when the peroxidase labeled antisera technique is used for mapping (236). The discovery of extensive telencephalic projections of these mesencephalic dopamine neurons opens new possibilities for possible influences of the brain stem on neocortical functions.

Tubero-infundibular System This system sends short dopaminergic axons from the arcuate nucleus (A12) in the hypothalamus into the external layer of the median eminence where they terminate in close proximity to portal blood

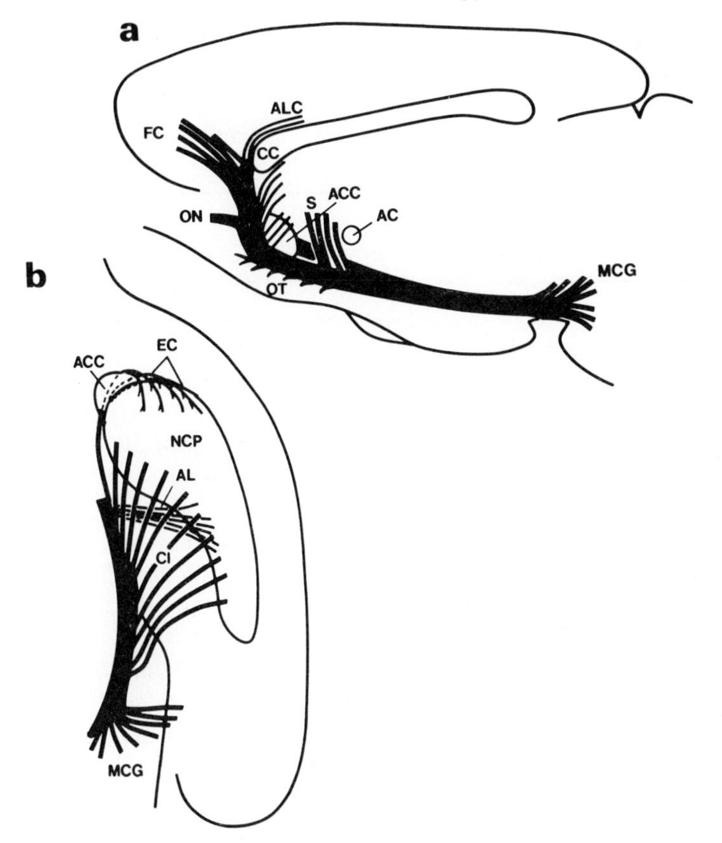

Figure 29. Extent of dopamine neuron systems in the brain as displayed by the glyoxylic acid fluorescence technique. *a,* schematic representation of the presumed dopaminergic mesolimbic and mesocortical systems in a sagittal projection. *b,* schematic representation in a horizontal plane of the presumed dopaminergic projections from the mesencephalic CA cell groups (*MCG*) to the nucleus caudatus-putamen (*NCP*). For abbreviations see Lindvall and Björklund (222). (From Lindvall and Björklund (222), by courtesy of *Acta Physiol. Scand.,* A.B. Grahns, Boktryckeri, Stockholm.)

vessels and to nerve endings that are thought to contain anterior pituitary releasing factors. Noradrenaline also appears to be involved in the regulation of pituitary function since the external layer of the median eminence receives additional innervation from a tuberohypophysial noradrenergic pathway that is derived from a branch of the ascending ventral noradrenergic bundle originating in the cell groups A1, A2, A5, and A7 in the pons and medulla. This pathway goes on to innervate the neuro-intermediate lobe of the pituitary (243). Interest in these pathways has increased recently because of the role of neurosecretory mechanisms in control of pituitary function and the known involvement of catecholamines in control of the pituitary-adrenal axis.

Central Adrenaline Pathways These have recently been mapped (232) by immunofluorescence techniques with antibodies against bovine adrenal phenylethanolamine-*N*-methyltransferase (PNMT), the enzyme responsible for converting noradrenaline to adrenaline. Since dopaminergic and noradrenergic neurons do not fluoresce with the PNMT antiserum, the assumption that is made is that PNMT-containing neurons represent adrenaline-containing neurons. In this way, adrenaline neurons have been localized in two groups of reticular nerve cell bodies (C1 and C2) in the medulla oblongata (Figures 30 and 31) whose neurons appear to have similar morphological characteristics to central noradrenaline neurons in that they send long ascending and descending fibers to the brainstem and spinal cord. These adrenaline-containing neurons project mainly to visceral afferent and efferent nuclei of the lower brainstem, to the locus coeruleus, and to certain nuclei in the hypothalamus and in the periventricular gray matter (Figure 30). In addition, PNMT positive axon bundles are observed in the reticular formation of the pons-medulla oblongata.

These immunohistochemical observations on the localization of PNMT to special reticular neurons provide the first morphological evidence that adrenaline neurons exist in the mammalian central nervous system. Their functional importance is not known but their distribution suggests that they could be important cellular participants in the control of such widespread physiological functions as food and water intake, body temperature, oxytocin and gonadotrophin secretion, sleep, wakefulness, respiration, and blood pressure.

GABA Neurons in the CNS Since the first report in the 1950s (244) that γ-aminobutyric acid (GABA) is present in large concentrations in the vertebrate central nervous system, this molecule has been studied extensively with a view to its postulated role as an inhibitory neurotransmitter. It now seems certain that it fulfils this role at the crustacean neuromuscular junction though the evidence for GABA being the major inhibitory transmitter in vertebrate nervous systems has been more difficult to obtain.

Autoradiographic Mapping As is the case for catecholamines, so too for GABA there appears to be a rapid uptake process that removes it from its extracellular site of action, thereby terminating its physiological effects (245). Autoradiographic studies in the central nervous system indicate that extracellular GABA is accumulated by certain types of neurons and also to a lesser extent by certain glial cells, whereas in the peripheral nervous system such rapid accumulation of exogenous GABA takes place only into glial cells (246).

So far as functional relationships between chemically characterized cell types are concerned, the cerebellum has been studied most fully. Autoradiography has been used to localize [^{3}H]GABA after stereotaxic injection into the cerebral and cerebellar cortices of rats which had been pretreated with amino-oxyacetic acid, a drug which prevents GABA breakdown (247). The isotope was found to be highly concentrated in stellate and basket cells in the cerebellar cortex and in stellate cells in the cerebral cortex (Figure 32, *a* and *b*). The Golgi cells also appeared to accumulate [^{3}H]GABA but the pyramidal cells showed only low activity. Surprisingly, Purkinje cells, which are the strongest candidates for a "GABA-neuron," showed only a low reaction; this apparent contradiction may be explained in part by differences between the intact brain and tissue slices in

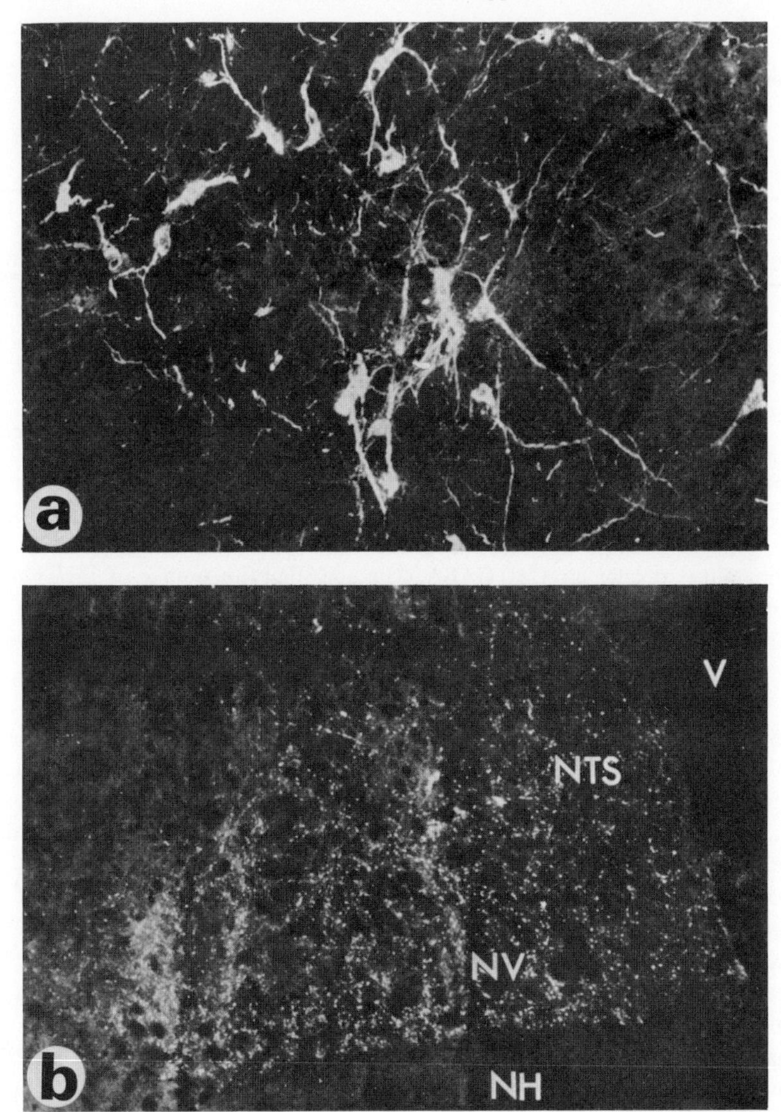

Figure 30. Immunofluorescence demonstration of presumed adrenaline neurons in the brain. *a,* immunofluorescence micrograph of the ventro-lateral reticular formation of the rostral medulla oblongata. Specific PNMT fluorescence is observed in oval and spindle-shaped nerve cell bodies and their processes (group C1). Note the extensive dendritic network. Magnification × 140. *b,* immunofluorescence micrograph of the nucleus dorsalis motorius nervi vagi (*NV*) and nucleus tractus solitarii (*NTS*). PNMT fluorescence is observed in nerve terminals of these nuclei. The density is highest in NV but there is a total lack of specific fluorescence in nucleus hypoglossus (*NH*). *V,* fourth ventricle. Magnification × 140. (From Hökfelt et al. (232) by courtesy of Elsevier, Amsterdam.)

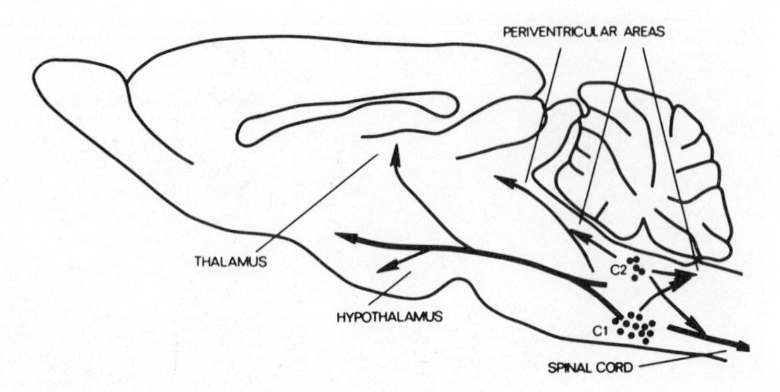

Figure 31. Schematic representation of adrenaline neuron pathways in the rat brain. The diagram shows the two PNMT positive cell groups (C1 and C2) and their hypothetical adrenergic projections. Both ascending and descending axons are seen giving rise to axon terminals in the brainstem and spinal cord. (From Hökfelt et al. (232) by courtesy of Elsevier, Amsterdam.)

their ability to take up [^{3}H]GABA. The specificity of uptake of [^{3}H]GABA into tissue slices and into tissues after injection in vivo, and its subsequent axonal transport (248), have consequently been the subject of some controversy (249, 250). Recently several alternative experimental systems, including the use of cultures of mammalian central nervous system tissue (246), have been investigated in which the problems of tissue damage and non-uniform labeling of tissues are minimized. Preliminary results with quantitative electron microscopic autoradiography of [^{3}H]GABA uptake into cultured cerebellum indicate that GABA neuron cell bodies, processes, and terminals accumulate label to the same extent. In 28-day cultures (obtained from dissociated cerebellar cells from 2-day-old rats), accumulation of GABA in glial cells accounted for only one-tenth of the total neuronal accumulation.

Immunohistochemical Mapping The morphological identity of GABA synapses has also been in dispute (246). Biochemical subfractionation studies combined with electron microscopy have shown that the enzyme L-glutamate decarboxylase (L-glutamate 1-carboxylyase; EC 4.1.1.15; GAD) that catalyzes the α-decarboxylation of L-glutamate to form GABA and CO_2 is particularly concentrated in the fraction enriched in nerve ending particles (synaptosomes). Eugene Roberts and colleagues (251, 252) have now purified the enzyme and prepared cross-species reactive antisera to it that have been used with immunoperoxidase techniques to localize the enzyme in rat cerebellum. The question as to whether asymmetrical or symmetrical synapses are characteristic of GABA neurons now appears to have been resolved by immunohistochemical observations with peroxidase-labeled anti-GAD, which indicate that GABA synapses may undergo morphological transformation during development. Two types of GAD-positive terminals were found: *rounded* terminals, resembling those described previously for Purkinje cell recurrent collaterals, which make synaptic contact with either the smooth surfaces of Purkinje cell bodies or their periso-

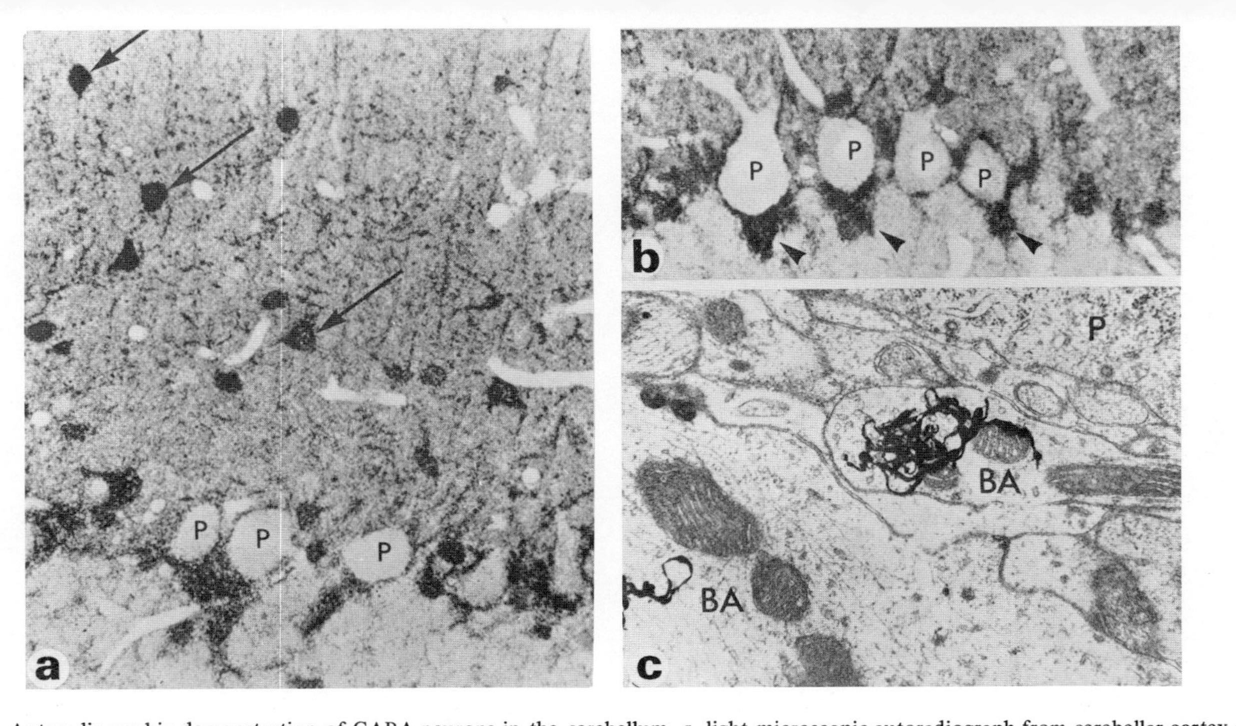

Figure 32. Autoradiographic demonstration of GABA-neurons in the cerebellum. *a,* light microscopic autoradiograph from cerebellar cortex after injection of [^{3}H]GABA. Many cell bodies (*arrows*) with a high activity are seen in the molecular layer. Note also dot and fiber-like accumulations. The Purkinje cells (*P*) have a low activity. Single cell bodies accumulating the isotope are seen in the granular layer. Magnification × 330. *b,* light microscopic autoradiograph from the Purkinje cell layer of the cerebellar cortex after injection of [^{3}H]GABA. A zone of high activity surrounds the Purkinje cell bodies (*P*) and this zone is expecially pronounced at the basal part (*arrow heads*) of the Purkinje cell. Magnification × 330. *c,* electron microscope autoradiograph from the Purkinje cell layer of the cerebellum after injection of [^{3}H]GABA. Grains are seen over nerve ending of basket cell axons (*BA*). *P,* peripheral part of a Purkinje cell body. Magnification × 22,500. (From Hökfelt and Ljungdahl (247), by courtesy of Springer-Verlag, Berlin.)

matic spines (Figure 33*a*), and *elongated* terminals (Figure 33, *b* and *c*), that resemble basket cell endings in the adult, and make synaptic contacts on the Purkinje somata but not on perisomatic spines. Both *rounded* and *elongated* GAD-positive terminals make symmetrical synaptic contacts and usually contain flattened or pleomorphic vesicles in contrast to the adjacent assymetrical GAD-negative climbing fiber endings that contain clear, rounded vesicles.

The histochemical distribution of glutamic acid decarboxylase, which is sparse in the Purkinje cell bodies but concentrated in discrete sites around the Purkinje cells and around the neurons of the deep cerebellar nuclei, is consistent with the large body of indirect biochemical, physiological, and morphological data that suggests an inhibitory synaptic role for GABA in neurons in the cerebellum. The evidence is consistent with the proposal that GABA is important in mediating the inhibitory actions of stellate and basket cells upon the Purkinje cells and of the Golgi type-II cells upon granule-cell dendrites, and between Purkinje cells and cells in the deep cerebellar nuclei. While these results are encouraging in that they support the role of GABA as a putative inhibitory transmitter in the cerebellum, further proof is required, and is presently being sought at two levels.

First, it is necessary to show that the histochemical reaction product can be localized to boutons of likely GABA neurons at the electron microscope level and that experimental manipulations known to modulate GABA levels produce a morphological change in these terminals consistent with a transmitter role for GABA.

Second, in accordance with the view that Purkinje cells use GABA as their neurotransmitter, it would be desirable to show that GABA and GAD are transported from the neuronal perikarya to the nerve terminals. McGeer, Hattori, and McGeer (253) have recently shown that anterograde axoplasmic transport of [^{3}H]GABA occurs from cell bodies in the cerebellar cortex to nerve endings in the deep cerebellar nuclei. In this study they injected [^{3}H]GABA into the cerebellar cortex of rats, and processed the tissue for autoradiography some time later. The results from both light and electron microscopic localization of [^{3}H]GABA suggested that four of the five cell types in the cerebellum (Purkinje cells, stellate cells, basket cells, and Golgi cells) may use GABA as their neurotransmitter—only the excitatory granule cells being non-GABA-containing neurons. These results are in good agreement with the immunohistochemical findings with anti-GAD sera. However, before the evidence can be considered conclusive much more needs to be learned about the specificity of uptake and transport of GABA in GABA- and non-GABA-containing neurons.

It may not be too optimistic to predict that the combination of the autoradiographic and immunohistochemical techniques will soon provide answers to many of the questions still remaining about the functional role of GABA in the cerebellum and also provide a more detailed chemical mapping of GABA neurons in the cerebellum and in other areas of the central nervous system.

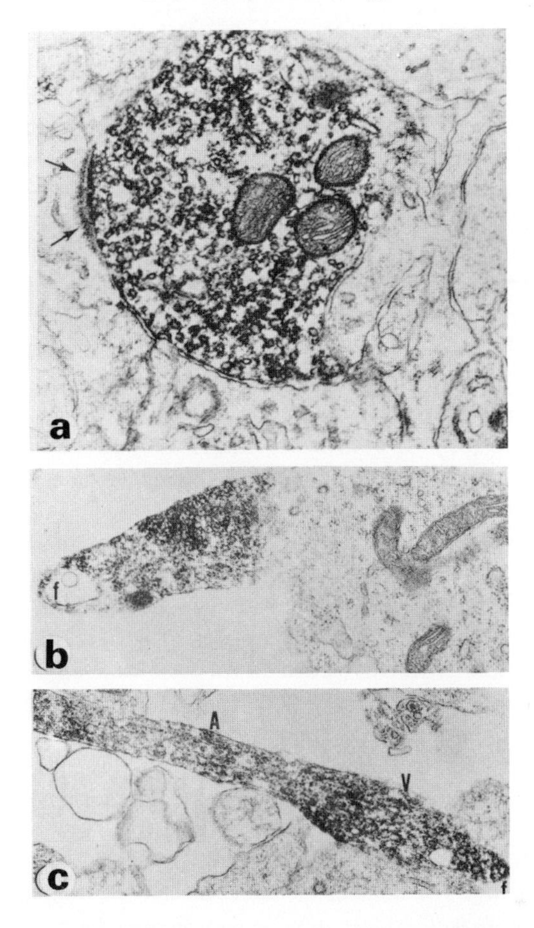

Figure 33. Immunohistochemical demonstration of presumed GABA terminals in the rat cerebellum. *a*, mature rounded GAD-containing synaptic terminals in the Purkinje cell layer, which synapse (*arrows*) on relatively immature dendritic profiles in the vicinity. Reaction product for GAD is associated with synaptic vesicle, mitochondrial and presynaptic junctional membranes. Other synaptic terminals in the field do not contain reaction product. 14 days postnatal × 24,160. *b*, GAD-containing process or foliopodium (*f*) extending from what appears to be a neuronal cell body. Reaction product is associated with small vesicles. 3 days postnatal. Purkinje cell layer × 16,020. *c*, long slender GAD-containing profile in which a more differentiated axonal region (*A*) expands into a varicose (*V*) enlargement. The leading tip of the varicosity may be a foliopodium (*f*). GAD-positive reaction product fills the entire profile and appears to be associated with small vesicles. 3 days postnatal. Internal granular layer × 20,460. (From McLaughlin et al. (340), by courtesy of Elsevier, Amsterdam.)

Neurosecretory Pathways Until recently the tracing of neurosecretory pathways relied on the classic light microscopic techniques for demonstration of the sulfur-rich neurosecretory material. These methods (chrome alum hematoxylin, aldehyde fuschsin, the various basophilic dyes, and pseudoisocyanin techniques that were used for demonstrating protein-bound cysteine and cystine) were found to lack specificity in that they reacted with the peptide hormones (e.g., oxytocin and vasopressin) but also equally well with their carrier proteins (neurophysins), and to some extent with a variety of other tissue components (254, 255). Nevertheless, these techniques and others such as Golgi staining (256), intracellular injection of procion yellow (257, 258) or cobalt chloride (259), and dark field illumination of stained (260, 261) or unstained living material (262) have been useful in identifying neurosecretory neurons in a wide range of species.

Retrograde transport of horseradish peroxidase has recently been used to trace neurosecretory neurons in the rat after injecting the tracer into the pituitary (263). After 2–3 days very little reaction product was visible in the neurosecretory axons, but their cells of origin, the neurons of the supraoptic nuclei, the magnocellular parts of the paraventricular nuclei, and the various accessory neurosecretory hypothalamic nuclei showed accumulation of the tracer within lysosome-like dense bodies (0.4–0.6 μm diameter) in the perikarya.

More sensitive and specific immunochemical techniques (264–267) (Figure 34) have now allowed the identification of the hormones (268) and their specific carrier proteins within separate neurosecretory neurons (269) (Figure 35, *a* and *b*) in the hypothalamus of mammalian species and, when combined with axonal ligation (270) (Figure 36), for the axonal transport of neurophysin to be studied. The increased sensitivity provided by these immunohistochemical techniques has permitted the visualization of neurophysins in neurosecretory cell bodies, axons and terminals extending throughout the hypothalamo-neurohypophysial system. Use of these techniques has led to the discovery (271) of a new neurophysin-containing neurosecretory pathway arising in the magnocellular hypothalamic nuclei and terminating around the portal circulation in the external infundibular zone of the median eminence stalk (265–266, 272–274) (Figure 35, *c* and *d*). The number of neurophysin-containing fibers detectable in this region is markedly altered by adrenalectomy and administration of adrenal steroids and by treatment with colchicine (276, 277).

Although the functional significance of this pathway is not known, it has been suggested (272, 275) that this pathway may be involved in control of adenohypophyseal function through modulation of the portal system blood flow and membrane permeability by vasopressin. Vasopressin is known to potentiate the release of other anterior hormone releasing factors and is associated with the rapid release of adrenocorticotropin under conditions of stress. At the present time much active research is directed towards determining whether neurophysins play a role in effecting any of these extraneuronal regulatory mechanisms.

Tracing of Neuronal Pathways with Exogenously Administered Tracers

In recent years, axonal transport has been exploited as a neuroanatomical method using a variety of tracers carried retrogradely and anterogradely. These include the use of radioactive amino acids for tracing the anterograde movement of *endogenous* transmitters (248) and proteins (214, 218, 278) and the administration of *exogenous* proteins such as horseradish peroxidase, to trace antero-

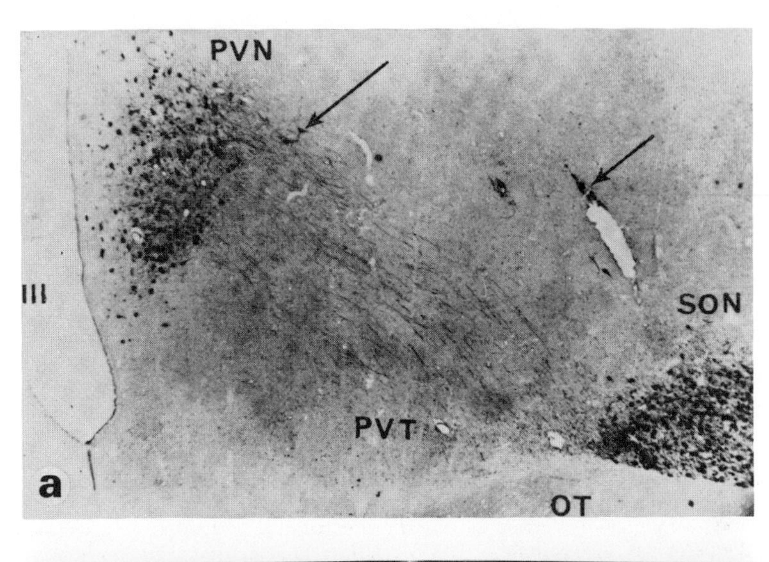

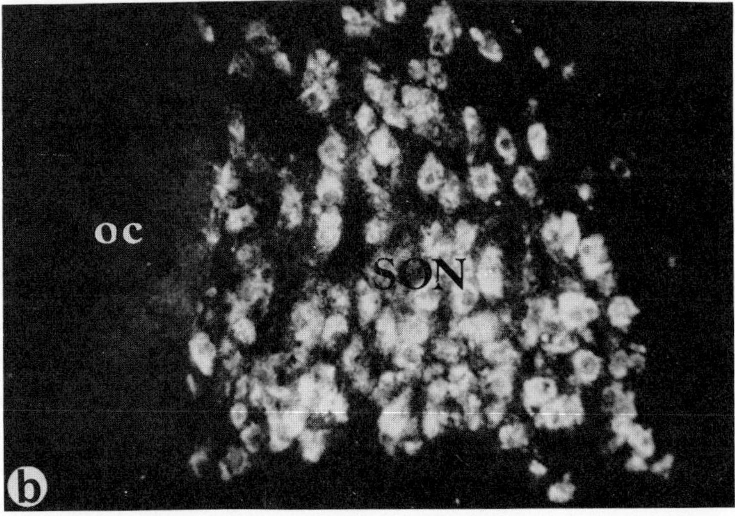

Figure 34. Immunohistochemical demonstration of neurosecretory neurons in the hypothalamus. *a,* transverse section of a rhesus monkey hypothalamus treated by the three-layer immunoperoxidase bridge method for the demonstration of neurophysin using antibody to bovine neurophysin I. Neurophysin is present in the paraventricular nucleus (*PVN*) and supraoptic nucleus (*SON*), concentrated near blood vessels (*arrows*) and in neurons of the paraventricular tract (*PVT*), optic tract (*OT*), and third ventricle (*III*). × 31. (From Zimmerman et al. (336) by courtesy of J.B. Lippincott Co., USA.) *b,* immunofluorescence demonstration of neurophysin in the perikarya of neurons in the rat supraoptic nucleus. The optic chiasma is on the left. (From Livett (264), by courtesy of the New York Academy of Sciences.)

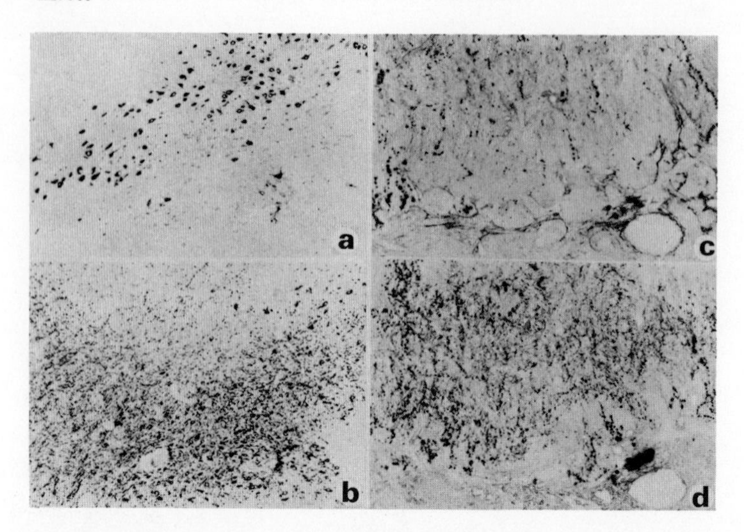

Figure 35. Immunohistochemical demonstration of neurophysin I and neurophysin II in separate neurosecretory neurons in the hypothalamus and median eminence. *a* and *b*, adjacent parasagittal sections of the supraoptic nucleus. Up, dorsal; down, ventral; apical is on the right. In *a* and *b* compare the dorsal region with the ventral region. *a*, section stained with the peroxidase-anti-peroxidase (PAP) bridge technique, using rabbit anti-bovine neurophysin I serum neutralized with neurophysin II. *b*, adjacent section stained with the PAP method, using rabbit anti-bovine neurophysin II serum neutralized with neurophysin I. In *a* note the selectively stained neurophysin I cells; in *b* the selectively stained neurophysin II cells. × 29. (From de Mey et al. (269), by courtesy of Springer-Verlag, Berlin.) *c* and *d*, adjacent serial sections of the distal part of the external region of the bovine median eminence. In *c* note the selectively stained neurophysin I fibers; in *d* the selectively stained neurophysin II fibers. PAP method using differentially adsorbed rabbit anti-bovine neurophysin I serum (*c*) or differentially adsorbed rabbit anti-bovine neurophysin II serum (*d*). Both × 90. (From de Mey et al. (274), by courtesy of Springer-Verlag, Berlin.)

grade (279) and retrograde (115, 263, 280–285) transport in a variety of neuronal systems. It has been of interest to compare the results obtained with these more recent techniques with those obtained from normally stained material, classic Golgi work (286), and experimental degeneration studies.

Radioautography of Labeled Endogenous Proteins The most widely used isotopes for tracing pathways by axonal transport have been L-[^{3}H]proline, or L-[^{3}H]leucine. A recent study, in which these isotopes have been used to trace the afferent connections of the accessory olfactory bulb in the mouse (287), illustrates some of the difficulties encountered with this technique. The isotopes may be introduced either under direct observation (as into the lumen of the vomeronasal organ) or by iontophoretic injection under stereotaxic guidance (as into the corticomedial amygdala) in a dose of 0.1–5 μCi. After allowing a suitable time for the labeled amino acids to be incorporated into proteins and transported to the terminals (from a few hours to 10 days, depending on the system), the animals are perfused with fixative and the brains processed for autoradiography.

The results obtained can depend on the dose of isotope administered, e.g., in studies on the afferent connections to the mouse accessory olfactory bulb, it was

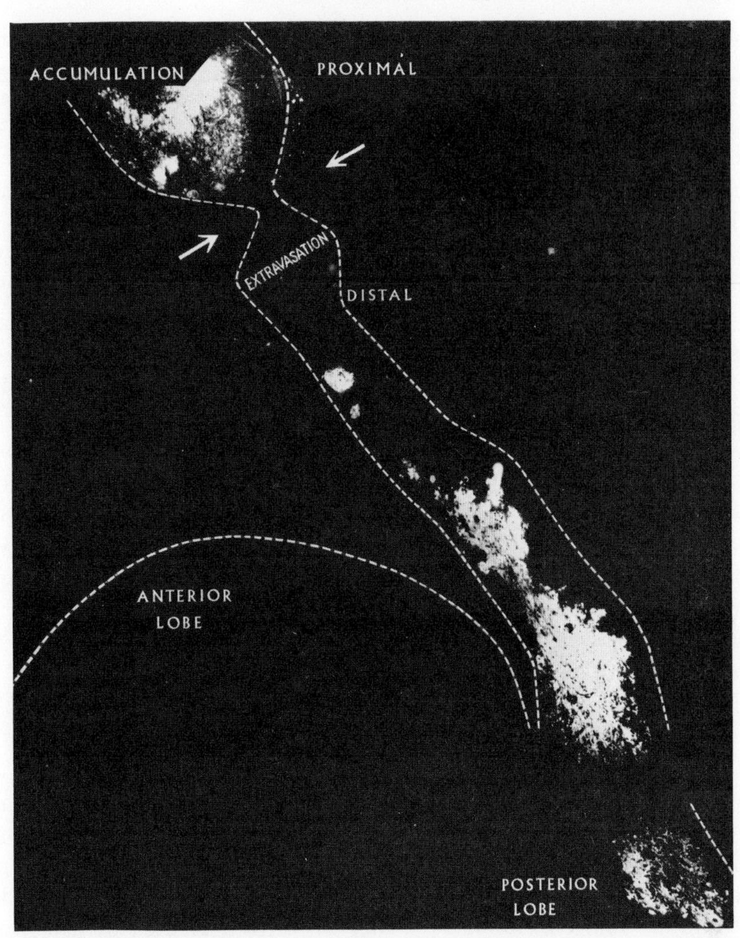

Figure 36. Immunofluorescence demonstration of axonal transport in neurosecretory neurons of the dog. Photomontage showing the accumulation of neurophysin-specific immunofluorescence proximal to a construction of the hypothalamo-neurohypophysial neurosecretory tract, 20 hours after operation. Specific fluorescence is absent from the axons immediately distal to the site of injury, while the terminals in the posterior lobe of the neurohypophysis show a decreased fluorescence. The results are interpreted as evidence for a rapid axonal transport of neurophysin from its site of synthesis in the cell bodies of the hypothalamus to the terminals in the posterior pituitary. (From Alvarez-Buylla et al. (270), by courtesy of Springer-Verlag, Berlin.)

found that by 5 days after iontophoretic injection of 0.05–0.1 μCi of L-[^{3}H]leucine into the corticomedial amygdala grains were confined to an area immediately overlying the deep granule cells, whereas with doses in excess of this (about 0.5 μCi) heavy labeling was apparent in a band over the internal plexiform layer, and with very high doses of isotope (5 μCi) over the external plexiform layer as well.

These observations raise the possibility of trans-synaptic transfer of the labeled isotopes (see sections "Trans-synaptic Migration" and "Mapping of

Neurons by Anterograde Axonal Transport" and "Trans-synaptic Transfer of [^{3}H]Adenosine and [^{3}H]Uridine"). With the higher levels of isotope (5 μCi, L-[^{3}H]leucine) injected iontophoretically into the corticomedial amygdala, it has been observed from the external plexiform layer as well as the granule cell and internal plexiform layers of the accessory olfactory bulb becomes labeled. To account for this observation it has been suggested that labeled material may have crossed *two* synapses, i.e., strial axon terminals to granule cells and granule cells to mitral cells. The exact chemical identity of the radioactive material transported and transferred across the synapse remains to be determined. Furthermore, it is not known whether this material has any functional role in the postsynaptic cells; however, it is clear from the above that the dose of isotope administered is an important determinant of the extent of labeling (and of mapping) of central neurons.

Another recent observation that may prove useful in tracing the origin of nerve ending components in the central nervous system is the finding (288) that colchicine (injected bitemporally into mice) blocks the accumulation in synaptosomes of radiolabeled soluble proteins transported to nerve endings only in limited areas of the brain. Radioautography when combined with this biochemical subfraction approach may be useful for in vivo studies concerning the origin of these nerve ending components.

Exogenous Tracers Neurons will take up exogenous proteins from the extracellular space. Once inside they may be transported 1) anterogradely, if taken up by the perikaryon or 2) retrogradely, if they have been taken up by axon terminals (115). Horseradish peroxidase (HRP) is well suited for these studies since it is a protein of molecular weight 40,000 and is therefore too large to diffuse through neuronal membranes in a nonspecific manner, but is small enough to be taken up by pinocytotic vesicles (see section "Trans-synaptic Migration"). A further advantage is the extremely high turnover number of this enzyme and the ability to localize the enzyme in tissue sections by demonstration of its insoluble polymeric reaction product after incubation of the sections in a medium containing Tris buffer, 3,3′-diaminobenzidine, and hydrogen peroxide (289).

$$H_2O_2 + 3,3'\text{-diaminobenzidine} \xrightarrow{\text{HRP}} \text{insoluble polymer} + H_2O$$

Retrograde Transport of HRP The cellular origin of axon terminal networks can be traced most effectively after stereotaxic injection of horseradish peroxidase into specific areas of the brain (117). The peroxidase, taken up into chemically heterogeneous populations of nerve endings, is transported retrogradely within the axons and may be visualized histochemically in the perikarya by the Graham and Karnovsky reaction (289) described above.

Recent studies with this technique in the central nervous system suggest that the degree of concentration (or coalescence) of HRP transported retrogradely is dependent on the density of the terminal axonal field available for incorporation of the enzyme at the injection site (290). Injection of HRP into the cortex of the frontal and parietal lobes of rhesus monkeys (Figure 37) results in retrograde transport of the enzyme to neurons in the nucleus basalis of the substantia innominata and in the hypothalamus, indicating that neurons in the basal forebrain areas send axons to the neocortex of these lobes (291). Such direct connections resemble those seen with the catecholamine fluorescence technique, where an ascending (57, 213, 218, 222, 234) monoaminergic pathway provides a direct link between cell bodies in the brainstem and terminals in the cortex. Taken together, these findings suggest that the various behavioral phenomena (sleep, food intake, and sexual activity) that can be elicited by stimulation of limbic basal forebrain areas may be brought about, in part, by way of direct connections to the cerebral cortex as well as by way of their descending pathways to the brainstem.

Although the biological role(s) of retrograde intra-axonal transport are not known, there are a number of regulatory processes operating in neurons, such as chromatolysis after axon injury (171), retrograde trans-synaptic changes (292),

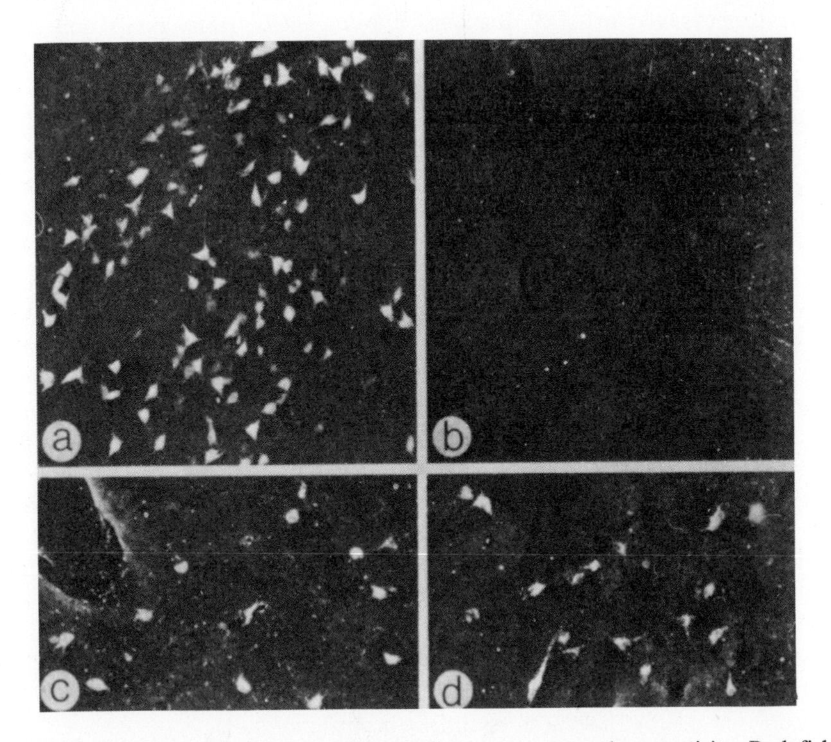

Figure 37. Use of horseradish peroxidase (HRP) to trace neuronal connectivity. Dark-field photomicrographs of HRP-positive neurons in basal forebrain areas after injection of the enzyme in frontal and parietal cortex. *a*, concentration of HRP-positive neurons (× 49) in medial part of substantia innominata. *b*, the same area (× 49) in control animal with saline injected in precentral gyrus. *c*, HRP-positive neurons (× 49) in caudal part of nucleus basalis after light counterstaining with cresyl violet. *d*, HRP-positive neurons (× 53) in lateral hypothalamus. (From Kievit and Kuypers (283) by courtesy of American Association for the Advancement of Science, Washington.)

acute glial reactions (293), and growth regulation according to the size of the peripheral field (294) that imply the existence of a retrograde transfer of information from the axons and terminals to the cell body.

Anterograde Transport of HRP The question of whether *physiological* anterograde transport of exogenous proteins (such as HRP) takes place has only recently been resolved through electron microscopic observation (279).

Anterograde transport of HRP escaped detection at the light microscope level because the enzyme is transported within fine tubular and vesicular structures below the resolution of the light microscope. The reciprocal connections between the caudoputaminal complex and the substantia nigra in the rat provided a convenient system for examining this question because the substantia nigra is organized into two anatomically distinct regions: 1) the pars compacta containing most, if not all, nigral neurons which project to the caudoputamen, and 2) the pars reticulata containing most of the afferent terminals to the substantia nigra from neurons in the caudoputaminal complex (Figure 38). Injection of HRP into the caudoputaminal complex therefore permitted observations to be made of HRP transported *retrogradely* from axons and axon terminals in the caudoputamen to the cell bodies of origin in the pars compacta of the substantia nigra, and of HRP transported *anterogradely* by neurons of the caudoputamen to terminals in the neuropil of the pars reticulata.

These and earlier studies (295–299) have shown that HRP is taken up into neuronal cell bodies by pinocytotic vesicles and eventually finds its way into lysosomes and multivesicular bodies. The more recent ultrastructural studies showed in addition that "a common characteristic of the localization of HRP at sites distant from the region of application was its appearance within cisternae of the agranular reticulum" (279). The presence of HRP within agranular reticulum in nerve terminals distant from cell bodies that had been exposed to HRP raises the possibility that the agranular reticulum may provide the normal physiological route for anterograde transport of proteins within neurons. Other evidence, reviewed above (section "Synthesis of Neuronal Components"), supports this view and provides evidence that elongated channels of agranular reticulum may extend the full length of the axon from the Golgi apparatus in the perikaryon to the terminals, and so provide channels for rapid anterograde transport of endogenous proteins. The electron microscopic observations of HRP within axonal and terminal agranular reticulum, after exposure of the cell bodies to low concentrations of enzyme, suggest that anterograde transport of HRP is a physiological rather than a pathological phenomenon. Anterograde transport of HRP may thus provide a new neuroanatomical tool for probing further into the dynamics of neurons having an essentially normal morphology.

Combined Use of Anterograde and Retrograde Tracers to Study Neuronal Connectivity The combined use of retrograde and anterograde tracers has been suggested as a possible way to study *reciprocal* neuronal connectivity (282). Anterograde and retrograde tracers injected together have been used to demonstrate reciprocal connectivity of thalamocortical and corticothalamic connections in the rat (300) after stereotaxically guided microinjection of a combined solution of HRP and [^{3}H]leucine into the *thalamus.* In larger animals, such as

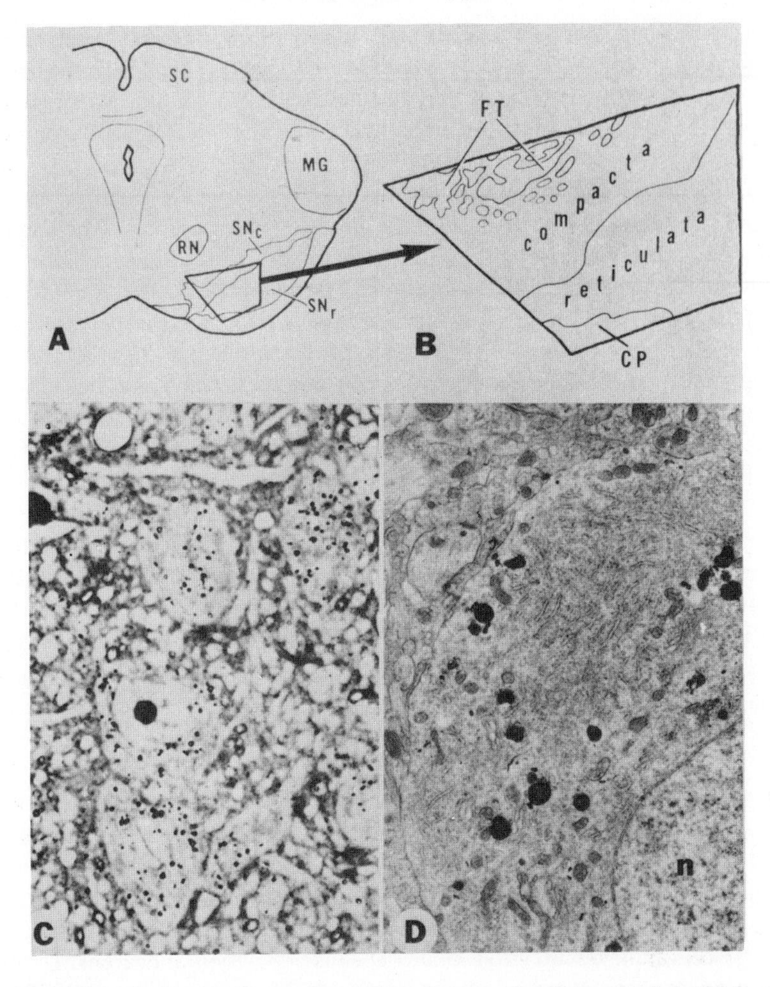

Figure 38. Light and electronmicroscopic tracing of neuronal pathways loaded with horseradish peroxidase. *A*, a transverse section through the midbrain. The region of the substantia nigra sectioned for electron microscopy is outlined. *SC*, superior colliculus; *MG*, medial geniculate; *RN*, red nucleus; SN_c, substantia nigra, pars compacta; SN_r, pars reticulata. *B*, a more detailed map of the region outlined in *A*. *CP*, cerebral peduncle; *FT*, fiber tracts. *C* shows the light microscopic appearance of *retrogradely* transported HRP in neurons of the pars compacta in 1-μm section. *D*, a low power electron micrograph of a *retrogradely* labeled neuron soma in the pars compacta. Note the normal appearance of the neuron and the localization of HRP within lysosomes, multivesicular bodies and the agranular reticulum. *n*, nucleus. Magnifications: *C*, × 1,510; *D*, × 7,270. (From Nauta et al. (279) by courtesy of Elsevier, Amsterdam.)

the squirrel monkey, these investigators (280) have demonstrated that *cortical* injections of the two tracers can be utilized to examine the reciprocal connectivity of *individual* thalamic nuclei within the cortex.

In this study, use was made of the retrograde tracer, horseradish peroxidase, and of the anterograde tracer [^{3}H]leucine in determining the reciprocity of neuronal connectivity of pulvinar and temporal lobes (superior temporal gyrus) in squirrel monkey (280). Both tracers were given stereotaxically as a combined solution (0.4–0.6 μl) for injection at a number of sites in the superior temporal gyrus, and after survival times of 12 hours to 3 days the tissues were processed to show HRP positive neurons and silver grains. In this way the topographical relationship between the injection sites in the superior temporal gyrus and the location of transported tracers in the posterior thalamus could be mapped. It was concluded that the connections between the superior temporal gyrus and pulvinar are topographical and reciprocal. However, it was also found that areas positively labeled with silver grains and HRP were not completely coextensive. To explain this finding it was proposed that this may be a result of an incomplete anatomical reciprocity of the corticothalamic and thalamocortical connections or it may be due to the difference of extent of diffusion of the two tracers at the injection site and/or differences in the quantity and rate of uptake and transport of the tracers.

This study confirmed earlier work on the existence of a topographical projection from the pulvinar to the temporal lobe, obtained after retrograde degeneration methods of the rhesus monkey (301), and extended this finding to another species (the squirrel monkey) in which anterograde degeneration studies on the corticothalamic projections from the temporal lobe had indicated only a general scheme for these connections. In addition, a convergent reciprocal relationship between widespread parts of the superior temporal gyrus related to the pulvinar was found, and in other experiments thalamic injections were used to yield useful information on reciprocal connectivity and cortical distribution of terminals and neurons. Such precise information about connectivity was not possible with the earlier anterograde and retrograde degeneration methods.

Combined Use of Immunohistochemical and Retrograde Tracers to Study Connectivity of Neurons Containing Specific Neurotransmitters A limitation of the radioactive and exogenous protein tracer methods mentioned above is that the tracers are taken up into a heterogeneous population of neurons. To obtain a precise chemical mapping of the brain, Hökfelt and colleagues (183) have recently applied the sensitive fluorescence and immunofluorescence histochemical techniques (for demonstrating particular neurotransmitters in the brain), together with the La Vail and La Vail (117) technique of stereotaxic administration of horseradish peroxidase. By use of these methods, pathways traced retrogradely with peroxidase are, in addition, characterized histochemically with regard to their neurotransmitters.

In this study HRP was injected into the head of the caudate nucleus of rats, and 24 hours later the animals were perfused with ice-cold formalin, and cryostat sections of the brain processed for immunohistochemistry. Antibodies to tyrosine hydroxylase were used to identify dopamine neurons by the indirect immunofluorescence procedure and, after recording the position of the fluorescent dopamine neurons with a fluorescence microscope, the sections were treated with diaminobenzidine for visualization of peroxidase by light microscopy. The fluorescence and light micrographs were then compared (Figure 39).

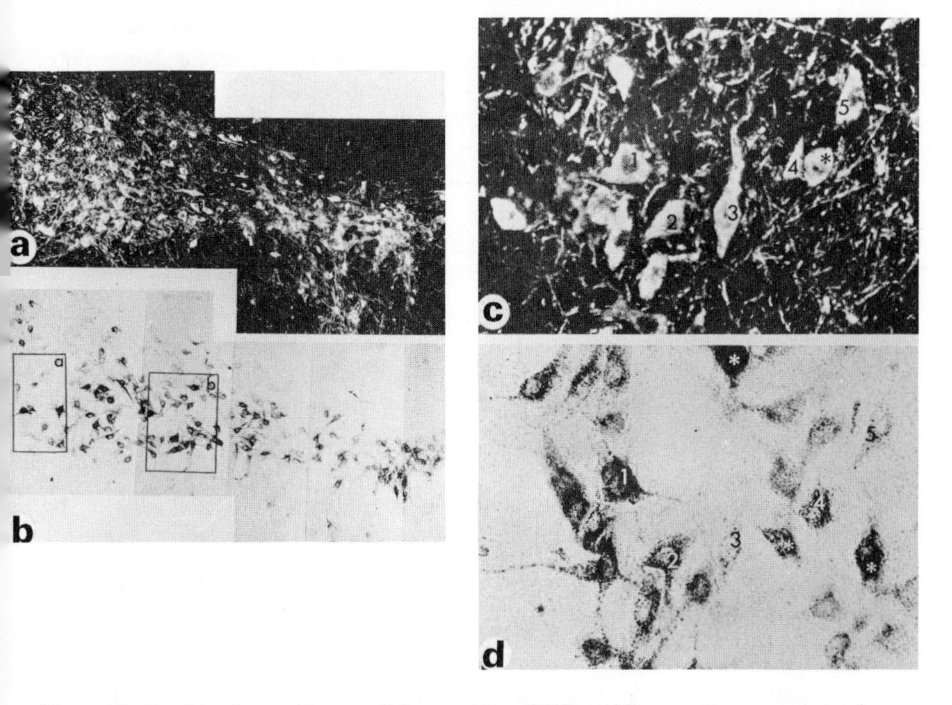

Figure 39. Combined use of horseradish peroxidase (HRP) and immunofluorescence tracing of tyrosine hydroxylase (TH) to study reciprocal connectivity of adrenergic neurons in the brain. Immunofluorescence micrograph (*a*) and light micrograph (*b*) of the same section of the substantia nigra of a rat injected with HRP into the head of the caudate nucleus. After incubation with antibodies to TH and photography (*a*), the section was processed for the demonstration of HRP (*b*). The distribution of TH and HRP positive cells is very similar. Note that in *a* both cell bodies and cell processes are strongly stained, whereas in *b* the HRP reaction is confined mainly to the cell bodies. Magnification × 70. *c* and *d*, higher magnifications of part of *a* and *b*, respectively (see area indicated with *b* within *b*). Most cells (1–5) contain both TH and HRP, whereas some cells are only TH positive (*black asterisks* in *c*) and others are only HRP positive (*white asterisks* in *d*). The weak appearance of some cells in micrograph (*d*) is due to the fact that these cells are slightly out of focus as a consequence of the thick sections. Magnification × 300. (From Ljungdahl et al. (183), by courtesy of Elsevier, Amsterdam.)

The distribution of the immunoreactive neurons closely paralleled that of dopamine cells projecting to the neostriatum as demonstrated previously with the formaldehyde fluorescence technique (200) and immunofluorescence (231). After treatment of the same sections with diaminobenzidine, the peroxidase-labeled cells were found in the same area, thus confirming the observations of Kuypers et al. (283) and Nauta et al. (302) who used larger injections of horseradish peroxidase. Most of the tyrosine hydroxylase positive cells also contained peroxidase showing that the majority of nerve endings arising from axons whose cell bodies are in the substantia nigra belong to dopamine neurons. These findings add further to the evidence for a nigrostriatal dopamine pathway and provide additional information about the extent of connectivity between these two regions.

This approach is not limited to the present combination of techniques and should be able to be applied to the tracing of other monoamine (adrenaline, noradrenaline, 5-hydroxytryptamine), GABA, and glycine cell bodies after combination of the appropriate formaldehyde fluorescence, immunofluorescence, or autoradiographic techniques, with peroxidase tracing.

Retrograde Transport of Nerve Growth Factor One of the few endogenous proteins for which physiological retrograde axonal transport has been shown to occur is nerve growth factor (303). Nerve growth factor (NGF) promotes the development and growth of the peripheral sympathetic nervous system (304) in neonatal animals. It is present in high concentrations in the mouse submaxillary gland from which it has been isolated and its structure determined (305) (Figure 40). The activity resides in a small basic polypeptide chain of 118 amino acids with a molecular weight of 13,259. Two such chains, when combined noncovalently in the dimeric structure of the native NGF, have a molecular weight of 26,518.

In *neonatal* animals, NGF administration enhances the differentiation of the terminal adrenergic neurons, this process being reflected by a selective induction of tyrosine hydroxylase and dopamine β-hydroxylase (306), the two enzymes involved in controlling the rate of synthesis of noradrenaline (58, 192, 210). NGF is required also for the maintenance of normal function of the peripheral sympathetic nervous system in *adult* animals (307, 308) as evidenced by the finding that removal of the submaxillary glands in adult mice causes a drop in the plasma and tissue concentrations of NGF which is followed by a reduction in the activity of all the enzymes involved in the synthesis of noradrenaline in both superior cervical and stellate ganglia (211, 308).

Soon after the initial suggestion by Hendry and Iversen (307) that the biological effect of NGF might be due to retrograde axonal transport of the protein from the nerve terminals, direct evidence for its retrograde transport was obtained (303). After injection of ^{125}I-NGF into the anterior eye chamber of rats and mice, it is taken up into adrenergic nerve terminals in the iris and transported at a relatively rapid rate (2.5 mm/hour) to the corresponding cell bodies in the superior cervical ganglion. This retrograde transport of NGF is sensitive to colchicine and is highly specific in that other proteins of similar molecular weight (e.g., cytochrome *c*) are not transported to a measurable extent, and even minor modifications to the NGF molecule (e.g., oxidation of the tryptophan residues) drastically reduce its retrograde transport (309). Since NGF on adrenergic perikarya in the superior cervical ganglion is not yet known, it mental conditions in which tetanus toxin is transported (310), it appears that the uptake and retrograde transport of NGF is not dependent on structural features common to all neurons.

Although the full extent of the biological effects of retrogradely transported NGF on adrenergic perikarya in the superior cervic ganglion is not yet known, it is known that these adrenergic cell bodies receive cholinergic input, and that the normal development of cholinergic nerve terminals within the superior cervical ganglion depends on the integrity of the sympathetic neurons with which they

Figure 40. Schematic representation of the amino acid sequence of the primary subunit of 2.5 S nerve growth factor from mouse submaxillary glands. (From Angeletti and Bradshaw (305) by courtesy of the National Academy of Sciences, Washington.)

synapse. Destruction of the adrenergic nerve cell bodies by use of an antiserum to NGF (311) or with 6-hydroxydopamine (58, 311) results in the failure of the normal development of the enzyme choline acetyltransferase, an enzyme located exclusively within cholinergic nerve terminals (311, 312). Axotomy of postganglionic adrenergic neurons in developing animals also prevents the normal development of choline acetyltransferase and presynaptic cholinergic nerve terminals on the superior cervical ganglion, and this effect can be reversed by NGF.

The physiological significance of retrograde transport of endogenous NGF is consequently of much interest to neurobiologists, and in a recent investigation Hendry (314) provided evidence that the arrival of NGF in the perikarya of developing adrenergic neurons may act as a trophic factor that ultimately determines the survival of ingrowing presynaptic cholinergic nerve terminals onto these adrenergic cell bodies by a retrograde trans-synaptic transfer of information. It is proposed that "the preganglionic cholinergic nerve terminals require adrenergic neurons to be in contact with their target cell before they can innervate them normally; however, the information the adrenergic cell derives from the periphery can be replaced by NGF" (314). Since it has been shown that NGF has no direct effect on any other cholinergic synapse (311, 315) it is concluded that the effects of NGF are on the adrenergic neurons directly and only indirectly on the cholinergic terminals. Whether the trans-synaptic information transmitted retrogradely to ingrowing cholinergic nerve terminals is in the form of a specific trophic molecule or is some consequence of the number or conformation of postsynaptic receptors is not known.

These observations on possible factors influencing synapse formation in the peripheral nervous system raise the question of whether similar mechanisms may operate in the central nervous system, especially at sites such as the locus coeruleus which appears to function as a central extension of the sympathetic chain (228). That retrograde transport of NGF is not restricted to afferent adrenergic neurons is shown by the fact that NGF is taken up by sensory nerve fibers in the rat forepaw and is transported retrogradely to the corresponding sensory dorsal root ganglia (316). The rate of retrograde transport of NGF in sensory neurons in the rat (316) is about 5 times faster (13 mm/hour) than in adrenergic neurons. The reason for this difference in rates is not clear, nor is it known why sensory neurons, though exhibiting the same high degree of selectivity for retrograde transport of NGF throughout life, only show responsiveness of their dorsal root ganglia within a restricted period of ontogenic development (304, 316).

Mapping of Neurons by Anterograde Axonal Transport and Trans-synaptic Transfer of [^{3}H]Adenosine and [^{3}H]Uridine Much attention has been focused recently on the axonal transport of RNA, which is reported to occur at a slow rate (see Jeffrey and Austin (6)). Of interest also is the rapid axonal transport of nucleotides, since these may, in addition to their role as precursors of RNA and DNA, supply the energy requirements of the neuron for dynamic processes such as axonal transport (7, 75) and transmitter release (317).

In a recent study, Schubert and Kreutzberg (114) mapped neurons in the visual corticothalamic projection of the rabbit by injecting [^{3}H]uridine or [^{3}H]adenosine into localized areas of the striate cortex. After 1, 2, and 3 days, the rabbits were killed by formalin perfusion and the brains processed for autoradiography. After 1 day, silver grains were localized at the injection site and over the myelinated fiber tracts leading to the lateral geniculate nucleus. By 3 days, the grain density in the lateral geniculate nucleus had increased and was localized over a strictly limited area of neuropil, indicating that the nucleoside derivatives had entered the axons and had been transported to the axon terminals by anterograde axonal transport. In addition, a large number of silver grains were located over the *postsynaptic* neurons, [^{3}H]uridine derivatives being more concentrated over the nerve cell nuclei, while [^{3}H]adenosine derivatives were found over the nuclei and cytoplasm.

The chemical nature of the transported radioactive material remains to be determined; however, preliminary studies have shown that the majority of the radioactive material transported into the target area was in a soluble form.

In view of the regulatory role of adenosine in the brain (see McIlwain (318)), these findings raise the possibility that nucleosides (or their derivatives) released from the terminals of one neuron may influence the metabolism of a second neuron or target cell. The demonstration that trans-synaptic transfer of these soluble materials occurs suggests that nucleosides and their derivatives should be considered as possible candidates for the elusive "trophic substance(s)" that mediate the effects of a neuron upon its effector cell.

Tracing of Selected Neurons with Cobalt Ions By injection of cobalt ions into a selected neuronal cell body, the full extent of its dendritic tree and axonal and terminal ramifications may be observed after precipitation as black cobalt sulfide (Figure 41). Terminals from other neurons which synapse all over the dendritic tree are not stained by this method. The technique can be used at the light or electron microscope level to trace connections made by the injected nerve cell.

In the 3 years since Pitman et al. (319) first introduced cobalt sulfide precipitation to identify nerve cell bodies and processes in the cockroach, the technique has been used successfully to study neuronal connectivity in several

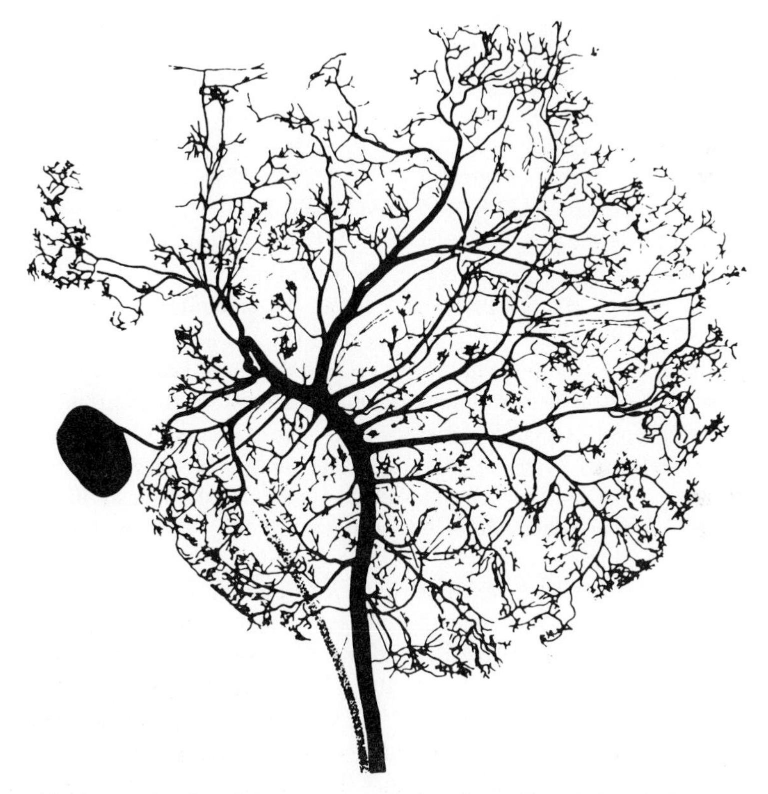

Figure 41. A nerve cell and its processes displayed by the cobalt sulfide precipitation technique. A nerve cell from the Australian plague locust, *Chortoicetes,* has been outlined in its entirety by injection with cobalt ions which are subsequently precipitated as the black sulfide. The nucleus is in the ovoid expansion to the left, and the axon, which is destined to form contacts with a muscle cell, continues on from the thickened shaft at the bottom. Terminals from other nerve cells end all over the branching dendritic tree. They are not stained by this method and can only be clearly seen by the electron microscope. (From Mark (106). Photo courtesy of Dr Mark Tyer, and Clarendon Press, Oxford.)

invertebrate nervous systems. More recently, the cobalt iontophoresis technique has been modified by Tyrer and Bell (320) and extended for use in studies of the vertebrate nervous system (321–323). A wide variety of vertebrates, including amphibians, reptiles, aves, and mammals, have been used with uniform success.

In this technique, the brain and spinal cord of the animal are removed following perfusion with saline, immersed in cold saline (4–5°C), and a suction electrode filled with 300 mM cobalt chloride placed over the cut end of the nerve trunk with the aid of a micromanipulator. The end of the nerve is gently sucked up into the electrode and cobalt ions are then subjected to electrophoresis (by means of a voltage divider) within the nerve fibers for periods of up to 48 hours. After iontophoresis, the brain is bathed in an ammonium sulfide solution which precipitates the cobalt as black cobalt sulfide. The course of the fibers can be traced after processing the tissue for light or electron microscopy.

The technique has been used in vitro to delineate neuronal pathways in the visual system and provide further information on the proposed retinohypothalamic projection in rats (322). It has also been used to trace afferent and efferent fiber pathways of cranial and spinal nerves, and fiber tracts within the hypothalamic-neurohypophyseal (259) and central nervous systems of vertebrates (323). In each case, where cobalt has been used to trace afferent fibers, the results obtained agreed with Nauta-Gygax and Fink-Heimer silver degeneration studies after sectioning of the nerves. As there is no indication that cobalt can cross nerve membranes or synapses when used in vitro it is concluded that the nerve fibers and cell bodies which contain the precipitate are the primary fibers and cell bodies of the particular nerve which was iontophoresed with the cobalt solution. The technique has the advantage that both afferent and efferent cells are labeled. A further advantage claimed (323) for cobalt iontophoresis is that it is more sensitive than horseradish peroxidase as a method for delineating the course of efferent fibers in the brain. The finding that both afferent and efferent fibers of vertebrate neurons can be traced simultaneously in vitro by cobalt iontophoresis indicates that this technique should see wide applicability in neuroanatomical tracing.

Mapping of Functional Pathways with [^{14}C]Deoxyglucose This method (324) is based on the fact that if sufficient time is allowed following an intravenous pulse of [^{14}C]deoxyglucose for the free carbon-14 deoxyglucose to have been cleared from the tissue, then the [^{14}C] concentration of the tissues in the central nervous system represents [^{14}C]deoxyglucose 6-phosphate and gives a measure of the rate of glucose consumption of the tissues. Altered functional activity alters metabolic activity and hence the uptake of [^{14}C]deoxyglucose in the tissues, so that [^{14}C]deoxyglucose can be used to map regions of the brain having an altered glucose utilization in response to alterations in local functional activity.

This property has been made use of for mapping of functional neuronal pathways in 1) the lumbar spinal cord of the rat after unilateral stimulation of the sciatic nerve; 2) the caudate nucleus, thalamus, putamen and globus pallidus of a monkey in which seizures were induced by application of penicillin to the ipsilateral motor cortex; 3) the lateral geniculate bodies, superior colliculi, striate cortex, and calcarine cortex of rats and monkeys which had undergone unilateral enucleation of the visual system; and 4) the central auditory components of the rat after unilateral occlusion of an auditory canal (Figure 42).

The advantage of this mapping technique over conventional mapping methods is that when combined with autoradiography it permits a survey of all the structures of the brain simultaneously. Improved structural resolution should be possible with this technique by replacement of [^{14}C]deoxyglucose with [^{3}H]deoxyglucose. This should permit high resolution radioautographs of thin sections and so yield even more information.

CONCLUSIONS

Over the last 25 years, a number of new techniques have been developed for studying neuronal connectivity. The introduction of the Nauta-Gygax (146) and

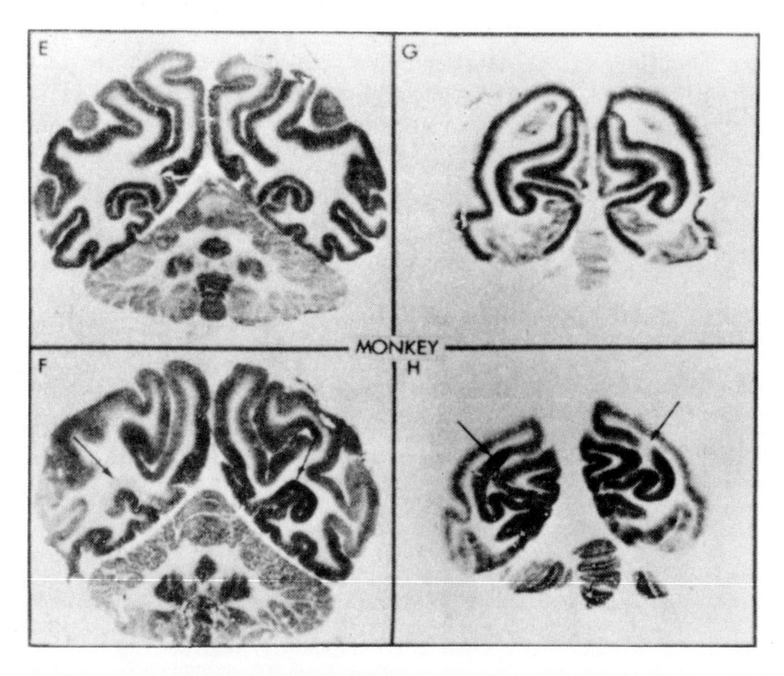

Figure 42. The use of [^{14}C]deoxyglucose to trace functional pathways in the brain. Effects of unilateral enucleation on [^{14}C]deoxyglucose uptake in components of the visual system in the rhesus monkey. Areas of asymmetry are seen only in regions with monocular input, namely, the rostral portion of the calcarine cortex (*F*) and the loci corresponding to the blind spots of the visual fields (*H*). Autoradiographs from corresponding sections of brain from a control monkey with both eyes intact exhibit no asymmetry in these regions (*E* and *G*). The section in (*H*) is only one of many serial sections demonstrating the loci of the blind spots, which appear to extend approximately 3.4 mm from front to back; the gross appearance of the section in (*G*) is somewhat different from that of the section in (*H*) because of a slight difference in the plane of sectioning, but it includes the portions of cortex containing the loci of the blind spots. (From Kennedy et al. (324), by courtesy of the American Association for the Advancement of Science, Washington.)

Fink-Heimer (157) silver methods for tracing anterograde degeneration enabled a more precise study of afferent and efferent connections of neuronal cell groups, while electron microscopy made possible the observation of degenerating boutons.

These studies have been complemented by techniques for demonstrating specific transmitters and proteins, by autoradiography and immunohistochemistry, and by procedures using horseradish peroxidase. In addition, the methods of axonal iontophoresis of various dyes and cobalt sulfide precipitation have contributed to our knowledge of neuronal connectivity in invertebrate and vertebrate nervous systems. Unlike the earlier silver methods which relied almost exclusively on degeneration techniques, these newer techniques are based on those dynamic properties of living neurons that enable them to transport tracer molecules within the axon (either anterogradely or retrogradely) to their final physiological location. An exciting recent development has been the introduction of labeled deoxyglucose for autoradiographic investigation of *functional* connections in the nervous system.

These contributions to neuroanatomical tracing techniques have come largely from biochemical and physiological investigations of axonal transport, studies which in themselves have let to a better appreciation of neuronal dynamics.

REFERENCES

1. Guth, L. (1974). Axonal regeneration and functional plasticity in the central nervous system. Exp. Neurol. 45:606.
2. Pomerat, C.M., Hendelman, W.J., Raiborn, C.W., and Massey, J.F. (1967). Dynamic activities of nervous tissue in vitro. *In* H. Hyden (ed.) The Neuron, p. 119. Elsevier Publishing Co., Amsterdam.
3. Barondes, S.H. (1969). Axoplasmic Transport. *In* A. Lajtha (ed.), Handbook of Neurochemistry, Vol. 2, p. 435. Plenum Press, New York.
4. Dahlström, A. (1971). Axoplasmic transport (with particular respect to adrenergic neurons). Phil. Trans. Roy. Soc. B. 261:325.
5. Grafstein, B. (1969). Axonal transport: communications between soma and synapse. *In* E. Costa and P. Greengard (eds.), Advances in Biochemical Psychopharmacology, Vol. 1, pp. 11. Raven Press, New York.
6. Jeffrey, P.L., and Austin, L. (1973). Axoplasmic transport. Prog. Neurobiol. 5:205.
7. Ochs, S. (1974). Systems of material transport in nerve fibers (axoplasmic transport) related to nerve function and trophic control. Ann. N.Y. Acad. Sci. 228:202.
8. Droz, B. (1973). Renewal of synaptic proteins. Brain Res. 62:383.
9. Droz, B., Di Giamberardino, L., and Koenig, H.L. (1974). Transports axonaux de macromolecules presynaptiques. Actualités Neurophysiologiques 10e:236.
10. Bennett, G., Di Giamberardino, L., Koenig, H.L., and Droz, B. (1973). Axonal migration of protein and glycoprotein to nerve endings.II. Radioautographic analysis of the renewal of glycoproteins in nerve endings of chicken ciliary ganglion after intracerebral injection of [^{3}H] fucose and [^{3}H] glucosamine. Brain Res. 60:129.
11. Taylor, A.C., and Weiss, P. (1965). Demonstration of axonal flow by the movement of tritium-labeled protein in mature optic nerve fibers. Proc. Natl. Acad. Sci. USA 54:1521.
12. Karlsson, J.L., and Sjöstrand, J. (1968). Transport of labeled proteins in the optic nerve and tract of the rabbit. Brain Res. 11:431.

13. Marko, P., and Cuenod, M. (1973). Contribution of nerve cell body to renewal of axonal and synaptic glycoproteins in the pigeon visual system. Brain Res. 62:273.
14. Marchisio, P.C., Gremo, F., and Sjöstrand, J. (1975). Axonal transport in embryonic neurons. The possibility of a proximo-distal axolemmal transport of glycoproteins. Brain Res. 85:281.
15. Austin, L., Bray, J.J., and Young, R.J. (1966). Transport of proteins and ribonucleic acid along nerve axons. J. Neurochem. 13:1267.
16. Bray, J.J., and Austin, L. (1966). Axoplasmic transport of ^{14}C-proteins at two rates in chicken sciatic nerve. Brain Res. 12:230.
17. Ochs, S., Johnson, J., and Ng, M.H. (1967). Protein incorporation and axoplasmic flow in motoneuron fibers following intra-cord injection of labeled leucine. J. Neurochem. 14:317.
18. Bradley, W.G., Murchison, D., and Day, M.J. (1971). The range of velocities of axoplasmic flow: a new approach, and its application to mice with genetically inherited spinal muscular atrophy. Brain Res. 35:185.
19. Austin, L., Komiya, Y., and Tang, B.Y. (1974). Axoplasmic flow of protein and phospholipid in dystrophic mice. *In* Recent Advances in Myology, Proc. III International Congress on Muscle Diseases, Newcastle-upon-Tyne, p. 224. Excerpta Medica Foundation, Amsterdam.
20. Weiss, P., and Holland, Y. (1967). Neuronal dynamics and axonal flow. II. The olfactory nerve as model test object. Proc. Natl. Acad. Sci. USA 57:258.
21. Gross, G.W., and Beidler, L.M. (1973). Fast axonal transport in the C-fibers of the garfish olfactory nerve. J. Neurobiol. 4:413.
22. Cancalon, P., and Beidler, L.M. (1975). Distribution along the axon and into the various subcellular fractions of molecules labeled with [^{3}H]leucine and rapidly transported in the garfish olfactory nerve. Brain Res. 89:225.
23. Ramirez, G., Levitan, I.B., and Mushynski, W.E. (1972). Highly purified synaptosomal membranes from rat brain. Incorporation of amino acids into membrane proteins in vitro. J. Biol. Chem. 247:5382.
24. Zelena, J. (1970). Ribosome-like particles in myelinated axons of the rat. Brain Res. 24:359.
25. Livett, B.G., Rostas, J.A.P., Jeffrey, P.L., and Austin, L. (1974). Antigenicity of isolated synaptosomal membranes. Exp. Neurol. 43:330.
26. Rostas, J.A.P., and Jeffrey, P.L. (1975). Restricted mobility of neuronal membrane antigens. Neurosci. Lett. 1:47.
27. Morgan, I.G., Zanetta, J.P., Breckenbridge, W.C., Vincendon, G., and Gombos, G. (1973). The chemical structure of synaptic membranes. Brain Res. 62:405.
28. Edds, M.V., Barkley, D.S., and Fambrough, D.M. (1972). Genesis of neuronal patterns. Neurosci. Res. Prog. Bull. 10:253.
29. Marchase, R.B., Barbera, A.J., and Roth, S. (1975). A molecular approach to retino-tectal specificity. CIBA Foundation Symposium. Cell Patterning, Vol. 29, p. 315. Elsevier, Amsterdam.
30. Nicholson, G.L. (1974). Interactions of lectins with animal cell surfaces. Int. Rev. Cytol. 39:89.
31. Hubbell, W.L., and McConnell, H.M. (1968). Spin label studies of the excitable membranes of nerve and muscle. Proc. Natl. Acad. Sci. USA 61:12.
32. Comoglio, P.M., and Filogamo, G. (1973). Plasma membrane fluidity and surface mobility of mouse C1300 neuroblastoma cells. J. Cell Sci. 13:415.
33. Keith, A.D., and Snipes, W. (1974). Viscosity of cellular protoplasm. Science 183:666.
34. Kadenback, B. (1967). Synthesis of mitochondrial proteins: demonstration of a transfer of proteins from microsomes into mitochondria. Biochim. Biophys. Acta 134:430.
35. Weiss, P.A., and Mayr, R. (1971). Neuronal organelles in neuroplasmic (axonal) flow. I. Mitochondria. Acta Neuropath (Berl.) Suppl. V.:187.
36. Di Giamberardino, L., Bennett, G., Koenig, H.L., and Droz, B. (1973). Axonal migration of protein and glycoprotein to nerve endings. III. Cell fraction analysis of chicken ciliary ganglion after intracerebral injection of labeled precursors of proteins and glycoproteins. Brain Res. 60:147.

37. Bray, J.J., Kon, C.M., and Breckenridge, B.M. (1971). Adenyl cyclase, cyclic nucleotide phosphodiesterase and axoplasmic flow. Brain Res. 26:385.
38. Lubinska, L., and Niemierko, S. (1971). Velocity and intensity of bidirectional migration of acetylcholinesterase in transected nerves. Brain Res. 27:329.
39. Grafstein, B. (1971). Transneuronal transfer of radioactivity in the central nervous system. Science 172:177.
40. Grafstein, B., Miller, J.A., Ledeen, R.W., Haley, J., and Specht, S.C. (1975). Axonal transport of phospholipid in goldfish optic system. Exp. Neurol. 46:261.
41. Austin, L., Lowry, H., Brown, J.G., and Carter, J.G. (1972). The turnover of protein in discrete areas of rat brain. Biochem. J. 126:351.
42. Kerkut, G.A., Shapira, A., and Walker, R.K. (1967). The transport of ^{14}C-labeled material from CNS to muscle along a nerve trunk. Comp. Biochem. Physiol. 23:729.
43. Keen, P., and McLean, W.G. (1974). The effect of nerve stimulation on the axonal transport of noradrenaline and dopamine-β-hydroxylase. Brit.J. Pharmacol. 52:527.
44. Dahlström, A., and Häggendal, J. (1969). Recovery of noradrenaline in adrenergic axons of rat sciatic nerves after reserpine treatment. J. Pharm. Pharmacol. 21:633.
45. Fibiger, H.C., and McGeer, E.G. (1973). Increased axoplasmic transport of ^{3}H-dopamine in nigroneostriatal neurons after reserpine. Life Sci. 13:1565.
46. Norström, A., and Sjöstrand, J. (1972). Effect of salt-loading, thirst and water-loading on transport and turnover of neurohypophysial proteins of the rat. J. Endocrinol. 52:87.
47. Norström, A., and Sjöstrand, J. (1972). Effect of suckling and parturition on axonal transport and turnover of neurohypophysial proteins of the rat. J. Endocrinol. 52:107.
48. Altman, J. (1966). Autoradiographic examination of behaviourally induced changes in the protein and nucleic acid metabolism of the brain. *In* J. Gaito (ed.), Macromolecules and Behavior, p. 103. North Holland Publishing Co., Amsterdam.
49. Lux, H.D., Schubert, P., Kreutzberg, G.W., and Globus, A. (1970). Excitation and axonal flow: autoradiographic study on motoneurones intracellularly injected with a ^{3}H-amino acid. Exp. Brain Res. 10:197.
50. Nörstrom, A., and Sjöstrand, J. (1971). Axonal transport of protein in the hypothalamo-neurohypophysial system of the rat. Acta Neuropath. (Berl.) Suppl. V:249.
51. Grainer, F., and Sloper, J.C. (1972). Overactivity of the hypothalamo-neurohypophysial neurosecretory system and the problem of the mechanisms of transporting neurosecretory material. VI Int. Symp. on Neurosecretion 10 Abst.
52. Almon, R.R., and McClure, W.O. (1974). The effect of nerve growth factor upon axoplasmic transport in sympathetic neurons of the mouse. Brain Res. 74:255.
53. Hier, D.B., Arnason, B.G.W., and Young, M. (1972). Studies on the mechanism of action of nerve growth factor. Proc. Natl. Acad. Sci. USA 69:2268.
54. Grafstein, B., Forman, D.S., and McEwan, B.S. (1972). Effects of temperature on axonal transport and turnover of protein in goldfish optic system. Exp. Neurol. 34:158.
55. Pleasure, D.E., Mishler, K.C., and Engel, C. (1969). Axonal transport of proteins in experimental neuropathies. Science 166:524.
56. Fibiger, H.C., Pudritz, R.E., McGeer, P.L., and McGeer, E.G. (1972). Axonal transport in nigrostriatal and nigrothalamic neurons: Effects of medial forebrain bundle lesions and 6-hydroxydopamine. J. Neurochem. 19:1697.
57. Ungerstedt, U. (1971). Stereotaxic mapping of monoamine pathways in the rat brain. Acta. Physiol. Scand. Suppl. 367:95.
58. Thoenen, H. (1972). Neuronally mediated enzyme induction in adrenergic neurons and adrenal chromaffin cells. Biochem. Soc. Symp. 36:3.
59. Sorimachi, M. (1975). Susceptibility of catecholaminergic cell bodies to 6-hydroxydopamine. Brain Res. 88:572.
60. Jacobowitz, D., and Kostrzewa, R. (1971). Selective action of 6-hydroxydopa on noradrenergic terminals: mapping of preterminal axons of the brain. Life Sci. 10:1329.
61. Baumgarten, H.G., and Lachenmayer, L. (1972). 5, 7-Dihydroxytryptamine: Improvement in chemical lesioning of indoleamine neurons in the mammalian brain. Z. Zellforsch. 135:399.

62. Baumgarten, H.G., and Lachenmayer, L. (1973). Falsche Übertragerstoffe Im Gehirn. Dtsch. Med. Wschr. 98:574.
63. Bradley, W.G., and Jaros, E. (1973). Axoplasmic flow in axonal neuropathies. II. Axoplasmic flow in mice with motor neuron disease and muscular dystrophy. Brain 96:247.
64. Komiya, Y., and Austin, L. (1974). Axoplasmic flow of protein in the sciatic nerve of normal and dystrophic mice. Exp. Neurol. 43:1.
65. Tang, B.Y., Komiya, Y., and Austin, L. (1974). Axoplasmic flow of phospholipids and cholesterol in the sciatic nerve of normal and dystrophic mice. Exp. Neurol. 43:13.
66. Karlsson, J.O., and Sjöstrand, J. (1971). Transport of microtubular protein in axons of retinal ganglion cells. J. Neurochem. 16:975.
67. James, K.A.C., and Austin, L. (1970). The binding in vitro of colchicine to axoplasmic proteins from chicken sciatic nerve. Biochem. J. 117:773.
68. Dahlström, A. (1968). Effect of colchicine on transport of amine storage granules in sympathetic nerves of rat. Eur. J. Pharmacol. 5:111.
69. Banks, P., Mayor, D., Mitchell, M., and Tomlinson, D. (1971). Studies on the translocation of noradrenaline-containing vesicles in post-ganglionic sympathetic neurons in vitro. Inhibition of movement by colchicine and vinblastine and evidence for the involvement of axonal microtubules. J. Physiol. (Lond.) 216:625.
70. Dahlström, A., and Häggendal, J. (1966). Studies on the transport and life-span of amine storage granules in a peripheral adrenergic neuron system. Acta. Physiol. Scand. 67:278.
71. Livett, B.G., Geffen, L.B., and Austin, L. (1968). Proximo-distal transport of [^{14}C] noradrenaline and protein in sympathetic nerves. J. Neurochem. 15:931.
72. Schmitt, F.O. (1968). Fibrous proteins–neuronal organelles. Proc. Natl. Acad. Sci. USA 60:1092.
73. Banks, P., Mayor, D., and Mrax, P. (1973). Metabolic aspects of the synthesis and intra-axonal transport of noradrenaline storage vesicles. J. Physiol. (Lond.) 229:383.
74. Ochs, S. (1971). Characteristics and a model for fast axoplasmic transport in nerve. J. Neurobiol. 2:331.
75. Ochs, S. (1972). Fast axoplasmic transport of materials in mammalian nerve and its integrative role. Ann. N.Y. Acad. Sci. 193:43.
76. Tennyson, V.M. (1974). *In* J.P. Dyck, P.K. Thomas, and E.H. Lambert (eds.), Peripheral Neuropathy. W.B. Saunders Co., Philadelphia [quoted in Ochs, 1974].
77. Khan, M.A., and Ochs, S. (1974). Magnesium or calcium activated ATPase in mammalian nerve. Brain Res. 81:413.
78. Wellington, B.W., Austin, L., and Jeffrey, P.L. Magnesium or calcium activated nucleotide triphosphatase in chicken sciatic nerve. In preparation.
79. Wellington, B.S., Livett, B.G., Jeffrey, P.L., and Austin, L. (1975). Biochemical and immunochemical studies on chick brain neurostenin. Neuroscience 1:22.
80. De Potter, P., De Schaepdryver, H., Moerman, E., and Smith, A.D. (1969). Evidence for the release of vesicle proteins together with noradrenaline upon stimulation of the splenic nerves. J. Physiol. (Lond.) 204:102.
81. Geffen, L.B., Livett, B.G., and Rush, R.A. (1969). Immunological localization of chromogranins in sheep sympathetic neurons and their release by nerve impulses. J. Physiol. (Lond.) 204:58.
82. Uttenthal, L.O., Livett, B.G., and Hope, D.B. (1971). Release of neurophysin together with vasopressin by a Ca^{++}-dependent mechanism. Phil. Trans. Roy. Soc. Lond. B. 261:379.
83. Ceccarelli, B., Hurlbut, W.O., and Mauro, A. (1972). Depletion of vesicles from frog neuromuscular junctions by prolonged tetanic stimulation. J. Cell. Biol. 54:30.
84. Heuser, T.S., and Reese, J.E. (1973). Evidence for recycling of synaptic vesicle membrane during transmitter release at the frog neuromuscular junction. J. Cell. Biol. 57:315.
85. Porter, C.W., and Barnard, E.A. (1975). The density of cholinergic receptors at the endplate postsynaptic membrane: ultrastructural studies in two mammalian species. J. Membrane Biol. 20:31.
86. Couteaux, R., and Pecot-Dechavassine, M. (1970). Vesicules synaptiques et poches au

niveau des zones actives de la junction neuromusculaire. Acad. Sci. (Paris) D. 271:2346.
87. Heuser, J.E., Reese, T.S., and Landis, D.M.D. (1974). Functional changes in frog neuromuscular junctions studied with freeze-fracture. J. Neurocytol. 3:109.
88. Hoss, W., and Abood, L.G. (1974). Fluidity in hydrophobic protein regions of synaptic membranes. Eur. J. Biochem. 50:177.
89. Livett, B.G., Fenwick, E.M., and Eadie, J. (1974). Ca^{2+}-Dependent redistribution of antigenic components during membrane fusion. Proc. Aust. Biochem. Soc. 7:57.
90. Howe, P.R.C., Fenwick, E.M., Rostas, J.A.P., and Livett, B.G. (1975). Further evidence for incorporation of synaptic vesicle proteins into synaptosomal plasma membranes. Proc. Aust. Biochem. Soc. 8:95.
91. Jeffrey, P.L., Fenwick, E.M., Howe, P.R.C., Rostas, J.A.P., and Livett, B.G. (1975). Chick brain synaptic membranes: Biochemical characterization and immunological evidence for common antigens in synaptic plasma membranes and vesicles. Neuroscience. In press.
92. Birks, R.I., Huxley, H.E., and Katz, B. (1960). The fine structure of the neuromuscular junction of the frog. J. Physiol. (Lond.) 150:134.
93. Thesleff, S., Vyskočil, F., and Ward, M.R. (1974). The action potential in end-plate and extrajunctional regions of rat skeletal muscle. Local postsynaptic sites of regenerative electrical response. Acta. Physiol. Scand. 91:196.
94. Axelsson, J., and Thesleff, S. (1959). A study of supersensitivity in denervated mammalian skeletal muscle. J. Physiol. (Lond.) 147:178.
95. Diamond, J., and Miledi, R. (1962). A study of fetal and newborn rat muscle fibers. J. Physiol. (Lond.) 162:393.
96. Lømø, T., and Rosenthal, J. (1972). Control of ACh sensitivity by muscle activity in the rat. J. Physiol. (Lond.) 221:493.
97. Drachman, D., and Witzke, F. (1972). Trophic regulation of acetylcholine sensitivity of electrical stimulation. Science 176:514.
98. Fischbach, G.D., and Cohen, S.A. (1973). The distribution of acetylcholine sensitivity over uninnervated and innervated muscle fibers grown in cell culture. Dev. Biol. 31:147.
99. Fishbach, G.D., and Robbins, N. (1971). Effect of chronic disuse of rat soleus neuro-muscular junctions on post-synaptic membrane. J. Neurophysiol. 34:562.
100. Miledi, R., and Potter, L. (1972). Acetylcholine receptors in muscle fibers. Nature 233:599.
101. Chang, C.C., Chen, T.F., and Chuang, S.T. (1973). N,O-di and N,N,O-tri-[^{3}H] acetyl α-bungarotoxins as specific labeling agents of cholinergic receptors. Brit. J. Pharmacol. 47:147.
102. Chang, C.C., and Huang, M.C. (1975). Turnover of junctional and extrajunctional acetylcholine receptors of the rat diaphragm. Nature 253:643.
103. Fambrough, D.M., Hartzell, H.C., Powell, J.A., Rash, J.E., and Joseph, N. (1974). On the differentiation and organization of the surface membrane of a postsynaptic cell. The skeletal muscle fiber. *In* M.V.L. Bennett (ed.), Synaptic Transmission and Neuronal Interaction, Vol. 28, 285 pp. Society of General Physiologists Series, Raven Press, New York.
104. Raisman, G., and Field, P.M. (1973). A quantitative investigation of the development of collateral reinnervation after partial deafferentation of the septal nuclei. Brain Res. 50:241.
105. Pfenninger, K.H., Bunge, M.B., and Bunge, R.P. (1974). Nerve growth cone plasmalemma–its structure and development. 8th. Int. Congress on Electron Microscopy, Canberra, 1974. *2:*234.
106. Mark, R. (1974). Memory and nerve cell connections. Criticisms and contributions from Developmental Neurophysiology. Clarendon Press, Oxford.
107. Globus, A., Lux, H.D., and Schubert, P. (1973). Transfer of amino acids between neuroglia cells and neurons in the leech ganglion. Exp. Neurol. 40:104.
108. Droz, B., Keonig, H.L., and Di Giamberardino, L. (1973). Axonal migration of protein and glycoprotein to nerve endings. I. Radioautographic analysis of the renewal of protein in nerve endings of chicken ciliary ganglion after intracerebral injection of [^{3}H] lysine. Brain Res. 60:93.

109. Korr, I.M., Wilkinson, P.N., and Chornock, F.W. (1967). Axonal delivery of neuroplasmic components to muscle cells. Science 155:342.
110. Korr, I.M., and Appeltauer, G.S.L. (1970). Continued studies on the axonal transport of nerve proteins to muscle. J. Amer. Osteopath. Assn. 69:1040.
111. Korr, I.M., and Appeltauer, G.S.L. (1971). Axonal transport of nerve cell proteins to muscle. Fed. Proc. 30:665.
112. Appeltauer, G.S., and Korr, I.M. (1975). Axonal delivery of soluble, insoluble and electrophoretic fractions of neuronal proteins to muscle. Exp. Neurol. 46:132.
113. Hendrickson, A.E. (1972). Electron microscopic distribution of axoplasmic transport. J. Comp. Neurol. 144:381.
114. Schubert, P., and Kreutzberg, G.W. (1974). Axonal transport of adenosine and uridine derivatives and transfer to post synaptic neurons. Brain Res. 76:526.
115. Kristensson, K., Olsson, Y., and Sjöstrand, J. (1971). Axonal uptake and retrograde transport of exogenous proteins in the hypoglossal nerve. Brain Res. 32:399.
116. Kristensson, K., and Olsson, Y. (1971). Retrograde axonal transport of protein. Brain Res. 29:363.
117. La Vail, J.H., and La Vail, M.M. (1972). Retrograde axonal transport in the central nervous system. Science 176:1416.
118. La Vail, M.M., and La Vail, J.H. (1975). Retrograde intra-axonal transport of horseradish peroxidase in retinal ganglion cells of the chick. Brain Res. 85:273.
119. Bunt, A.H. (1969). Formation of coated and "synaptic" vesicles within neurosecretory axon terminals of the crustacean sinus gland. J. Ultrastruc. Res. 28:411.
120. Heuser, J.E., and Reese, T.S. (1974). Morphology of synaptic vesicle discharge and reformation at the frog neuromuscular junction. *In* M.V.L. Bennett (ed.), Synaptic Transmission and Neuronal Interaction–Society of General Physiologists Series Vol. 28, p. 59. Raven Press, New York.
121. Holtzman, E., Freeman, A.R., and Kashner, L.A. (1971). Stimulation dependent alterations in peroxidase uptake at lobster neuromuscular junctions. Science 173:733.
122. Pysh, J.J., and Wiley, R.G. (1974). Synaptic vesicle depletion and recovery in cat sympathetic ganglia electrically stimulated. J. Cell. Biol. 60:365.
123. La Vail, J.H., and La Vail, M.M. (1974). The retrograde intra-axonal transport of horseradish peroxidase in the chick visual system: a light and electron microscopic study. J. Comp. Neurol. 157:303.
124. Bunt, A.H., and Lund, R.D. (1974). Blockage by vinblastine of ortho- and retrograde axonal transport in retinal ganglion cells. Anat. Rec. 178:507.
125. Kristensson, K., and Sjöstrand, J. (1972). Retrograde transport of protein tracer in the rabbit hypoglossal nerve during regeneration. Brain Res. 45:175.
126. Ukena, T.E., and Berlin, R.D. (1972). Effect of colchicine and vinblastine on the topographic separation of membrane junctions. J. Exp. Med. 136:1.
127. Paulson, J.C., and McClure, W.O. (1974). Microtubules and axoplasmic transport. Brain Res. 73:333.
128. Holtzman, E., Teichberg, S., Abrahams, S.J., Citkowitz, E., Crain, S.M., Kawai, N., and Peterson, E. (1973). Notes on synaptic vesicles and related structures, endoplasmic reticulum, lysosomes and peroxisomes in nervous tissue and adrenal medulla. J. Histochem. Cytochem. 21:349.
129. Singer, M. (1974). Neurotrophic control of limb regeneration in the newt. Ann. N.Y. Acad. Sci. 228:308.
130. Clark, G. (ed.) (1973). "Staining Procedures"–used by the Biological Stain Commission. 3rd Ed. The Williams and Wilkins Co., Baltimore.
131. Mallory, F.B. (1938). Pathological Technique. W.B. Saunders Co., Philadelphia.
132. Heidenhain, M. (1915). Über die Mallorysche Bindegewebsfärbung mit Karmin und Azokarmin als Vorfarben. Ztschr. f. Wissensch Mikr. 32:361.
133. Nonidez, J.F. (1939). Studies on the innervation of the heart. I. Amer. J. Anat. 65:361.
134. Holmes, W. (1947). The peripheral nerve biopsy. *In* S.C. Dyke (ed.), Recent Advances in Clinical Pathology. p. 402. Churchill, London.
135. Collonnier, M. (1964). The tangential organization of the visual cortex. J. Anat. 98:327.
136. Cox, W.H. (1891). Imprägnation des centralen Nervensystems mit Quecksilbersalzen. Arch. Mikrosk. Anat. 37:16.

137. Stell, W.K. (1965). Correlation of retinal cytoarchitecture and ultrastructure in Golgi preparations. Anat. Rec. 153:389.
138. Nauta, W.J.H., and Ebbesson, S.O.E. (eds.) (1970). Contemporary Research Methods in Neuroanatomy. Springer, Berlin, Heidelberg and New York.
139. Kolb, H. (1970). Organization of the outer plexiform layer of the primate retina: electron microscopy of Golgi-impregnated cells. Phil. Trans. Roy. Soc. B 258:261.
140. Nissl, F. (1885). Über die Untersuchungsmethoden der Grosshirnrinde. Neurol. Centralbl. 4:500.
141. Nissl, F. (1894). Über eine neue Untersuchungsmethode des centralorgans speciell zür Festellung der Localization der Nervenzellen. Neurol. Centralbl. 13:507.
142. Sheehan, D.C., and Hrapchak, B.B. (1973). Theory and practice of histotechnology. The C.V. Mosby Co., St. Louis.
143. McConnell, C.H. (1932). The development of the ectodermal nerve net in the buds of Hydra. Quart. J. Microsc. Sci. 75:495.
144. Bodian, D. (1936). A new method for staining nerve fibers and nerve endings in mounted paraffin sections. Anat. Rec. 65:89.
145. Romanes, G.J. (1950). The staining of nerve fibers in paraffin sections with silver. J. Anat. 84:104.
146. Nauta, W.J.H., and Gygax, P.A. (1954). Silver impregnation of degenerating axons in the central nervous system; a modified technique. Stain Technol. 29:91.
147. De Myer, W. (1958). Impregnation of axons and terminal buttons in routine, paraffin or frozen sections of central and peripheral nervous tissue: Adaptation of Hortega's silver carbonate method for neurofibrils. Amer. J. Clin. Path. 29:449.
148. Ráliš, H.M., Beesley, R.A., and Ráliš, Z.A. (1973). Techniques in Neurohistology. Butterworths, London.
149. Bielschowsky, M. (1908). Eine Modifikation meines Silberimprägnations verfahrens zur Darstellung der Neurofibrillen. J. Psychol. Neurol. 12:135.
150. Von Braunmuhl, A. (1957). Alterserkrankungen des Zentralnervensystems Senile Involution, Senile Demenz, Alzheimersche Krankheit. *In* O. Lubarsch and F. Hinke (eds.), Hdbel. d.Spez. Pathol. Vol. XIII, IA, 337 pp. Springer, Heidelberg.
151. Pal, J. (1886). Ein Beitrag zur Nervenfärbetechnik. Wien.med.Jahrb. N.F. 1:619. Abstracted in Z.Wissen. Mikr. 4:92.
152. Weigert, K. (1885). Eine Vergesserung der Haematoxylin Blutaugensalzmethode für das Centralnervensystem. Fortschr. d. Med. 3:236.
153. Weil, A. (1928). A rapid method for staining myelin sheaths. Arch.Neurol.Psychiat. (Lond.) 20:392.
154. Berube, G.R., Powers, M.M., and Clark, G. (1965). Iron hematoxylin chelates. I. The Weil staining bath. Stain Technol. 40:53.
155. Eager, R.P. (1973). A method for degenerating axons. Cited in Ralis, H.M., Beesley, R.A., and Ralis, Z.A., Techniques in Neurohistology, 99. Butterworth, London.
156. Hjorth-Simonsen, A. (1970). Fink-Heimer silver impregnation of degenerating axons and terminals in mounted cryostat sections of fresh and fixed brains. Stain Technol. 45:199.
157. Fink, R.R., and Heimer, L. (1967). Two methods for selective silver impregnation of degenerating axons and their synaptic endings in the central nervous system. Brain Res. 4:369.
158. Guillery, R.W., Shirra, B., and Webster, K.E. (1961). Differential impregnation of degenerating nerve fibers in paraffin-embedded material. Stain Technol. 36:9.
159. Glees, P. (1946). Terminal degeneration within the central nervous system as studied by a new silver method. J.Neuropathol.Exp.Neurol. 5:54.
160. Nauta, W.J.H., and Gygax, P.A. (1951). Silver impregnation of degenerating axon terminals in the central nervous system: (1) technic (2) chemical notes. Stain Technol. 26:5.
161. Guillery, R.W. (1965). Some electron microscopical observations of degenerative changes in central synapses. Progr. Brain Res. 14:57.
162. Gudden, B. (1870). Experimentaluntersuchungen über das peripherische und zentrale Nervensystem. Arch.Psychiat. Nervenkrankh. 2:693.
163. Leiberman, A.R. (1974). Some factors affecting retrograde neuronal responses to axonal lesions. *In* R. Bellairs and E.G. Gray (eds.), Essays on the Nervous System, p. 71. Clarendon Press, Oxford.

164. Lavelle, A. (1973). Levels of maturation and reactions to injury during neuronal development. Progr.Brain Res. 40:161.
165. Brattgard, S.O., Edström, J.E., and Hyden, H. (1957). The chemical changes in regenerating neurons. J. Neurochem. 1:316.
166. Watson, W.E. (1965). An autoradiographic study of the incorporation of nucleic acid precursors by neurones and glia during nerve regeneration. J.Physiol. (Lond.) 180:741.
167. Watson, W.E. (1968). Observations on the nucleolar and total cell body nucleic acid of injured nerve cells. J. Physiol. (Lond.) 196:655.
168. Gunning, P.W., Kaye, P.L., and Austin, L. The synthesis of rapidly-labeled RNA in the nodose ganglion and vagus nerve following nerve injury. Exp. Neurol. In press.
169. Adams, D.H., and Fox, M.E. (1969). Some studies on rat brain microsomes in relation to growth and development. Brain Res. 12:157.
170. Pannese, E. (1963). Investigations on the ultrastructural changes of the spinal ganglion neurons in the course of axon regeneration and cell hypertrophy. I. Changes during axon regeneration. Z.Zellforsch 60:711.
171. Cragg, B.G. (1970). What is the signal for chromatolysis? Brain Res. 23:1.
172. Brodal, A. (1940). Modification of Gudden method for study of cerebral localization. Arch.Neurol.Psychiatr.Chic. 43:46.
173. Grant, G. (1970). Neuronal changes central to the site of axon transection. A method for the identification of retrograde changes in perikarya, dendrites and axons by silver impregnation. *In* W.J.H. Nauta and S.O.E. Ebbeson (eds.), Contemporary Research Methods in Neuroanatomy, 173 pp. Springer Verlag, Berlin, Heidelberg, New York.
174. Erulkar, S.D., Nichols, C.W., Popp, M.B., and Koelle, G.B. (1968). Renshaw elements: localization and acetylcholinesterase content. J.Histochem. 16:128.
175. Stretton, A.O.W., and Kravitz, E.A. (1968). Neuronal geometry: determination with a technique of intracellular dye injection. Science 162:132.
176. Falck, B., Hillarp, N-Å, Thieme, G., and Torp, A. (1962). Fluorescence of catecholamines and related compounds condensed with formaldehyde. J.Histochem.Cytochem. 10:348.
177. Jonsson, G. (1973). Quantitation of biogenic monoamines demonstrated with the formaldehyde fluorescence method. *In* A.A. Thaer and M. Sernetz (eds.), Fluorescence Techniques in Cell Biology, p. 191. Springer-Verlag, Berlin.
178. Björklund, A., and Falck, B. (1973). Cytofluorometry of biogenic monoamines in the Falck-Hillarp method. Structural identification by spectral analysis. *In* A.A. Thaer and M. Sernetz (eds.), Fluorescence Techniques in Cell Biology, p. 171. Springer-Verlag, Berlin.
179. Eränkö, O., and Raisanen, L. (1966). Demonstration of catecholamines in adrenergic nerve fibers by fixation in aqueous formaldehyde solutions and fluorescence microscopy. J. Histochem. Cytochem. 14:690.
180. Sakharova, A.V., and Sakharov, D.A. (1971). Visualization of intraneuronal monoamines by treatment with formaldehyde solution. *In* O. Eranko (ed.), Histochemistry of nervous transmission, Progr.Brain Res. 34:100. Elsevier, Amsterdam, London, and New York.
181. Laties, A.M., Lund, R., and Jacobowitz, D. (1967). A simplified method for the histochemical localization of cardiac catecholamine-containing nerve fibers. J. Histochem. Cytochem. 15:535.
182. Hökfelt, T., and Ljungdahl, A. (1972). Modification of the Falck-Hillarp formaldehyde fluorescence method using the vibratome: simple, rapid and sensitive localization of catecholamines in sections of unfixed or formalin fixed brain tissue. Histochemie 29:325.
183. Ljungdahl, A., Hökfelt, T., Goldstein, M., and Park, D. (1975). Retrograde peroxidase tracing of neurons combined with transmitter histochemistry. Brain Res. 84:313.
184. Björklund, A., Falck, B., and Håkanson, R. (1968). Histochemical demonstration of tryptamine. Properties of the formaldehyde-induced fluorophores of tryptamine and related indole compounds in models. Acta Physiol. Scand. Suppl. 318:1.
185. Björklund, A., and Falck, B. (1969). Histochemical characterization of a tryptamine-like substance stored in cells of the mammalian adenohypophysis. Acta Physiol. Scand. 77:475.
186. Björklund, A., and Stenevi, U. (1970). Acid catalysis of the formaldehyde condensa-

tion reaction for sensitive histochemical demonstration of tryptamines and 3-methoxylated phenylethylamines. I. Model experiments. J.Histochem. Cytochem. 18:794.
187. Björklund, A., Nobin, A., and Stenevi, U. (1971). Acid catalysis of the formaldehyde condensation reaction for a sensitive histochemical demonstration of tryptamines and 3-methoxylated phenlethylamines. 2. Characterization of amine fluorophores and application to tissues. J. Histochem. Cytochem. 19:286.
188. Angelakos, E.T., and King, M.P. (1967). Demonstration of nerve terminals containing adrenaline by a new histochemical technique. Nature 213:393.
189. Juhlin, L., and Shelley, W.B. (1966). Detection of histamine by a new fluorescent *o*-phtalaldehyde stain. J. Histochem. Cytochem. 14:525.
190. Dahlström, A., and Fuxe, K. (1964). A method for the demonstration of adrenergic nerve fibers in peripheral nerves. Z.Zellforschung 62:602.
191. Eränkö, O., and Härkönen, M. (1965). Effect of axon division on the distribution of noradrenaline and acetylcholinesterase in sympathetic neurons of the rat. Acta Physiol. Scand. 63:411.
192. Geffen, L.B., and Livett, B.G. (1971). Synaptic vesicles in sympathetic neurons. Physiol. Rev. 51:98.
193. Kapeller, K., and Mayor, D. (1967). The accumulation of noradrenaline in constricted sympathetic nerves as studied by fluorescence and electron microscopy. Proc. Roy. Soc. Lond. B 167:282.
194. Geffen, L.B., and Ostberg, A. (1969). Distribution of granular vesicles in normal and constricted sympathetic neurons. J. Physiol. (Lond.) 204:583.
195. Laduron, P., and Belpaire, F. (1968). Transport of noradrenaline and dopamine β-hydroxylase in sympathetic nerves. Life Sci. 7:1.
196. Livett, B.G., Geffen, L.B., and Rush, R.A. (1968). Immunohistochemical evidence for the transport of dopamine-β-hydroxylase and a catecholamine binding protein in sympathetic nerves. Biochem. Pharmacol. 18:923.
197. Brimijoin, S. (1972). Transport and turnover of dopamine-β-hydroxylase (EC 1.14.2.1) in sympathetic nerves of the rat. J. Neurochem. 19:2183.
198. Coyle, J.T., and Wooten, G.F. (1972). Rapid axonal transport of tyrosine hydroxylase and dopamine-β-hydroxylase. Brain Res. 44:701.
199. Norberg, K.A., Risley, P.L., and Ungerstedt, U. (1971). Adrenergic innervation of the male reproductive ducts in some mammals. II. Effects of vasectomy and castration. Experientia 23:392.
200. Andén, N.E., Fuxe, K., Hamberger, B., and Hökfelt, T. (1966). A quantitative study on the nigro-neostriatal dopamine neuron system in rat. Acta. Physiol. Scand. 67:306.
201. Kobayashi, R.M., Palkovit, M., Kopin, I.J., and Jacobowitz, D.M. (1974). Biochemical mapping of noradrenergic nerves arising from the rat locus coeruleus. Brain Res. 77:269.
202. Kapeller, K., and Mayor, D. (1969). An electron microscopic study of the early changes proximal to a constriction in sympathetic nerves. Proc. Roy. Soc. Lond. 172:39.
203. Livett, B.G. (1973). Histochemical visualization of peripheral and central adrenergic neurons. Brit. Med. Bull. 29:93.
204. Geffen, L.B., and Rush, R.A. (1968). Transport of noradrenaline in sympathetic nerves and the effect of nerve impulses on its contribution to transmitter stores. J. Neurochem. 15:925.
205. Folkow, B. (1952). Impulse frequency in sympathetic vasomotor fibers correlated with the release and elimination of the transmitter. Acta Physiol. Scand. 25:49.
206. Celander, O. (1954). The range of control exercised by the sympathico-adrenal system. Acta Physiol. Scand. Suppl. 116:32.
207. De Potter, W.P. (1971). Noradrenaline storage particles in splenic nerve. Phil.Trans. Roy.Soc.B 261:313.
208. Geffen, L.B. Biochemical and histochemical methods of tracing transmitter-specific neuronal molecules (with special reference to adrenergic neurons). Prog. Brain Res. In press.
209. Weiner, N. (1970). Regulation of norepinephrine synthesis. Annu.Rev.Pharmacol. 10:273.
210. Molinoff, P.B., and Axelrod, J. (1971). Biochemistry of catecholamines. Annu.Rev. Biochem. 40:465.

211. Thoenen, H., Hendry, I.A., Stoeckel, K., Paravicini, U., and Oesch, F. (1974). Regulation of enzyme synthesis by neuronal activity and nerve growth factor. *In* K. Fuxe, L. Olson, and Y. Zotterman (eds.), Dynamics of Degeneration and Growth in Neurons, p. 315. Pergamon Press, Oxford.
212. Thoenen, H., Tranzer, J.P., and Häusler, G. (1970). Chemical sympathectomy with 6-hydroxydopamine. *In* H.J. Schumann and G. Kroneberg (eds.), New aspects of storage and release mechanisms of catecholamines, p. 130. Proc. Bayer Symp. II held at Grosse Ledder near Cologne, October 9–12, 1969. Springer, Berlin.
213. Ungerstedt, U. (1971). Stereotaxic mapping of monoamine pathways in the rat brain. Acta Physiol. Scand. Suppl. 367:1.
214. Cowan, W.M., Gottlieb, D.I., Hendrickson, A.I., Price, J.L., and Woolsey, T.A. (1972). The autoradiographic demonstration of axonal connections in the central nervous system. Brain Res. 37:2.
215. McGeer, E.G., Searle, E., and Fibiger, H.C. (1975). Chemical specificity of dopamine transport in the nigrostriatal tract. J.Neurochem. 24:283.
216. McGeer, P.L., Fibiger, H.C., Hattori, T., Singh, V.K., McGeer, E.G., and Maler, L. (1974). Biochemical neuroanatomy of the basal ganglia. *In* R.D. Myers and R.R. Drucker-Colin (eds.), Neurohumoral Coding of Brain Function, Plenum, New York. Adv.Behav.Biol. 10:27.
217. Fibiger, H.C., McGeer, E.G., and Atmadja, S. (1973). Axoplasmic transport of dopamine in nigro-striatal neurons. J. Neurochem. 21:373.
218. Pickel, V.M., Segal, M., and Bloom, F.E. (1974). A radiographic study of the efferent pathways of the nucleus locus coeruleus. J. Comp. Neurol. 155:15.
219. Lindvall, O., Björklund, A., Hökfelt, T., and Ljungdahl, A. (1973). Application of the glyoxylic acid method to vibratome sections for the improved visualization of central catecholamine neurons. Histochemie 35:31.
220. Björklund, A., Falck, B., and Stenevi, U. (1974). Microspectrofluoremetric characterization of monoamines in the central nervous system: evidence for a new neuronal monoamine-like compound. Prog. Brain. Res. 34:63.
221. Axelsson, S., Björklund, A., Falck, B., Lindvall, O., and Svensson, L.-A. (1973). Glyoxylic acid condensation: A new fluorescence method for the histochemical demonstration of biogenic monoamines. Acta Physiol. Scand. 87:57.
222. Lindvall, O., and Björklund, A. (1974). The organization of the ascending catecholamine neuron systems in the rat brain. As revealed by the glyoxylic acid fluorescence method. Acta Physiol. Scand. Suppl. 412:1.
223. Livett, B.G., Geffen, L.B., and Rush, R.A. (1969). Immunohistochemical evidence for the transport of dopamine β-hydroxylase and a catecholamine binding protein in sympathetic nerves. Biochem. Pharmacol. 15:923.
224. Hartman, B.K., and Udenfriend, S. (1970). Immunofluorescent localization of dopamine β-hydroxylase in tissues. Molec.Pharmacol. 6: 85.
225. Geffen, L.B., Livett, B.G., and Rush, R.A. (1970). Immunohistochemical localization of chromogranins in sheep sympathetic neurons and their release by nerve impulses. *In* H.J. Schumann and G. Kroneberg (eds.), New Aspects of Storage and Release Mechanisms of Catecholamines, p. 58. Springer Verlag, Berlin.
226. Cheah, T.B., and Geffen, L.B. (1970). Immunofluorescent staining of dopamine β-hydroxylase in the central nervous system. Proc.Aust.Physiol.Pharmacol.Soc. I:22.
227. Fuxe, K., Goldstein, M., Hökfelt, T., and Joh, T.H. (1972). Cellular localization of dopamine β-hydroxylase and phenylethanolamine N-methyl transferase as revealed by immunohistochemistry. *In* O. Eranko (ed.), Histochemistry of Nervous Transmission, Prog. in Brain Res. 34:127. Elsevier, Amsterdam.
228. Hartman, B.H., Zide, D., and Udenfriend, S. (1972). The use of dopamine β-hydroxylase as a marker for the central noradrenergic nervous system in rat brain. Proc. Natl. Acad. Sci. USA 69:2722.
229. Pickel, V.M., Joh, T.H., Field, P.M., Becker, C.G., and Reiss, D.J. (1975). Cellular localization of tyrosine hydroxylase by immunohistochemistry. J.Histochem. Cytochem. 23: 1.
230. Goldstein, M., Fuxe, K., and Hökfelt, T. (1972). Characterization and tissue localization of catecholamine synthesizing enzymes. Pharmacol. Rev. 24:293.
231. Hökfelt, T., Fuxe, K., and Goldstein, M. (1973). Immunohistochemical studies on monoamine containing cell systems. Brain Res. 62:461.

232. Hökfelt, T., Fuxe, K., Goldstein, M., and Johansson, O. (1974). Immunohistochemical evidence for the existence of adrenaline neurons in the rat brain. Brain Res. 66:235.
233. Olson, M., and Fuxe, K. (1971). On the projections from the locus coeruleus noradrenaline neurons: the cerebellar innervation. Brain Res. 28:165.
234. Tohyama, M., Maeda, T., and Shimizu, N. (1974). Detailed noradrenaline pathways of locus coeruleus neuron to the cerebral cortex with use of 6-hydroxydopa. Brain Res. 79:139.
235. Kostrzewa, R., and Jacobowitz, D. (1973). Acute effects of 6-hydroxydopa on central monoaminergic neurons. Eur. J. Pharmacol. 21:70.
236. Pickel, V.M., Joh, T.H., and Reiss, D.J. (1975). Immunohistochemical localization of tyrosine hydroxylase in brain by light and electron microscopy. Brain Res. 85:295.
237. Hartman, B.K. (1973). Immunofluorescence of dopamine β-hydroxylase application of improved methodology to the localization of the peripheral and central noradrenergic nervous system. J. Histochem. Cytochem. 21:312.
238. Björklund, A., and Stenevi, U. (1971). Growth of central catecholamine neurons into smooth muscle grafts in the rat mesencephalon. Brain Res. 31:1.
239. Hamilton, B.L. (1973). Projections of the nuclei of the periaqueductal gray matter in the cat. J.Comp.Neurol. 152:45.
240. Hornykiewicz, O. (1973). Dopamine in the basal ganglia–its role and therapeutic implications (including the clinical use of L-dopa). Brit. Med. Bull. 29:172.
241. Fuxe, K., and Hökfelt, T. (1971). Histochemical fluorescence detection of changes in central monoamine neurons provoked by drugs acting on the CNS. Triangle 10:73.
242. Thierry, A.M., Stinus, L., Blanc, G., and Glowinski, J. (1973). Some evidence for the existence of dopaminergic neurons in the rat cortex. Brain Res. 50:230.
243. Björklund, A., Falck, B., Hromek, F., Owman, C., and West, K.A. (1970). Identification and terminal distribution of the tubero-hypophyseal monoamine fiber systems in the rat by means of stereotaxic and microspectrofluorimetric techniques. Brain Res. 17:1.
244. Roberts, E., and Frankel, S. (1950). γ-Aminobutyric acid. Fed. Proc. 9:219.
245. Iversen, L.L. (1971). Role of transmitter uptake mechanisms in synaptic neurotransmission. Brit. J. Pharmacol. 41:571.
246. Burry, R.W., and Lasher, R.S. (1975). Uptake of GABA in dispersed cell cultures of postnatal rat cerebellum: an electron microscope autoradiographic study. Brain Res. 88:582.
247. Hökfelt, T., and Ljungdahl, A. (1972). Autoradiographic identification of cerebral and cerebellar cortical neurons accumulating labeled gamma-aminobutyric acid (^{3}H-GABA). Exp. Brain Res. 14:354.
248. Schon, F., and Iversen, L.L. (1974). The use of autoradiographic techniques for the identification and mapping of transmitter-specific neurons in the brain. Life Sci. 15:157.
249. Iversen, L. (1975). High affinity uptake of neurotransmitter amino acids. Nature 253:481.
250. Levi, G., and Raiteri, M. (1975). High affinity uptake of neurotransmitter amino acids. Nature 253:481.
251. Saito, K., Barber, R., Wu, J-Y, Matsuda, T., Roberts, E., and Vaughn, J.E. (1974). Immunohistochemical localization of glutamate decarboxylase in rat cerebellum. Proc.Natl.Acad.Sci. USA 71:269.
252. McLaughlin, B.J., Wood, J.G., Saito, K., Barber, R., Vaughn, J.E., Roberts, E., and Wu, J. (1974). The fine structural localization of glutamate decarboxylase in synaptic terminals of rodent cerebellum. Brain Res. 76:377.
253. McGeer, P.L., Hattori, T., and McGeer, E.G. (1975). Chemical and autoradiographic analysis of γ-aminobutyric acid transport in Purkinje cells of the cerebellum. Exp. Neurol. 47:26.
254. Sloper, J.C. (1972). The validity of current concepts of hypothalamo-neurohypophysial neurosecretion. Prog. Brain Res. 38:123.
255. Watkins, W.B. (1975). Immunohistochemical demonstration of neurophysin in the hypothalamo-neurohypophysial system. Int.Rev.Cytol. 41:241.
256. Leontovich, T.A. (1969). The neurons of the magnocellular neurosecretory nuclei of the dog hypothalamus. A Golgi study. J.Hirnforsch. 11:499.

257. Hayward, J.N. (1974). Physiological and morphological identification of hypothalamic magnocellular neuroendocrine cells in goldfish preoptic nucleus. J. Physiol. (Lond.) 239:103.
258. Sakharov, D.A., and Salanki, J. (1971). Study of neurosecretory cells of *Helix pomatia* by intracellular dye injection. Experientia (Basel) 27:655.
259. Winlow, W., and Kandel, E.R. The morphology of identified neurons in the abdominal ganglion of *Aplysia californica.* In preparation. (Quoted in Swindale and Benjamin (261)).
260. Donev, S. (1970). Occurrence of the neurosecretory substance during the embryonic development of the guinea pig. Z.Zellforsch. 104:517.
261. Swindale, N.V., and Benjamin, P.R. (1975). Dark field illumination of material stained for neurosecretion. Brain Res. 89:175.
262. Maynard, D.M. (1961). Thoracic neurosecretory structures in brachyura. II. Secretory neurons. Gen.Comp.Endocr. 1:237.
263. Sherlock, D.A., Field, P.M., and Raisman, G. (1975). Retrograde transport of horseradish peroxidase in the magnocellular neurosecretory system of the rat. Brain Res. 88:403.
264. Livett, B.G. (1975). Immunochemical studies on the storage and axonal transport of neurophysins in the hypothalamo-neurohypophyseal system. Ann. N.Y. Acad. Sci. 248:112.
265. Zimmerman, E.A., Defendini, R., Sokol, H.W., and Robinson, A.G. (1975). The distribution of neurophysin-secreting pathways in the mammalian brain: Light microscopic studies using the immunoperoxidase technique. Ann.N.Y. Acad.Sci. 248:92.
266. Watkins, W.B. (1975). Neurosecretory neurons in the hypothalamus and median eminence of the dog and sheep as revealed by immunohistochemical methods. Ann.N.Y. Acad.Sci. 248:134.
267. Pelletier, G., Leclerc, R., Labrie, F., and Puviani, R. (1974). Electron microscope immunohistochemical localization of neurophysin in the rat hypothalamus and pituitary. Molec.Cell.Endocrinol. 1:157.
268. Leclerc, R., and Pelletier, G. (1974). Electron microscope immunohistochemical localization of vasopressin in the hypothalamus and neurohypophysis of the normal and Brattleboro rat. Amer. J. Anat. 140:583.
269. De Mey, J., Vandesande, F., and Dierickx, K. (1974). Identification of neurophysin producing cells. II. Identification of the neurophysin I and the neurophysin II producing neurons in the bovine hypothalamus. Cell Tissue Res. 153:531.
270. Alvarez-Buylla, R., Livett, B.G., Uttenthal, L.O., Hope, D.B., and Milton, S.H. (1973). Immunochemical evidence for the transport of neurophysin in the hypothalamo-neurohypophysial system of the dog. Z. Zellforsch. 137:435.
271. Livett, B.G., and Parry, H.B. (1971). Accumulation of neurophysin in the median eminence and the cerebellum of sheep with natural scrapie. Brit. J. Pharmacol. 43:423.
272. Parry, H.B., and Livett, B.G. (1973). A new hypothalamic pathway to the median eminence containing neurophysin and its hypertrophy in sheep with natural scrapie. Nature 242:63.
273. Parry, H.B., and Livett, B.G. (1976). Neurophysin in the brain and pituitary gland of normal and scrapie-affected sheep. I. Its localization in the hypothalamus and neurohypophysis with particular reference to a new neurosecretory pathway to the median eminence. Neuroscience Vol. 1. In press.
274. De Mey, J., Dierickx, K., and Vandesande, F. (1975). Immunohistochemical demonstration of neurophysin I- and neurophysin II-containing nerve fibers in the external region of the bovine median eminence. Cell Tissue Res. 157:517.
275. Livett, B.G., and Parry, H.B. (1972). The distribution of vasopressin and neurophysin in the hypothalamo-distal-neurohypophysial and hypothalamo-infundibular neurosecretory systems of normal and scrapie-affected sheep. J. Physiol. (Lond.) 230:20P.
276. Vandesande, F., De Mey, J., and Dierickx, K. (1974). Identification of neurophysin producing cells. I. The origin of neurophysin-like substance containing nerve fibers of the external region of the median eminence of the rat. Cell Tissue Res. 151:187.
277. Watkins, W.B., Schwabedal, O., and Bock, R. (1974). Immunohistochemical demonstration of a CRF-associated neurophysin in the external zone of the rat median eminence. Cell Tissue Res. 152:411.

278. Höllander, H. (1974). Projections from striate cortex to the diencephalon in the squirrel monkey (*Saimiri sciureus*). A light microscopic autoradiographic study following intracortical injections of [^{3}H] leucine. J. Comp. Neurol. 155:425.
279. Nauta, H.J.W., Kaiserman-Abramof, I.R., and Lasek, R.J. (1975). Electron microscopic observations of horseradish peroxidase transported from the caudoputamen to the substantia nigra in the rat: Possible involvement of the agranular reticulum. Brain Res. 85:373.
280. Trojanowski, J.Q., and Jacobson, S. (1975). A combined horseradish peroxidase-autoradiographic investigation of reciprocal connections between superior temporal gyrus and pulvinar in squirrel monkey. Brain Res. 85:347.
281. Jacobson, S., and Trojanowski, J.Q. (1974). The cells of origin of the corpus callosum in rat, cat and rhesus monkey. Brain Res. 74:149.
282. Graybiel, A.M., and Devor, M. (1974). A microelectrophoretic delivery technique for use with horseradish peroxidase. Brain Res. 68:167.
283. Kuypers, H.G.J.M., Kievit, J., and Groen-Klevant, A.C. (1974). Retrograde axonal transport of horseradish peroxidase in rat forebrain. Brain Res. 67:211.
284. Lasek, R., Joseph, B.S., and Whitlock, D.G. (1968). Evaluation of a radioautographic neuroanatomical tracing method. Brain Res. 8:319.
285. La Vail, J.H., Winston, K.R., and Tish, A. (1973). A method based on retrograde intra-axonal transport of protein for identification of cell bodies of origin of axons terminating within the C.N.S. Brain Res. 58:470.
286. Ramon y Cajal, S. (1911). Reprinted C.S.I.C. Madrid, 1955. Histologie du Systême Nerveux L'Homme et des Vertêbrés.
287. Barber, P.C., and Field, P.M. (1975). Autoradiographic demonstration of afferent connections of the accessory olfactory bulb in the mouse. Brain Res. 85:201.
288. Crothers, S.D., and McCluer, R.H. (1975). Effect of colchicine on the delayed appearance of labeled protein into synaptosomal soluble proteins. J. Neurochem. 24:209.
289. Graham, R.C., and Karnovsky, M.J. (1966). The early stages of absorption of injected horseradish peroxidase in the proximal tubules of mouse kidney: ultrastructural cytochemistry by a new technique. J. Histochem. Cytochem. 14:291.
290. Jones, E.G. (1975). Possible determinants of the degree of retrograde neuronal labeling with horseradish peroxidase. Brain Res. 85:249.
291. Kievit, J., and Kuypers, H.G.J.M. (1975). Subcortical afferents to the frontal lobe in the rhesus monkey studied by means of retrograde horseradish peroxidase transport. Brain Res. 85:261.
292. Cowan, W.M. (1970). Anterograde and retrograde transneuronal degeneration in the central and peripheral nervous system. *In* W.J.H. Nauta and S.O.E. Ebberson (eds.), Contemporary Research Methods in Neuronanatomy, pp. 217–251. Springer Verlag, New York.
293. Sjöstrand, J. (1965). Proliferative changes in glial cells during nerve regeneration. Z. Zellforsch. 68:481.
294. Prestige, M.C. (1970). Differentiation, degeneration and the role of the periphery: quantitative considerations. *In* F.D. Schmitt (ed.), The Neurosciences Second Study Program. pp. 73–82. Rockefeller Univ. Press, New York.
295. Becker, N.H., Hirand, A., and Zimmerman, H.M. (1968). Observations of the distribution of exogenous peroxidase in the rat cerebrum. J. Neuropathol. Exp. Neurol. 27:439.
296. Holtzman, E. (1971). Cytochemical studies of protein transport in the nervous system. Phil. Trans. Roy. Soc. B 261:401.
297. Holtzman, E., and Peterson, J. (1969). Uptake of protein by mammalian neurons. J. Cell. Biol. 40:863.
298. Nagasawa, J., Douglas, W.W., and Schultz, R.A. (1971). Micropinocytotic origin of coated and smooth microvesicles (synaptic vesicles) in neurosecretory terminals of posterior pituitary glands demonstrated by incorporation of horseradish peroxidase. Nature 232:341.
299. Sellinger, O.Z., and Petiet, P.D. (1973). Horseradish peroxidase uptake *in vivo* by neuronal and glial lysosomes. Exp. Neurol. 38:370.
300. Jacobson, S., and Trojanowski, J.Q. (1975). Corticothalamic neurons and thalamo-

cortical terminal field: an investigation in rat using horseradish peroxidase and autoradiography. Brain Res. 85:385.
301. Locke, S. (1960). The projection of the medial pulvinar of the macaque. J. Comp. Neurol. 115:155.
302. Nauta, H.J.W., Pritz, M.B., and Lasek, R.J. (1974). Afferents to the rat caudoputamen studied with horseradish peroxidase. An evaluation of a retrograde neuroanatomical research method. Brain Res. 67:219.
303. Hendry, I.A., Stoeckel, K., Thoenen, H., and Iversen, L.L. (1974). Retrograde axonal transport of the nerve growth factor. Brain Res. 68:103.
304. Levi-Montalcini, R., and Angeletti, P.U. (1969). Nerve growth factor. Physiol. Rev. 48:534.
305. Angeletti, R.H., and Bradshaw, R.A. (1971). Nerve growth factor from mouse submaxillary gland. Proc. Natl. Acad. Sci. USA 68:2417.
306. Thoenen, H., Angeletti, P.U., Levi-Montalcini, R., and Kettler, R. (1971). Selective induction by nerve growth factor of tyrosine hydroxylase and dopamine β-hydroxylase in the rat superior cervical ganglia. Proc. Natl. Acad. Sci. USA 68:1598.
307. Hendry, I.A., and Iversen, L.L. (1973). Changes in tissue and plasma concentrations of nerve growth factor following removal of the submaxillary glands in adult mice and their effects on the sympathetic nervous system. Nature 243:500.
308. Hendry, I.A., and Thoenen, H. (1974). Changes of enzyme pattern in the sympathetic nervous system of adult mice after submaxillary gland removal. Response to exogenous nerve growth factor. J. Neurochem. 22:999.
309. Stoeckel, K., Paravicini, U., and Thoenen, H. (1974). Specificity of the retrograde axonal transport of nerve growth factor. Brain Res. 76:413.
310. Stoeckel, K., and Thoenen, H. (1975). Retrograde axonal transport of nerve growth factor: specificity and biological importance. Brain Res. 85:337.
311. Black, I.B., Hendry, I.A., and Iversen, L.L. (1972). The role of post-synaptic neurons in the biochemical maturation of pre-synaptic cholinergic nerve terminals in a mouse sympathetic ganglion. J. Physiol. (Lond.) 221:149.
312. Black, I.B., Hendry, I.A., and Iversen, L.L. (1971). Trans-synaptic regulation of growth and development of adrenergic neurons in a mouse sympathetic ganglion. Brain Res. 34:229.
313. Hebb, L.O., and Waites, G.M.H. (1956). Choline acetylase in antero- and retrograde degeneration of a cholinergic nerve. J. Physiol. (Lond.) 132:667.
314. Hendry, I.A. (1975). The retrograde trans-synaptic control of the development of cholinergic terminals in sympathetic ganglia. Brain Res. 86:483.
315. Thoenen, H., Saner, A., Angeletti, P.U., and Levi-Montalcini, R. (1972). Increased activity of choline acetyltransferase in sympathetic ganglia after prolonged administration of nerve growth factor. Nature New Biol. 236:26.
316. Stoeckel, K., Schwab, M., and Thoenen, H. (1975). Specificity of retrograde transport of nerve growth factor (NGF) in sensory neurons: A biochemical and morphological study. Brain Res. 89:1.
317. Berl, S., Puszkin, S., and Nicklas, W.J. (1973). Actomyosin-like protein in brain. Science 179:441.
318. McIlwain, H. (1973). Adenosine in neurohumoral and regulatory roles in the brain. *In* E. Genazzani and H. Herken (eds.), Central nervous system–studies on metabolic regulation and function, pp. 3–11. Springer, Berlin.
319. Pitman, R.M., Tweedle, C.D., and Cohen, M.J. (1972). Branching of central neurons; intracellular cobalt injection for light and electron microscopy. Science 176:412.
320. Tyrer, N.M., and Bell, E.M. (1974). The intensification of cobalt-filled neuron profiles using a modification of Timm's sulphide–silver method. Brain Res. 73:151.
321. Llinás, R. (1973). Procion yellow and cobalt as tools for the study of structure-function relationships in vertebrate central nervous systems. *In* S.B. Kater and C. Nicholson (eds.), Intracellular staining in neurobiology, p. 211. Springer, New York.
322. Mason, C.A. (1975). Delineation of the rat visual system by the axonal iontophoresis–cobalt sulphide precipitation technique. Brain Res. 85:287.
323. Fuller, P.M., and Prior, D.J. (1975). Cobalt iontophoresis techniques for tracing afferent and efferent connections in the vertebrate CNS. Brain Res. 88:211.
324. Kennedy, C., Rosiers, Des. M.H., Jehle, J.W., Reivich, M., Sharpe, F., and Sokoloff, L.

(1975). Mapping of functional neural pathways by autoradiographic survey of local metabolic rate with [^{14}C] deoxyglucose. Science 187:850.

325. Paravicini, U., Stoeckel, K., and Thoenen, H. (1975). Biological importance of retrograde axonal transport of nerve growth factor in adrenergic neurons. Brain Res. 84:279.
326. Brodal, A. (1973). Anterograde and retrograde degeneration of neurons in the central nervous system. *In* W. Haymaker and R.D. Adams (eds.), Histology and histopathology of the nervous system. Charles C. Thomas, Springfield, Ill.
327. Romanes, G. (1946). Motor localization and the effects of nerve injury on the ventral horn cells of the spinal cord. J. Anat. 80:117.
328. Hess, A. (1957). The experimental embryology of the foetal nervous system. Biol. Rev. 32:231.
329. La Velle, A., and Sechrist, J.W. (1970). Immature and mature reaction patterns in neurons after axon section. Anat. Rec. 166:335.
330. Torvik, A. (1972). Phagocytosis of nerve cells during retrograde degeneration. An electron microscope study. J. Neuropath. Exp. Neurol. 31:132.
331. Torvik, A., and Skjorten, F. (1971). Electron microscopic observations of nerve cell regeneration and degeneration after axon lesions. Acta Neuropath. Berl. 17:248.
332. Birks, R.I. (1974). The relationship of transmitter release and storage to fine structure in a sympathetic ganglion. J. Neurocytol. 3:133.
333. Hillarp, N-Å. (1959). The construction and functional organization of the autonomic innervation apparatus. Acta Physiol. Scand. 46. Suppl. 157:1.
334. Hillarp, N-Å. (1946). Structure of the synapse and the peripheral innervation apparatus of the autonomic nervous system. Acta Anat. 2. Suppl. 4:1.
335. Bennett, T., Cobb, J.L.S., and Malmfors, T. (1973). Fluorescence histochemical and ultrastructural observations on the effects of intravenous injections of vinblastine on noradrenergic nerves. Z.Zellforsch. 141:517.
336. Zimmerman, E.A., Hsu, K.C., Robinson, A.G., Carmel, P.W., Frantz, A.G., and Tannenbaum, M. (1973). Studies of neurophysin secreting neurons with immunoperoxidase techniques employing antibody to bovine neurophysin. I. Light microscopic findings in monkey and bovine tissues. Endocrinol. 92:931.
337. Geuze, J.J., and Kramer, M.F. (1974). Function of coated membranes and multivesicular bodies during membrane regulation in stimulated exocrine pancreas cells. Cell Tiss. Res. 156:1.
338. Drachman, D.B. (ed.) (1974). Trophic functions of the neuron. Ann. N.Y. Acad. Sci. 228:1.
339. Harris, A.J. (1974). Inductive functions of the nervous system. Annu. Rev. Physiol. 36:251.
340. McLaughlin, B.J., Wood, J.G., Saito, K., Roberts, E., and Wu, J-Y. (1975). The fine structural localization of glutamate decarboxylase in developing axonal processes and presynaptic terminals of rodent cerebellum. Brain Res. 85:355.
341. Vacca, L.L., Rosario, S-L., Zimmerman, E.A., Tomashefsky, P., Ng, P-Y., and Hsu, K.C. (1975). Application of immunoperoxidase techniques to localize horseradish peroxidase-tracer in the central nervous system. J. Histochem. Cytochem. 23:208.

International Review of Physiology
Neurophysiology II, Volume 10
Edited by Robert Porter
 University Park Press Baltimore

3
Long-Term Regulation in the Vertebrate Peripheral Nervous System

D. PURVES
Washington University School of Medicine, St. Louis, Missouri

---

INTRODUCTION

The ability of excitable cells to influence one another in ways other than the usual transmitter induced conductance change at synapses has long been a source of interest to neurobiologists. Certainly the enthusiasm of many workers in this field stems from the suspicion that understanding the long-term changes that occur as a result of denervation and subsequent re-innervation, or the removal of the neuron's peripheral target, may provide clues to the ways in which animals learn and remember, abilities which would seem to require some kind of more or less permanent alteration in the nervous system. While it may seem a long way from the effects of denervating a muscle to higher functions in the central nervous system, it is likely that the mechanisms by which cells influence one another over long periods in the peripheral nervous system are similar to central interactions.

Just as the neuromuscular junction of vertebrates has served as a useful system in which to study the fundamental rules of synaptic transmission (see Refs. 1 and 2 for excellent reviews of this subject), so the advantages of anatomical simplicity, ease of intracellular recording from the postsynaptic cell, and knowledge of the neurotransmitter have made the neuromuscular junction a valuable preparation for studying long-term interactions between cells. For similar reasons, the vertebrate autonomic system also provides many preparations in which long-term interactions between neurons can be studied. The anatomical, physiological, and pharmacological complexity of the central nervous system continues to make analogous experiments more difficult (although by no means impossible) in brain and spinal cord.

This review will not give an exhaustive account of experiments relevant to the ways in which nerve cells (and nerve and muscle cells) interact over periods of days, weeks, or months, but rather will focus on a limited number of topics in this field which at the moment seem especially exciting, and to a certain extent controversial. While opinion on what these topics are would no doubt vary widely, perhaps most would agree that three areas in particular have caught the interest of a broad range of workers in this field: 1) the effects of denervation, especially the mechanisms by which denervation changes are brought about; 2) the effects of axon interruption on neuronal properties, especially the influence of the "target organ" on the nerve cells which innervate it; and 3) the nature of re-innervation, especially the mechanisms by which re-innervation can be specific.

A recent review by Harris (3) ranges more widely over the field of long-term interactions between excitable cells and is recommended for further background.

DENERVATION CHANGES AND THEIR MECHANISM

Perhaps the best studied long-term interaction between excitable cells is the series of changes which follows denervation. A major interest has been the underlying mechanism of the dramatic alterations in the properties of muscle fibers which follow motor nerve section. At present there is controversy over whether the cause of denervation effects is the absence of a chemical mediator (trophic factor) which is normally released by motor nerve endings, or whether the effects are due simply to the relative lack of postsynaptic cell activity. Since the mechanism, whatever it turns out to be, has broad implications for neuronal interactions, the current debate is an important one.

Changes That Follow Denervation of Vertebrate Skeletal Muscle

For more than a century, it has been known that denervation has profound effects on vertebrate skeletal muscle fibers (for reviews of the older literature see Refs. 4 and 5).

The best studied of these changes is an increase in the surface area of denervated fast-twitch muscle fibers that is sensitive to the neurotransmitter, acetylcholine (ACh). The most widely used measure of ACh sensitivity in recent years has been the membrane depolarization produced by a known amount of ACh applied locally by ejecting it in small, electrically controlled pulses from a drug-filled pipette (a technique called microiontophoresis). In normally innervated fast-twitch muscle fibers the ACh receptor region is localized to the subsynaptic membrane (the end-plate region) (6–13). Indeed, recent studies using differential interference contrast microscopy, and improved iontophoretic methods, indicate that receptor sensitivity is very sharply circumscribed, falling off to 3% or less of the end-plate value within as little as 10–15 μm of the edge of the synaptic specialization (14–18). The precise localization of ACh receptors has also been beautifully demonstrated by the application of immunofluorescent labeled α-bungarotoxin (19) (which binds specifically to ACh receptors; see Refs. 40 and 41), as well as by an immunoperoxidase stain of α-bungarotoxin (20) and autoradiographically (21, 22). In the vicinity of the end-plate in some muscle fibers there appears to be a surround of membrane with a much lower but still appreciable density of receptors (11–13, 17, 23). Receptors in the surround have been reported to have different physiological properties (24), although in a more recent study (18) some of these findings could not be repeated. In slow-twitch muscle fibers, such as those of the rat soleus muscle, and in lower vertebrate slow muscle fibers (such as those of the frog iliofibularis), the entire surface of the muscle fiber is slightly sensitive to transmitter (25–27). Also, in frog twitch muscle fibers there is considerable sensitivity in the myotendinous region (28), as well as very slight generalized sensitivity in occasional fibers (13). These provisos, however, should not obscure the central fact of a sharp localization of ACh sensitivity to the end-plate region of all normally innervated fibers. In spite of this restriction of sensitivity, junctional receptors appear to be turning over at a slow rate, probably with a half-time of about a week in both the frog (29) and the rat (30, 31) (see also Ref. 36).

When a muscle loses its motor innervation, there is a marked increase in the degree to which extrajunctional membrane responds to the iontophoretic application of ACh, a phenomenon generally called denervation supersensitivity. In mammalian fast-twitch muscle fibers the change is detectable within a matter of hours, and reaches a maximum in about a week, by which time the entire surface of the muscle fiber is nearly as sensitive to transmitter as the normal end-plate region (10, 13, 26, 32). In frog muscle fibers the change occurs more slowly, reaching a maximum only after 2 weeks or more (11, 27). The end-plate region of frog twitch muscle fibers remains more sensitive than the rest of the fiber (13), and has even been reported to increase somewhat in sensitivity (13, 27). At the mammalian end-plate, however, there is considerable loss of receptors after long-term denervation as shown by labeled bungarotoxin binding studies (33). Since the development of supersensitivity (and extrajunctional α-bungarotoxin binding) can be blocked by the administration of inhibitors of protein synthesis in the period immediately following denervation, it seems likely that increased sensitivity represents incorporation of newly synthesized receptors into extrajunctional membrane (34–37). *De novo* synthesis of extrajunctional ACh receptors in denervated muscle has recently been demonstrated directly by receptor isolation after exposure of organ cultured muscle to [^{35}S]methionine (38). Other mechanisms, such as exposure of occult receptors, have been proposed, but have no experimental support. The possibility of receptors diffusing from the end-plate region is ruled out by the observation that supersensitivity develops in segments of muscle experimentally isolated from the end-plate region (39). The discovery of the specific and irreversible receptor blocking agent α-bungarotoxin by Lee and his co-workers (40) has provided a more reliable tool than the iontophoretic method for the study of extrajunctional receptor incorporation. For example, by iontophoretic measurements the extrajunctional region of denervated mammalian muscle can appear to be about as sensitive to ACh as the end-plate, yet bungarotoxin binding studies show that the density of receptors extrajunctionally does not approach that at the end-plate (21, 41–43). In addition, there is a clear difference in the rate of receptor turnover in junctional and extrajunctional regions, extrajunctional receptor turning over at least several times faster than subsynaptic receptor (29–31, 44). There are also differences in extrajunctional receptors with respect to their response to curare (41, 45, 46), the duration of the "elementary event" as determined by analysis of the ACh noise spectrum (47), and possibly the ACh reversal potential (48). Whether or not this means that significant molecular differences exist between receptors in the two regions is not fully known; recent results indicate that purified junctional and extrajunctional receptors are similar but not identical in their chemical and physical properties (49, 50).

The presence of denervation supersensitivity in other types of muscle has been more difficult to evaluate. For vertebrate smooth muscle there is no doubt that a variety of effector organs are more sensitive to their normal transmitter after denervation (5, 51). However, the basis of such supersensitivity may be more complex than that in denervated skeletal muscle. For example, since the action of norepinephrine on smooth muscle is in large measure terminated through specific re-uptake by the nerve terminal, absence of the nerves might in itself contribute to the apparent supersensitivity (51). The fact that supersensitivity of the nictitating membrane develops in two phases with different pharmacological responses supports the view that supersensitivity of smooth muscle has both a pre- and a postsynaptic component (52). This feature, along with the anatomical complexity of a distributed and morphologically ill-defined innervation, and the technical problems of working with smaller smooth muscle cells, has resulted in a less precise understanding of supersensitivity of smooth muscle fibers transmitter sensitivity is also normally restricted to the synaptic of supersensitivity in invertebrate muscle fibers. It is clear that in invertebrate muscle fibers transmitter sensitivity is also normally restricted to the synaptic region (53–55). However, the events following denervation in some invertebrate species are different from those occurring in vertebrates. In crustaceans, for example, cutting a motor nerve leads only to gradual degeneration of the distal axons (56–59), and even many months after such a procedure (when the nerve has finally disappeared) there is no evidence of supersensitivity (60). Thus Cannon's "Law of Denervation" (that supersensitivity is a universal occurrence after denervation) (5) has at least this restriction. Denervation supersensitivity has, however, been shown to occur in denervated insect muscle (61). These findings in invertebrates raise a broader and recurring question, namely, the extent to which one can generalize from the kinds of long-term changes found in a particular animal. Although one is used to thinking of basic mechanisms in the nervous system being common to a wide segment of the phylogenetic spectrum, there appear to be major differences among different types of animals regarding the phenomena under discussion here.

A second important change that follows denervation of mammalian muscle is the onset of spontaneous electric activity (fibrillation) (4). An innervated mammalian skeletal muscle fiber normally gives an action potential only in response to the release of transmitter which brings its membrane to threshold. In contrast, action potentials begin to occur spontaneously within a few days of denervation (4, 62–64). Such potentials often occur at high rates (see for example Refs. 64–66) and probably continue for as long as the muscle remains denervated, in some cases for up to a year or more (4, 66). This impressive degree of autogenic activity arises from local fluctuations in membrane potential (63, 64, 67–70); the cause of these potential changes is not known with certainty, but they depend on an increase in membrane conductance to Na^+ which is distinct from the Na^+ conductance increase induced by ACh. Thus local potential fluctuations underlying spontaneous activity are blocked by reducing Na^+ in the bathing medium or by tetrodotoxin, but are unaffected by curare (62, 69, 71). This alteration in the Na^+ conductance properties of the membrane could be based on the incorporation of a different type of Na^+ channel or some modification of the structure or efficacy of existing voltage sensitive Na^+ channels. The onset of fibrillation can also be blocked by inhibitors of protein synthesis (37, 72), suggesting, but not proving, that new channels are synthesized and incorporated into the denervated membrane. Fibrillation does not occur in denervated frog muscle (R. Miledi, personal communication).

Other changes in the Na^+ conductance mechanism of mammalian muscle fibers induced by denervation are a loss of sensitivity of the action potential to the blocking action of tetrodotoxin (TTX) (73, 74), and a reduction in the rate of rise of the action potential (75, 76). Again qualification must be made concerning the species, since the effect is not seen in avian (77) or amphibian muscle fibers (27, 76). In mammalian muscle, however, within several days of interrupting the nerve supply, increasingly large doses of TTX are required to abolish the Na^+ conductance increase associated with the rising phase of the action potential. Within a week (and thereafter) about 100–1000 times more drug is required to abolish or markedly reduce the regenerative response to depolarization than in normal muscle (73, 74). The basis of these effects is unclear. The fact that the development of TTX resistance, as well as the slowed rate of rise, can be prevented by inhibitors of protein synthesis (35) again suggests that an altered channel is incorporated into the muscle fiber membrane, or at least that protein synthesis is required for the expression of the altered channel characteristics. As is the case for fibrillation, these phenomena are not directly related to the newly incorporated ACh receptors (76).

Other electrical changes which do not depend on a voltage sensitive sodium conductance mechanism also occur in denervated muscle fibers. In both amphibians and mammals there is a marked increase in the input resistance of fibers after denervation, as well as changes in specific membrane capacitance and internal resistance (26, 32, 78, 79) (however, see Ref. 27). Although the diameter of denervated fibers is ultimately reduced because of atrophy (see below), the increase in input resistance cannot be accounted for solely on this basis, and has been shown to represent an increase in specific membrane resistance (26, 32, 78). In mammalian muscles, but not in the frog, there is also a fall in resting potential that begins within hours of denervation and reaches its full extent in about a week (26, 32, 78, 80). Since membrane resistance depends largely on potassium channels and on non-specific leakage channels (81, 82), denervation presumably affects these channels as well as Na^+ channels. Denervated muscle also undergoes an abnormal form of contracture in response to ACh or increased K^+ in the bathing fluid which is associated with increased Ca^{2+} permeability (83).

Obvious changes in muscle structure and biochemistry also occur following denervation. The most conspicuous sign of structural change is muscle atrophy, which at the ultrastructural level is due to a progressive destruction of the fiber's contractile elements resulting in a decrease in fiber diameter (84, 85) (see also Ref. 86). As might be expected, the mechanics of contraction are also altered: if adult or developing mammalian muscles are deprived of their innervation, the speed of the contractile response is slowed, much more so in fast-twitch than slow-twitch muscle fibers (87–89) (see also Refs. 90 and 91). These changes have been extensively studied in experiments in which fast- and slow-twitch muscles are cross-reinnervated (i.e., where a fast muscle is reinnervated with the motor nerve to a slow muscle, and vice versa) (see for example Refs. 88 and 89), and the results have provided strong evidence for the influence of nerve on muscle properties (see below). Some histochemical properties (which largely reflect enzyme patterns in muscle fibers) are also altered as a consequence of denervation (92), as are many aspects of muscle metabolism (see Refs. 86 and 93 for extensive treatments of this topic).

Finally, one of the most important and least understood changes of denervated muscle fibers is a renewed ability to receive synaptic contacts (94–96).

Although normal muscle fibers resist "hyperinnervation" from foreign nerves (and presumably from their own nerve fibers as well), denervated fibers accept contacts from a wide variety of sources including other motor nerves (95, 96), preganglionic autonomic fibers (97–99), and possibly even sensory nerves (100) (these findings are further discussed under "Specific Re-innervation of Muscle Fibers").

To summarize, within a few days of denervation many of the major functional properties of mammalian muscle fibers are affected. Since the altered properties are based on a wide variety of molecular mechanisms (ranging from membrane ionophores to contractile proteins), it is unlikely (but not impossible) that the effects of denervation are due to a single change in the genome of the muscle cell. Similar changes occur in amphibian and avian skeletal muscle fibers, but often more slowly, and to a less dramatic degree. In invertebrates, denervation has been less fully studied, but changes appear less marked, and in some instances may not occur at all.

Upon prompt re-innervation, muscle fiber properties begin to revert to normal (101–103).

Changes That Follow Denervation of Neurons

The interest in denervation changes in muscle fibers has been due partly to the idea that similar changes might occur in neurons deprived of their synaptic input. Neurons are generally more difficult to study than muscle fibers: unlike the vertebrate neuromuscular junction, neuronal innervation is usually widely distributed on the soma and dendrites, impalements with microelectrodes frequently cause more damage and are less stable, and the interaction of the transmitter with postsynaptic receptors is often more complex. In spite of these limitations, a number of studies have examined the effects of denervation on neurons.

Surprisingly, supersensitivity of neurons after denervation was described in the 1930s, long before much was known of the basis of this phenomenon in muscle fibers (104–106) (see also Ref. 5 for references to older literature). This knowledge came largely from the work of Cannon and his collaborators, who were concerned to show that pharmacological supersensitivity of muscle after denervation was a manifestation of a much more general change that occurred in denervated (or partially denervated) autonomic ganglion cells, spinal neurons, and even neurons within the cortex (5). Many of their experiments were carried out on the superior cervical ganglion of cats, using as an assay the response of the nictitating membrane to close arterial injection of acetylcholine, or direct local ACh application to the superior cervical ganglion. Cannon's group found that contraction of the nictitating membrane was greatly accentuated following section of the cervical sympathetic trunk (which provides the preganglionic fibers innervating the ganglion) (104, 106) (see also Ref. 107), or after partial denervation achieved by cutting some thoracic rami (105). This was not entirely a result of supersensitivity of the smooth muscle of the end-organ, since the nictitating membrane's response to postganglionic nerve stimulation or injected norepinephrine was only slightly greater than normal. The mechanism of ganglionic supersensitivity, however, remained unclear. In a series of more recent experiments Kuffler, McMahan, Dennis, and Harris (108–111) (see also Ref. 112), examined supersensitivity of denervated neurons with more modern techniques. An obstacle to studying the surface of sensitivity of neurons is that, since they are small and multiply innervated, it is difficult to distinguish a synaptic from an extrasynaptic region in a living preparation. A major advance was the application of differential interference contrast microscopy (Nomarski optics) (113) which allows visualization of living neuronal synapses in a few favorable systems. Two such preparations have been described in some detail, the parasympathetic ganglion of the atrial septum of the frog (108), and the cardiac parasympathetic ganglion of the mudpuppy (114) (see also Refs. 115 and 116). Both these autonomic ganglia have the advantage that the neurons are embedded in a thin sheet of tissue; thus with Nomarski optics one can see single cells and some of the synapses on them. The ability to see synapses permits accurate placement of a pipette containing transmitter (ACh) at synaptic and extrasynaptic regions to test ACh sensitivity, in more or less the same way that such experiments are carried out on muscle fibers. In normally innervated ganglion cells the synaptic regions are about 2–10 times more sensitive to transmitter than random (presumably extrasynaptic) application sites (110, 117). The reason for the small difference between synaptic and extrasynaptic regions (compared to muscle) may be largely technical: the synapses are close enough together (average separation on frog ganglion cells is about 8 μm) so that a nominal extrajunctional location of the pipette may allow transmitter to diffuse to a nearby synaptic region (110). Thus it is uncertain if in autonomic neurons transmitter localization is very sharp, or whether there is significant extrasynaptic sensitivity.

The cardiac ganglion of both the frog and the mudpuppy can be denervated (although denervation is only partial in the mudpuppy) by cutting the vagus nerves near their exit from the skull (111, 117). Within a few days, synapses disappear from the surface of frog parasympathetic ganglion cells (108) and are greatly diminished on ganglion cells in the mudpuppy (which also receive other intrinsic synapses) (114, 118). When ACh sensitivity is tested after denervation, the level typical of normal synaptic regions is encountered at virtually every point on the cell surface (111, 117). Since acetylcholinesterase is highly localized at synaptic regions on these neurons (108), a change in this enzyme would be unlikely to cause this result. Although these experiments represent a limited sample, it seems probable that denervation supersensitivity will prove a general phenomenon in vertebrate neurons. As in muscle fibers, the basis of supersensitivity is presumably incorporation of new receptors into extrasynaptic membrane, although this has not yet been shown. A mammalian autonomic ganglion, the submandibular ganglion of the rat, in which the neurons occur in a thin connective tissue sheet, has also been briefy described (108, 119); thus it should be possible to study the spread of ACh sensitivity in mammalian neurons as well. It is interesting that partially denervated neurons which are demonstrably supersensitive continue to give a normal response to stimulation of residual terminals, as judged by evoked and spontaneously occurring synaptic potentials (117). This implies that, in this instance, supersensitivity has little effect on remaining terminals.

Other effects of denervation on neurons are less well documented, and in some cases have simply not been studied, again because of the generally greater difficulties and ambiguities involved in experiments on neurons. Spontaneous activity of denervated sympathetic neurons has been described, and has been suggested to be analogous to the fibrillation of denervated muscle fibers (120–122). These experiments were carried out with early electrophysiological techniques and the observations could not be entirely confirmed in a more recent study, although an occasional sympathetic ganglion denervated 1–8 weeks previously showed signs of spontaneous activity (123). In the inferior mesenteric ganglion of the rabbit, McLennan and Pascoe (124) show clear records of spontaneous activity following denervation, although these nerve cells are normally silent. The underlying mechanism is not known. In spite of these examples, spontaneous activity has not been noted in other studies (often using intracellular recording) of denervated autonomic neurons in amphibians (111, 117, 125), avians (126), and mammals (127, 128). It is clear, then, that spontaneous activity is an inconstant phenomenon following denervation of peripheral neurons, despite earlier reports to the contrary. Some evidence for increased spontaneous activity of denervated central neurons has also been presented (129). Electrical properties of peripheral nerve cells have been found to change relatively little after denervation. Input resistance rises in frog peripheral neurons after denervation (125), but this has not been found in denervated sympathetic ganglia of mammals (128). Resting potentials, although difficult to measure in autonomic neurons with any accuracy, have generally not been noted to be much different in normal and denervated neurons (111, 117, 118, 125, 127, 128). The reduced effect of tetrodotoxin in blocking action potentials, which is so striking in mammalian muscle fibers, has not yet been studied in a mammalian neuronal preparation.

Peripheral neurons appear to undergo relatively minor anatomical changes following denervation. For example, Hamlyn (130) found little histological change in mammalian autonomic ganglion cells several months after denervation, other than mild cell shrinkage and some loss of the intensity of Nissl staining. There was no apparent degeneration of neurons. The suggestion that extensive morphological changes (other than synapse loss) do not occur in denervated vertebrate peripheral neurons has been confirmed by a number of more recent studies at both the light and electron miscroscopic level (125, 128, 131–134). One interesting finding is that the postsynaptic membrane specialization appears to remain intact on autonomic nerve cells as it does at the muscle end-plate, perhaps serving a localizing function during re-innervation (see below) (131, 132, 134). Denervation of neurons, as muscle fibers, induces sprouting of nearby intact presynaptic elements (135, 136).

This relative indifference to innervation is not universal among neurons, since within the central nervous system anatomical changes are often more extensive. While a discussion of this literature is beyond the scope of this article, severe degeneration (usually referred to as anterograde transneuronal degeneration) occurs at many sites in the brain following "denervation" (especially in young animals) (see Ref. 137 for a review). One of the best studied systems is the lateral geniculate nucleus in cats and primates (138, 139); enucleation results in a severe degenerative reaction of many lateral geniculate neurons within a few days in monkeys (139), and a few months in cats (138). Indeed, visual deprivation alone in young cats leads to morphological and functional changes in the lateral geniculate body (140, 141). While it is difficult to generalize from these findings, it seems fair to say that the morphological effects of denervation on vertebrate neurons in adult animals range from slight in the peripheral nervous system, to moderate or even severe at some sites in the central nervous system.

Denervation makes nerve cells, like muscle fibers, receptive to foreign innervation (the ability of denervated autonomic neurons to receive a variety of foreign synapses is further described under "Specific Re-innervation of Neurons"). Because of the anatomical complexity of neuronal innervation it is unclear whether true hyperinnervation can occur. For example, a denervated muscle fiber will receive synapses in extrajunctional regions (94–96), and such foreign innervation will be retained indefinitely in mammalian muscle even when the original nerve fibers have re-innervated the end-plate region (142–145). Mammalian autonomic neurons can also retain dual foreign and native innervation indefinitely (146), but it is not known whether this represents true hyperinnervation or (more probably) foreign synapses occupying sites previously contacted by native fibers. The ability of central neurons to receive abnormal contacts after partial denervation has also been extensively studied in the last few years and has been discussed elsewhere (147, 148).

Biochemical studies of peripheral neurons have also demonstrated enzymatic changes following denervation. For instance, a number of recent experiments have examined transynaptic regulation (regulation mediated by preganglionic fibers) of the enzyme tyrosine hydroxylase (and other cell functions) in developing sympathetic ganglia (134, 149–151). Such an interaction might be an important regulatory mechanism in the adult autonomic nervous system as well, and some evidence for this has been presented (134, 152). However, the general question might be raised of whether the evidence to date indicates a major trophic role for innervating fibers of nerve cells, analogous to the trophic influence of nerve on muscle fibers. In the light of what is known, it appears that a neuron is less critically dependent on the synaptic contacts it receives than is a vertebrate muscle fiber.

Mechanism of Denervation Effects

Experiments aimed at determining the mechanism by which innervating fibers influence the properties of postsynaptic cells have generally been carried out at the neuromuscular junction. It seems likely, however, that to the extent neurons are trophically affected by the fibers that innervate them, similar mechanisms may operate. The major alternatives that have been considered are that denervation changes are due to inactivity of the deprived postsynaptic cell or that they are due to the absence of a trophic factor.

Evidence for a Trophic Factor At the neuromuscular junction, the prevailing view has been that many, if not all, of the effects of denervation stem from the absence of some factor (factors) normally released by nerve fibers (see for example Ref. 153). According to this view, aspects of muscle fiber function which are normally not expressed (presumably because of a direct or indirect action of the trophic substance on the genome of the muscle fiber) are freed from repression when the nerve fiber and its chemical messenger are removed (11, 154). This proposal, generally called the trophic factor hypothesis, has a good deal of experimental evidence to recommend it. Most of the relevant experiments were carried out at the frog neuromuscular junction, many of them by Miledi and his collaborators whose work has sharply focused this issue.

One of the most frequently cited experiments supporting the trophic factor hypothesis is the partial denervation of frog sartorius muscle fibers (11). Many fibers in this muscle have two separate end-plates, thus permitting an experiment that distinguishes between a denervation effect due simply to the lack of activity, and an effect independent of activity, presumably due to a trophic factor. Since the nerve which supplies the muscle splits into several major branches, it is possible to cut some of these; fibers which receive two nerve endings are thus partially denervated, the nerve degenerating at one end-plate while remaining intact at the other. The remaining nerve would continue to impose a normal pattern of activity on the muscle fibers due to the high safety factor for neuromuscular transmission. When a denervated end-plate region was found in a fiber which was also innervated by an intact synapse (activation of which initiated an action potential), the ACh sensitivity of the two regions was compared. In every case there was a definite increase in ACh sensitivity in the region of the denervated end-plate, although the sensitivity of the same fiber was normal elsewhere. There was also a small increase in the input resistance measured in the vicinity of the denervated end-plate. Since fibers could still be normally activated by transmitter release from the intact nerve ending, the interpretation given these findings was that supersensitivity had developed in the presence of a normal degree of activity. If muscle fiber activity did not play a role, then a chemically mediated influence seemed likely, and Miledi suggested (11, 154) that denervation changes might be the consequence of removal of a normally secreted neural factor. An alternative interpretation of this result, for which there is some evidence, is that degenerating nerve fibers produce a substance which induces a local increase in extrajunctional sensitivity (155–157) (see below).

A second experiment carried out by Miledi (101) was a study, not of the onset of supersensitivity, but rather its decline when denervated muscle fibers became re-innervated by the outgrowing nerve. When a frog skeletal muscle is denervated (again the sartorius was used), the end-plate region can still be found because of the local release of quanta of ACh by Schwann cells (101, 158–160). Although the mechanism of this phenomenon is not entirely understood (160), it serves as a useful indicator of the end-plate region. Schwann cell ACh release is characterized by an amplitude distribution of miniature potentials which is skewed towards the amplifier noise level, with a mean smaller than normal miniature end-plate potentials, and a relatively low frequency of occurrence (101, 160). The arrival of the re-innervating nerve fiber is thus signaled by a shift in the amplitude distribution of miniatures from abnormal Schwann cell release towards the more normal pattern, and an increase in frequency (101, 161, 162). Thus it is possible to know with reasonable certainty whether a given fiber is still denervated or has been recently re-innervated. In at least some fibers there is a period following re-innervation during which there is little or no functional synaptic transmission (101), apparently due to failure of the action potential to invade the newly formed terminal (161). When ACh sensitivity was tested during this "non-transmitting" period in a small number of selected fibers, it was found that extrajunctional ACh sensitivity had returned to normal levels (101). Thus the change in transmitter sensitivity could not be explained by a renewal of muscle fiber activity since transmission was absent; this again favored the idea of a chemical mediator. A similar result has been obtained in a study of re-innervation of avian muscle (163). Analogous experiments are more difficult to perform in mammalian muscles since the period of paralysis after re-innervation is much shorter, particularly if re-innervation occurs at the site of the former end-plate (161, 164, 165). In the mouse, however, Tonge (165) has reported that extrajunctional ACh sensitivity remains high until functional transmission is restored, and has suggested that activity may be more important in controlling ACh sensitivity in mammals than in frog or fowl.

A third line of evidence supporting the trophic factor hypothesis, which is less direct, concerns the time of onset of denervation effects as a function of the length of the nerve stump remaining distal to the point of transection. The initial experiment carried out by Luco and Eyzaguirre (166), and later by Emmelin and Malm (167), measured the time of onset of increased responsiveness of mammalian muscle fibers to ACh when the nerve was cut either near the muscle surface or several centimeters away. They found that the longer the nerve stump, the slower the onset of supersensitivity. Since there would be no activity in the denervated muscle no matter what the length of the nerve stump (fibrillation developing only after several days), the interpretation of these results was that the longer nerve degenerated more slowly and thus provided a larger amount of trophic factor, which could somewhat delay the onset of denervation changes. This view is given additional support by the finding that after severing the axon the ultrastructural changes of degeneration in nerve terminals, and cessation of quantal release, are delayed if the cut is made at some distance from the ending (158, 168). Similar results were obtained using the onset of fibrillation or the decrease in resting potential as indicators of denervation change (80, 166). More recently this experiment has been repeated examining another denervation effect, the development of resistance to the blocking action of tetrodotoxin (169). The outcome was in accord with the previous experiments: the longer the stump, the greater the delay in the development of action potential resistance to TTX. Since the lag was about 2 hours/cm of nerve stump in these experiments, a rate corresponding roughly to that of fast axoplasmic transport, it was suggested that axoplasmic transport was the mechanism by which a factor synthesized in the perikaryon reached the terminal.

Finally, agents which might interfere with the transport or release of a possible trophic factor from nerve terminals have been used in a number of additional experiments which bear on the validity of the trophic factor hypothesis. An intriguing finding initially made some years ago was that the toxin produced by the bacterium *Clostridium botulinum* prevents ACh release from nerve terminals (170, 171). Electrophysiological studies of Thesleff (172) (see also Ref. 173) further showed that chronic application of botulinum toxin (BTX) to mammalian muscle resulted in a spread of ACh sensitivity characteristic of denervation, without any obvious change in nerve terminal morphology. Other experiments have demonstrated that chronic application of BTX produces additional denervation effects including the development of spontaneous activity (fibrillation) (172), the ability to accept foreign synapses in extrajunctional regions (142), and axonal sprouting and the formation of new end-plates (174–176). Thesleff's initial interpretation of his results (172) was that the trophic substance might well be ACh itself, a view that is to some extent still current (see for example Ref. 117). This possibility now seems less likely due to the observation that ACh sensitivity develops in organ cultured muscle even in the presence of high concentrations of ACh (11) and that denervation changes develop in the presence of normal spontaneous transmitter release when impulses are blocked by a local anesthetic (178) (see also Ref. 179). Furthermore, when drugs which interfere with axoplasmic transport (e.g., colchicine) are applied to motor nerves (180, 181) denervation changes occur in the presence of apparently normal impulse traffic and ACh release (these experiments are further discussed below).

An alternative interpretation of the BTX results is that in addition to blocking ACh release, the toxin also blocks the release (or uptake) of a trophic factor. Some evidence supporting this interpretation comes from the experiments of Bray and Harris (182) who examined rat diaphragms in which synaptic transmission was partially blocked by botulinum toxin. Bray and Harris (182) found that 5–10 days after the onset of local BTX intoxication they could impale fibers in which action potentials were still elicited by nerve stimulation (several workers have shown that BTX usually causes less than a complete block of spontaneous (183, 184) and evoked release (176)). Some of these fibers showed extrajunctional supersensitivity to iontophoretic ACh application (however, see also Ref. 176). Conversely, after two weeks or more (i.e., during recovery from the BTX block) some fibers failed to respond to nerve stimulation but had little or no extrajunctional sensitivity, a result somewhat similar to the experiments of Miledi (101) during nerve regeneration in frog muscle. Recovery from BTX effects appears to depend largely on the formation of new end-plates (165, 175, 182). Bray and Harris (182) concluded that since supersensitivity can develop in the presence of muscle activity, and recede in its absence, these effects are not due to activity level but to BTX blockade of trophic factor release from nerve. The idea that an additional action of BTX is to block release of trophic factor is bolstered by experiments which show that proteins carried by fast axoplasmic transport accumulate to abnormally high levels in nerve terminals blocked by BTX (182). Although one might object that the state of activity of muscle fibers before the acute experiments is not known, the results provide additional evidence in support of the trophic factor hypothesis.

A further experiment using BTX was carried out by Miledi and Spitzer (185), and provides additional evidence for a trophic factor released by nerve terminals which operates independently of ACh secretion and muscle contraction. Slow muscle fibers in the frog are normally incapable of producing an action potential (186, 187); however, about 2 weeks after such fibers are denervated, regenerative spikes can be elicited by direct depolarization (188). Miledi and Spitzer (185) injected BTX into a muscle containing slow fibers (the iliofibularis) and found that although transmission was effectively blocked, the ability to respond regeneratively to depolarization failed to develop even after many weeks. The simplest explanation of this result would be that BTX, while blocking ACh release (and muscle activity), had not interfered with the release of a trophic factor responsible for suppressing action potentials in slow fibers. This is a different interpretation from that proposed by Bray and Harris (182) for their results in rat diaphragm; they suggest that BTX blocks release of a trophic factor more effectively than it blocks ACh release.

Drugs such as colchicine and vinblastine, which interfere with axoplasmic transport (among other effects) (see for example Refs. 189–191), make possible other experiments which might dissociate synaptic activation of muscle from a possible trophic factor effect of nerve on muscle. Application of these agents at appropriate concentrations to motor nerves for periods of a few days does not interfere with the conduction of action potentials, and therefore does not paralyze muscle fibers (180, 181, 192, 193). Such drugs might, however, be expected to interfere with the transport of a trophic factor from the cell body to nerve terminals. Thus the finding that colchicine application leads to the development of supersensitivity (and other changes associated with denervation), without disrupting synaptic transmission, was initially construed as strong evidence for a trophic factor mechanism (180, 181). However, this interpretation has recently been brought into question by the observation that systemic administration of colchicine leads to changes in limb muscles like those seen after denervation (192). This suggests that colchicine has a direct effect on muscle membrane, independent of any neuronal effects, which mimics the postsynaptic changes caused by denervation. This possibility is made more plausible by the fact that other agents such as muscle injury (39), or a piece of degenerating nerve (155, 156), have been shown to have such an effect on muscle fibers (see also Ref. 157). Recent experiments of Lømo (194) have indeed shown that colchicine can effect muscle fibers independently of any nerve mediated influence (see also Ref. 195), although these results have been disputed (193). The observations reported by Kaufman et al. (193) seem to indicate that under some conditions colchicine and vinblastine can cause denervation–like changes in the absence of a significant direct effect on muscle; however, Lømo and Westgaard (157) have raised serious objections to their approach. Thus the mechanism by which colchicine applied to motor nerves affects muscle fiber properties is somewhat uncertain.

Evidence for Muscle Fiber Inactivity as the Cause of Denervation Effects Although one can disagree with individual results, taken together the experiments described in the preceding section might be considered strong enough to have settled the question of whether activity of the muscle fibers plays any role in denervation effects. However, in the late 1960s Jones and Vrbová (196, 197) revived the activity/trophic factor controversy by briefly reporting that surprisingly short periods of stimulation of denervated mammalian muscles (either directly or via the degenerating nerve) could delay or reverse the development of ACh supersensitivity (see also Refs. 198 and 199). In addition, a number of other results (some already touched on) indicated that activity might be more important than generally thought. For example, in muscles of mice suffering from "heriditary motor end-plate disease" (which is characterized by a block of neuromuscular transmission) fibers show many of the signs of denervation, even though the nerve terminals appear quite normal and release packets of transmitter spontaneously; the defect in this disease is probably a failure of the action potential to invade the nerve terminal (179). The simplest explanation of the presence of extrajunctional ACh sensitivity and fibrillation in these muscle fibers is that lack of activity is the causal agent, as there is no obvious reason why a possible trophic substance should not also be released normally. This result suggested a different interpretation of the earlier experiments on the effect of BTX (172): since denervation-like changes developed in the presence of normal (if blocked) nerve terminals, it could be that lack of activity was the cause, rather than the absence of ACh or some other trophic factor. Other workers had looked directly at the role of activity and concluded that its effects were slight: denervation changes were not produced (or produced only to a small extent) by immobilizing muscles (200, 201), or isolating segments of spinal cord (202, 203). In these experiments, however, there was room for equivocation. For example, immobilization prevents only some aspects of contraction (shortening against tension), and either procedure may not adequately reduce activity.

Given such uncertainties, the work of Lømo and Rosenthal (178), which directly demonstrated that activity per se has important effects on mammalian (rat) muscle, has rightly received much attention. First, they confirmed that denervation effects could develop in the presence of intact nerve endings, as long as activity was abolished. This was done by the application of a cuff containing local anesthetic to a motor nerve which was blocked in this way for many days (see also Refs. 204 and 205). Under these conditions, impulse traffic down the nerve is abolished, although most terminals remain functionally intact, as demonstrated by stimulation of the nerve distal to the block. Denervation effects developed with their usual time course (although see Ref. 206). Next they implanted electrodes to directly stimulate muscles (soleus and extensor digitorum longus) to see if such stimulation could prevent the onset of extrajunctional ACh sensitivity, or reverse it once sensitivity had become established. This is the more critical experiment since the results with local anesthesia might have been due to an effect of the agents used on axoplasmic transport. The result was that activity was remarkably effective in reducing extrajunctional sensitivity, although there was no apparent decline in the sensitivity of the end-plate region itself (178). The effect of direct activity on extrajunctional sensitivity was quickly confirmed in other mammalian muscle preparations both in vivo (207) and in vitro (208), as well as in chick myotubes in culture (209). An important extension of these results was to prevent activity in innervated muscle by chronic curarization or treatment with α-bungarotoxin (210). Rats were paralyzed with these drugs for several days, and denervation changes looked for in muscle fibers of the diaphragm. After 3 days, clear signs of denervation were present, including detectable ACh sensitivity, changes in membrane electrical properties, and tetrodotoxin resistance. Axoplasmic flow, as measured by cholinesterase accumulation above a ligature, was normal (210). Conceptually, these experiments are similar to the use of an anesthetic cuff, but have the advantage that the block is at the level of the postsynaptic receptors (211). Thus the experiments avoid an objection to the cuff experiments, that local anesthetics might affect axoplasmic transport and thus influence the result. The outcome of chronic curarization experiments also makes it unlikely that activity is needed simply as a means of releasing a trophic factor, since denervation effects developed in spite of normal activity in the phrenic nerve (210).

One aspect of these results which is at first puzzling is the persistence of high ACh sensitivity in the face of muscle fiber fibrillation. It is known that denervated muscle fibers in mammals can be spontaneously active at average rates as great as or greater than those used in the direct stimulation experiments (see for example Refs. 64, 67, 208), and that such activity can continue for many months (4). This degree of activation would thus be expected to have an effect similar to that of imposed activity, namely, a marked reduction of extrajunctional sensitivity. The probable answer to this problem was provided by experiments in which single fibrillating fibers in organ cultured rat diaphragm strips were examined over prolonged periods (about 12–48 hours) (208). It was found that at any one time only about one-third of the fibers in a denervated muscle is spontaneously active, single fibers fibrillating in cyclical fashion with an active period of 21 hours on average, and a subsequent silent period of about 2–4 days. When all of the fibers in a muscle strip were directly stimulated for 24 hours the number of fibrillating fibers fell nearly to zero; conversely, when all the muscle fibers were rendered quiescent for 3 days by action potential blockade with TTX, then the percentage of active fibers approached 100% when the block was removed. Thus the reason for the cyclical activity of single fibers appears to be that activity is self-regulating. Such regulation would explain the co-existence of high levels of activity and supersensitivity: spontaneous activity in individual fibers inhibits itself before much reduction in extrajunctional ACh sensitivity can occur.

Lømo and Westgaard (157, 212) have further strengthened the argument that the activity level of muscle fibers exerts a major influence on extrajunctional ACh sensitivity by examining the effects of different patterns of activity on sensitivity. They showed that different amounts of muscle activity affect extrajunctional ACh sensitivity in a graded fashion, and that the same total amount of imposed activity is much more effective in reducing sensitivity when it is presented in high frequency bursts than in a continuous, low frequency pattern. The fact that bursts of activity could significantly reduce sensitivity even when separated by several hours might explain why immobilization of limbs (200, 201), or surgical isolation of motoneurons (202, 203), produces little or no supersensitivity. The mechanism by which activity reduces extrajunctional ACh sensitivity is not known; the fact that true slow muscle fibers become supersensitive after denervation (27) suggests that action potentials per se are not a necessary part of the controlling influence of activity, but this might not be true for twitch fibers. Thus the question of whether electrical activity or contraction (or both) are necessary for preventing or reducing supersensitivity remains unanswered.

In addition to its effect on extrajunctional sensitivity, activity also influences other properties of denervated muscle. As noted, fibrillation is also regulated by activity. Since fibrillation is initiated by a Na^+ conductance increase different from the conductance increase which follows the interaction of ACh and its receptor, this membrane property must also be affected (69). There is also good evidence that direct activation of a denervated rat muscle can prevent hyperinnervation by an implanted nerve, although re-innervation by the native nerve returning to the end-plate region can still occur (213) (see below). Westgaard (79) has further found that the altered electrical properties of denervated mammalian skeletal muscle fibers can be returned to normal by direct stimulation. Thus specific membrane resistance, internal resistance, and membrane capacitance were brought within the range found in innervated fibers by 2 weeks of stimulation. A similar return to normal after direct stimulation has been reported for the decreased resting potentials seen in denervated muscle (212). Since these aspects of muscle fiber function depend not on ACh receptors but on other membrane channels (as well as the internal structure of muscle fibers or, in the case of receptivity to innervation, an unknown aspect of the membrane), the implication is that activity affects at least several constituents of the muscle fiber.

Finally, changes in the contractile properties of mammalian skeletal muscle following denervation or cross-innervation also appear to be largely dependent on activity. Such an effect was first reported by Salmons and Vrbová (198), who found that innervated fast-twitch muscle underwent a slowing of the time course of contraction when stimulated at rates similar to the pattern of slow-twitch activation. More recently Lømo and his collaborators (214) have extended this work by showing that *denervated* slow-twitch muscle fibers acquire some of the properties of fast-twitch fibers when they are stimulated with high frequency bursts (probably comparable to the normal pattern of fast-twitch activation), but are unchanged by activity patterns similar to the normal pattern of firing of motor nerve fibers innervating slow-twitch muscles (see also Ref. 215). Thus Lømo et. al. suggest that the basis of changes in contractile properties of cross-reinnervated muscles may simply result from altered activity pattern, rather than from different trophic factors secreted by motor neurons to fast and slow muscle fibers, as was suggested originally (88).

Trophic Factor or Activity? In summary, there are experiments which strongly support the proposition that the cause of many denervation effects is a lack of activity (or more generally that muscle membrane properties are a function of activity level), as well as strong evidence to support the trophic factor hypothesis. On the one hand, the original experiments of Miledi (11, 101) and other more recent experiments (see for example Refs. 182 and 185) show that at least some of the phenomena which follow denervation of skeletal muscle are due to the loss of some interaction of nerve and muscle which is independent of synaptic activation. While the nature of this interaction remains somewhat speculative, the absence of a normally secreted trophic factor is certainly a logical mechanism. On the other hand, the experiments of Lømo and Rosenthal (178) and others leave no doubt that activity has an important influence on the ACh sensitivity of the muscle fiber membrane (196, 196, 207–210), as well as the ability to fibrillate (208), the nature of the contractile response (198, 214), the electrical properties of muscle fibers (79, 212), and the ability of denervated fibers to become hyperinnervated (213).

There are several ways in which these differing results might be partially reconciled. First, it is possible that different mechanisms operate in amphibians and mammals, a chemical mediator being the primary cause of denervation effects in the frog, and activity playing a more important role in mammals. While this explanation is unattractive in principle, the responses of frog and rat muscle to denervation differ in time course, magnitude, and even the nature of the changes undergone; it therefore might not be surprising to find some difference in mechanism as well. This possibility could be examined by looking at the effects of imposed activity on denervated frog muscle, and conversely looking for further evidence of trophic factor mediation in mammalian nerve-muscle interaction. Experiments using colchicine represent just such an effort, but have not been decisive for the reasons mentioned. The recent experiments of Bray and Harris (182) and Miledi and Spitzer (185) represent another approach along these lines, and the former offer stronger support for the idea that a trophic factor also plays a part at the mammalian neuromuscular junction. Second, some of the results supporting a trophic factor mechanism might have an alternative explanation if degenerating nerve terminals release a substance which has a direct local effect on muscle. According to this view (see for example Refs. 157 and 212) the direct action of a degenerating terminal on muscle would override the effect of activity, and give the result found after partial denervation (11). Conceivably, the regulating effect of nerve stump length could also be explained in this way (see also Ref. 216). Indeed, there is considerable experimental support for this idea, for if a degenerating nerve is simply placed in contact with a normal muscle, the membrane in the vicinity of the dying nerve becomes supersensitive (155, 156). The effects of degenerating nerve are thus similar to injury, which also has the ability to cause a local increase in ACh sensitivity (39). However, neither a local effect of dying nerve nor activity pattern would explain the ability of a regenerating nerve to affect the muscle fiber it contacts prior to the onset of functional transmission during the course of re-innervation (101, 182), or the ability of BTX to block induction of the action potential mechanism in denervated slow muscle (185). Perhaps the most prudent view in the light of the evidence to date is that both activity and a trophic factor are important in controlling muscle fiber properties, one or the other dominating depending on the preparation and the circumstances. In spite of this somewhat neutral conclusion, it is probably fair to say that the thrust of many recent experiments is towards a greater role for activity in vertebrate nerve-muscle interactions, and a correspondingly diminished role for chemical mediators. The question has reasonably been asked whether all the major phenomena of denervation might be due solely to lack of activity (157). While such a general conclusion may be as yet unwarranted, the onus is increasingly on those who favor a trophic factor mechanism as the basis for most denervation changes in muscle.

Other Trophic Interactions Whatever the resolution of this controversy concerning nerve-muscle interaction, one should hasten to add that abundant evidence exists for trophic factor mechanisms (or at least non-activity mediated interactions) operating elsewhere in the peripheral nervous system, as well as in other aspects of nerve-muscle interaction. For instance, the properties of the synaptic region itself (in contrast to extrajunctional membrane) are not obviously affected by the pattern or level of activity (see for example Refs. 30, 31, 33, 157, 178, and 212). The maintenance of high end-plate sensitivity can hardly be attributed to an ongoing influence of nerve since it persists in denervated muscle; however, the initial restriction of sensitivity to the end-plate zone during development must depend on a neural influence which is unlikely to be related solely to activity (see also "Specificity of Peripheral Re-innervation"). Similarly, the induction of cholinesterase in the synaptic region (217), as well as the elaborate postsynaptic membrane infolding (218) which occurs during the initial formation of the end-plate, must be based on a chemical interaction of some kind (see also Refs. 219 and 220). In general, it seems that the regulation of junctional and extrajunctional properties is quite different, and that activity affects primarily the latter.

Other instances in the peripheral nervous system where apparently chemical (trophic factor) interactions exist might also be mentioned. Some classes of taste receptor disappear within a few days of sectioning of the primary axon, and reappear when the peripheral nerve regenerates (see for example Ref. 221). It is hard to imagine how such an interaction could occur without some sort of chemical mediation. Another example comes from the examination of the effects of colchicine on sprouting of sensory nerves (222). Cutting a spinal nerve in a salamander causes compensatory sprouting of sensory fibers running in adjacent nerves, with consequent enlargement of the receptive field of nearby nerves. Sprouting of adjacent nerves was also induced by colchicine application to a spinal nerve. Conversely, compensatory sprouting of nearby nerves in response to section of a single spinal nerve could be prevented by colchicine application to those nerves. In no case was there evidence of nerve degeneration. The interpretation given these experiments is that the balance of peripheral fields between different nerves depends on a continuous supply of trophic factor carried peripherally by axoplasmic flow; if, for whatever reason, the supply is reduced, then nearby neurons will sprout (222). Similarly, in ontogeny, it is known that many nerve cells die, probably because they fail to make "appropriate" peripheral contacts (223–227). Conversely, nerveless limbs are atrophic and otherwise abnormal (see for example Ref. 228). A recent review by Hamburger (228) provides an especially lucid and witty account of this field. In the adult nervous system as well, there is a growing body of evidence which suggests that appropriate peripheral contacts are important in maintaining a neuron's properties (the subject of the subsequent section). None of these results can be explained by activity in any simple way.

The conclusion to be drawn from these several lines of evidence is that chemically mediated (trophic) interactions are probably widespread in the nervous system. If it turns out that activity, rather than a trophic factor, is the primary controlling agent in nerve-muscular interactions, the mechanism at the neuromuscular junction may still be the exception, rather than the rule, in long-term interactions between excitable cells.

EFFECTS OF AXON INTERRUPTION ON NEURONAL PROPERTIES

One of the points made in the preceding section was that relatively little is known about the extent to which the fibers innervating a neuron influence it over long periods of time. There is, however, at least one other possible source of trophic influence for neurons: since neurons send axons to contact another cell or cells, it is conceivable that the target of innervation (or the absence of a target) will determine the nerve cell's properties.

For many years the question of what happens to a neuron when its axon is interrupted (injured) has been intensively studied from both anatomical and physiological points of view. The motivation for many earlier studies appears to have been to determine how a neuron reacts to injury, in part to gain some insight into the ability of peripheral neurons (or central neurons which project peripherally) to recover following injury. The result has been a wealth of information about what happens to many classes of neurons after axotomy.

A neuron whose axon has been injured undergoes a series of anatomical changes generally known as chromatolysis. These changes have been extensively discussed elsewhere, at the level of both the light and the electron microscope (see for example Refs. 229–231). Briefly, the prominent light microscopic changes include a breaking up of Nissl bodies and a consequent decrease in cytoplasmic basophilia (hence the name chromatolysis), movement of the nucleus to an eccentric position in the cell body, an increase in perikaryal volume, nucleolar swelling, and in some cases later degeneration of nerve cells. At the ultrastructural level, the prominent features are (in the rat superior cervical ganglion as described by Matthews and Raisman (230) and Matthews (231)) a decrease and dispersion of the rough endoplasmic reticulum (corresponding to chromatolysis at the light microscopic level), a reduction in the number of large dense core vesicles frequently seen in these cells, and an increase in cytoplasmic dense bodies and autophagic vacuoles. Most of these characteristics revert to normal over a period of several weeks as injured axons regenerate (231). This initial spectrum of changes has been interpreted by Matthews and Raisman (230) as reflecting a shift in the metabolism of the cell away from those processes concerned with transmitter synthesis and receptor functions, representing a period of neuronal involution. A further anatomical change that has been reported after axon interruption is a retraction and then re-expansion of the dendritic tree of affected neurons (232, 233), although the death of many neurons following axotomy (see below) makes these apparent changes difficult to interpret. There seems little doubt, however, that the processes of the injured neurons undergo dramatic changes in geometry including axonal sprouting (231), and the appearance of dendritic swellings seen by injection of horse-radish peroxidase that, at the ultrastructural level, contain organelles characteristic of rapid membrane turnover (234). This is in contrast to invertebrate neurons which show little or no change after axotomy (235).

Corresponding to these morphological indicators of the altered demands on the cell's metabolic machinery are a number of biochemical changes in axonally injured neurons, which have been extensively studied by Watson in the hypoglossal nucleus of the rat (236–240) (see also Ref. 241). After interruption of the hypoglossal nerve, Watson found that affected neurons in the brainstem undergo an increase in total and nucleolar RNA content, and that there is a net shift of RNA from the nucleus to the cytoplasm, presumably secondary to increased synthesis of ribosomes by nucleoli (236, 237). There is also an intense glial cell reaction to injury (236, 240, 242–245). These changes subside during axon regeneration, and have been used to study the nature of the signal which initiates chromatolysis (see below).

The most dramatic reaction to axon interruption is cell death. Neuronal death after such an injury occurs in many types of neurons which ramify within the central nervous system of higher vertebrates (see Refs. 229 and 246 for a discussion of this topic), but has been less widely reported in mature peripheral neurons, or adult neurons located centrally and projecting peripherally. However, cell loss within one to a few weeks occurs when axons are interrupted in autonomic ganglia (234, 247, 248) (see also Ref. 229), in brainstem nuclei (236, 243, 244, 249), and spinal motoneurons (250, 251). Cell death has been reported to occur more swiftly in younger animals (250). In at least some preparations nearly all the neurons may die if their axons are prevented from regaining contact with their normal target (234). This might explain the often greater degree of degeneration after axon injury of central neurons (229) which have little or no ability to regenerate (see for example Refs. 252–254).

In addition to these structural changes, the synaptic function of terminals which contact an injured nerve cell is altered. The original observation of this phenomenon was made by Acheson, Lee and Morison (255) who found that the reflex discharge of respiratory motoneurons was markedly decreased after section of the phrenic nerve, but gradually recovered over a period of several weeks as the cut nerve regenerated. Their suggestion that this might represent a synaptic defect has been subsequently confirmed in a number of other preparations. In autonomic neurons (234, 256–260) and spinal motoneurons (261–265) the events following axon interruption are similar: within a few days of injury by crush, or by nerve section, the excitatory postsynaptic potential (e.p.s.p.) recorded in the injured neuron declines sharply, often failing to bring the cell to threshold. This synaptic depression presumably accounts for the original observation of Acheson and his co-workers (255). When the injured axons are allowed to regenerate, depression of synaptic transmission is largely reversed over a period of weeks (234, 257, 263). In the superior cervical ganglion the amplitudes of synaptic potentials recorded in ganglion cells about 2 months after postganglionic nerve crush are indistinguishable from those of control neurons (234). In addition to synaptic depression, dendrites of both spinal motoneurons (263, 266) and autonomic ganglion cells (234) develop the ability to generate action potentials after axotomy. The ability of normally non-regenerative membrane to give rise to action potentials has also been reported in some invertebrate neurons after axotomy or colchicine application (267, 268) (as well as in amphibian slow muscle fibers after denervation (188); see above). This change, as well as an altered safety factor for antidromic invasion of the initial segment (234, 263, 266), suggests additional disturbances of the membrane properties of the injured cell.

Synaptic depression aftter axotomy might have a number of causes associated with one or more of the steps in the process of chemical synaptic transmission. Although some early evidence favored a decrease in the ability of the postsynaptic cell to respond to transmitter (256, 264), a number of more recent studies are in general agreement that loss of synaptic contacts is the major cause of synaptic depression after axon injury in mammals. The number of synapses on injured neurons are reduced in the hypoglossal nucleus (269–272), the facial nucleus of the rat and mouse (242, 243), spinal motoneurons (unpublished results of D.E. Hillman, cited in Ref. 265), and the superior cervical ganglion of the rat (260) and guinea pig (234). The events in these preparations are roughly similar, in so far as they can be compared. During the first few days following axon injury, the number of synapses observed in electron microscopic sections of autonomic ganglia in which most of the neurons have undergone axon interruption declines to about 30% of normal levels (234, 260). Synaptic disjunction and electrophysiological depression of transmission occur with roughly the same time course (234, 260). A careful study of the morphological events associated with synaptic disjunction in the rat superior cervical ganglion has shown that, during the first few days, presynaptic terminals are often unapposed by a postsynaptic element and instead enveloped by satellite cell processes (260). Such profiles are thought to represent synapses at which terminals have disengaged from principal cell contact. In both the rat and guinea pig (234, 260) (see also Ref. 272) postsynaptic thickenings, which presumably represent vacated synaptic sites, rapidly disappear. This finding is in contrast to the persistence of postsynaptic thickenings after denervation (131, 132, 134), where they may provide a marker for the return of terminals to the same locus (134). In both rat and guinea pig there is some evidence for the involution of the presynaptic terminals in addition to disjunction, since vesicle-filled profiles do not increase but rather decline in electron microscopic sections after axotomy (234, 260). Thus the effects of axotomy are not limited to the injured neurons. Synaptic loss is, however, limited to those neurons whose axons have been interrupted; when only one of several postganglionic branches is injured, neurons sending axons into intact branches show a normal intracellularly recorded response to preganglionic stimulation, even though adjacent injured neurons show marked synaptic depression (234). The mechanism of this loss of synaptic contacts remains unclear, although active glial cell interposition has been suggested by several workers (see for example Refs. 242, 243, 245, 260, 269). The way in which axon interruption leads to disjunction of boutons on the injured cell is a question of obvious interest.

The nature of the signal (or signals) which gives rise to synaptic disjunction, and other changes associated with axotomy, is not known with certainty. In general, the range of possibilities can be divided into those mechanisms concerned with axonal injury per se and the attempt to recover from it (e.g., loss of the bulk of the neuronal axoplasm, mobilization of the neuron's resources for axon regeneration), and mechanisms concerned with the disruption of a normally present informational (trophic) link between a neuron and its target. The view that the changes following axon interruption are primarily a response of the nerve cell to injury (see Ref. 273 for a more extensive discussion of this) is perhaps the simpler explanation, but has received firm evidence to support it only recently. The problem had been in the first place to devise reasonably quantitative assays for some aspect of the axotomy reaction, and, second, to devise experiments which would differentiate an injury response from a response due to the presence or absence of functional peripheral contact. Both these problems were partially overcome for some measures of the axotomy reaction in a series of experiments on the rat hypoglossal nucleus carried out by Watson (236–240). Watson first undertook to quantitate the metabolic responses during chromatolysis and axon regeneration, and then used these as indices in a variety of experimental situations. In a study of the incorporation of labeled precursors of RNA with autoradiographic and microanalytic techniques in cells after axotomy, injured neurons were found to incorporate precursors, and to transfer RNA from nucleus to cytoplasm, at a faster than normal rate within 48 hours of injury (236). As axons began to regenerate, however, the rate of incorporation and transfer fell and remained low. Accompanying these changes in neuronal metabolism was an increase in the number of perineuronal microglial cells (236). These findings were extended by measuring the nucleic acid content of neurons and their nucleoli by ultraviolet absorption microspectrometry (237). Knowing the time course and magnitude of these metabolic reactions, it was possible to examine whether they reflected removal of a peripheral trophic influence, or a response to the injury itself. If the hypoglossed nerve was cut or crushed at varying distances from the cell body, the nucleic acid changes also varied, being greater for proximal lesions; if the changes depended on peripheral contact they might be expected to vary in an all or none fashion.

Moreover if the hypoglossal nerve was kept from re-innervating the tongue (or presumably any other muscle) by implanting its cut end in the neck, then a second nerve injury caused nucleic acid changes similar to those occurring after the initial injury. Since in this case the nerve already lacked a target, the absence of normal peripheral contact was probably not essential to changes measured. Finally, the eventual decrease in RNA synthesis did not depend on renewed contact since it was not prevented when nerve regeneration was impeded (237). These results suggested that these particular aspects of the reaction to axotomy have little or nothing to do with the presence or absence of a peripheral trophic interaction. In order to clarify whether loss of axoplasm is the aspect of injury underlying nucleic acid changes after axotomy, Watson examined the effects on hypoglossal motoneurons of botulinum toxin injected into the tongue (238). He found that, indeed, similar RNA changes occurred after BTX blockade. Since no loss of axoplasm was involved, the interpretation of this result was that sprouting of the distal axon (known to occur after BTX induced paralysis–see above) was the critical aspect of the response to injury that determined the changes in neuronal metabolism. This idea gained further support from the observation that other conditions which induced axonal growth or sprouting (such as the stimulus provided by denervated muscle) also caused nucleic acid changes in hypoglossal motor neurons similar to those which follow axotomy (239). However, some aspects of these later experiments suggested that factors in addition to injury and axonal sprouting might also be involved in producing some of the effects of axotomy. For instance, injection of BTX into the tongue does not cause the initial microglial proliferation characteristic of axotomy (238), although later astrocytic changes can be induced (240). When the hypoglossal nerve was implanted into the innervated sternomastoid muscle, a metabolic reaction characteristic of axotomy occurred only when the sternomastoid muscle was deprived of its normal innervation (as might be expected on the basis of the axonal sprouting hypothesis) (239). However, if BTX was injected locally at the same time as cutting the native nerve to the sternomastoid, then the reaction of the hypoglossal nerve cells did not occur (239). This procedure should not interfere with axon sprouting, which has been shown to be induced by BTX application (174, 175), but would initially prevent the implanted hypoglossal nerve from forming junctional contacts with the sternomastoid (see Ref. 142). The results of these experiments are difficult to bring together into a unitary view of the underlying cause of axotomy changes. Their major import is perhaps the demonstration that, while injury per se clearly underlies some aspects of the neuronal response to axotomy, a similar metabolic response can occur in the absence of actual axon interruption. This latter type of response appears to be closely tied to the state of the distal axon, perhaps to the presence or absence of functional peripheral contact.

A number of other experiments directed primarily at the function and morphology synaptic contacts on neurons after their axons have been interrupted provide evidence for the view that synaptic changes after axotomy depend on peripheral contact. This was first suggested by the observation that the severity of functional loss (unlike the severity of the chromatolytic response) bears no relation to the level of phrenic nerve section (255). Pilar and Landmesser (258) further found that both the light microscopic changes of chromatolysis and depression of ganglionic synaptic transmission could be reproduced by brief application of colchicine to the postganglionic nerves of the avian ciliary ganglion. Colchicine's primary action under these conditions is assumed to be interference with normal axoplasmic transport (189–191). A colchicine-induced loss of synapses from affected neurons has been observed with intracellular recordings and electron microscopy in mammalian sympathetic neurons after similar postganglionic nerve treatment (274). Morphological evidence of bouton loss after colchicine application to the hypoglossal nerve has also been reported (Ref. 271, and R.E. Cull, personal communication). Since colchicine applied in this way would not be expected to cause axonal sprouting or interfere directly with peripheral adrenergic synaptic transmission (see, however, Refs. 275 and 276), disruption of the neuron's transport system seems the most likely cause of these effects. However, this does not necessarily mean that a peripheral trophic interaction is essential in ganglionic synaptic maintenance. Indeed, Pilar and Landmesser (258) favored the interpretation that disruption of the circulation of a substance produced by the neuron (rather than a factor produced peripherally) was involved in the effects they observed. While the ability of colchicine to mimic the effects of axotomy implies the importance of axoplasmic transport mechanisms in producing these effects, whether anterograde, retrograde, or some more general aspect of transport is critical remains to be demonstrated.

The fact that functional recovery eventually occurs after axotomy (see above) indicates that preganglionic fibers whose terminals have come away from injured neurons ultimately re-innervate them more or less completely. This provides an additional means of assessing the importance of peripheral contact in causing the synaptic effects of axon interruption. One can ask whether ganglionic transmission is restored at some predetermined time, or whether re-innervation of neurons whose axons have been cut (or crushed) depends on renewed peripheral contact. If postganglionic axons of neurons in the mammalian superior cervical ganglion are crushed (thus allowing prompt regeneration), e.p.s.p.s recorded in ganglion cells in response to maximal preganglionic nerve stimulation undergo a sharp reduction in amplitude during the succeeding few days which is largely due to preganglionic terminals coming off the dendritic tree (234). After about 1.5–2 months, however, all affected neurons once more have e.p.s.p.s of normal amplitude. The time course of recovery is roughly the same as the re-establishment of peripheral synaptic connections, which can first be detected after about 1 month. If, however, postganglionic axons are ligated, thus preventing them from regaining their peripheral target, then e.p.s.p.s recorded in surviving ganglion cells in response to preganglionic stimulation show no signs of recovery for at least 3 months (234). Similar results have been reported in anatomical studies of boutons on hypoglossal motoneurons during recovery from axotomy (271, 272). Using the zinc iodide-osmium method to stain boutons at the light microscope level, Cull (271) found that the number of boutons fell after crushing the hypoglossal nerve, but returned to normal in about 1 month. However, when the nerve was implanted into the normal sternomastoid muscle, so that peripheral re-innervation probably failed to occur, then the number of boutons continued to fall for at least 3 months. Restoration of boutons on hypoglossal motor neurons during recovery from axotomy has been confirmed at the electron microscopic level by Sumner (272).

Cell death after axon interruption (see above) provides a further measure of the importance of the periphery in neuronal survival. Such control, as already mentioned, is known to occur in vertebrate ontogeny: during development, neurons which lack a peripheral target or are unable to reach it, differentiate, migrate, and grow axons normally, but degenerate in large numbers when they fail to make appropriate peripheral connections (see for example Refs. 223, 225, 226; for reviews see Refs. 224 and 227). In the course of examining the effects of axotomy it had been observed that neuronal death is less pronounced after a crush injury than after nerve section (see for example Ref. 236), which presumably makes it somewhat harder for axons to find their way back to the periphery. Cell death has recently been examined in more detail in the mammalian superior cervical ganglion after both crush and ligation of postganglionic nerves (234). While a substantial number of cells degenerate after nerve crush, cell death is more widespread after ligation and eventually appears to affect nearly all neurons whose axons are enclosed by the ligature. One interpretation of this result is that in the adult, as in developing animals, contact with a target is essential for neuronal survival.

Some recent experiments by Kuno and his colleagues (277, 278) have provided evidence for the influence of peripheral contact in determining, in this case, the properties of spinal motoneurons. Motor neurons to mammalian fast- and slow-twitch muscles have somewhat different electrophysiological characteristics (in the duration of the afterpotential, conduction velocity, and overshoot amplitude); these differences are largely lost following motor nerve section (277). The restoration of these differential properties appears to depend on re-innervation, although not necessarily functional re-innervation (278). The specific properties of fast and slow motoneurons, however, were regained without regard to which type of muscle they re-innervated (278). A further striking example of the apparent influence of muscle on the properties of motor axon terminals in the lobster has been reported by Frank (279), although this particular sort of interaction might be peculiar to invertebrates. Although all excitatory synapses on muscle fibers of the accessory flexor muscle of the lobster walking leg are made by a single axon, the characteristics of individual junctions vary greatly with respect to facilitation of transmitter release. However, the facilitation of junctions of any one muscle fiber is similar, implying an influence of the muscle fiber on nerve endings (279).

While many of these experiments suggest a trophic influence of the peripheral target on neuronal properties, the nature of such a link remains obscure. There are, however, several promising pharmacological approaches to the problem. In cholinergic systems botulinum toxin has already been used to examine the interaction between skeletal muscle and hypoglossal motor neuron properties (see for example Refs. 239, 240, and 280). The results of Bray and Harris (182) suggest that the toxin may, in addition to blocking ACh release, block the transfer of trophic agents between pre- and postsynaptic cells. Although this suggestion awaits confirmation, BTX may provide a powerful tool for the further analysis of the interaction between neurons and their targets. The phenomenon described by Watson (239) of local BTX application preventing the metabolic response of hypoglossal neurons in some situations might be an example of this novel action of BTX blocking a centripetal trophic effect. In the sympathetic nervous system, the drug 6-hydroxydopamine, which selectively destroys the terminals of adrenergic neurons (281, 282), might allow another means of analyzing the mechanism of axotomy effects. If functional adrenergic terminals are essential for the maintenance of intraganglionic synapses and survival of sympathetic neurons, then their selective destruction might be expected to reproduce the effects of axotomy. In the sympathetic system there is a strong candidate for a trophic factor operating between neurons and their target. The protein nerve growth factor (NGF) (283) clearly plays a trophic role in development, since a course of antiserum to NGF administered to young animals results in nearly complete destruction of the peripheral sympathetic system (284). Little is known about NGF's biologic importance in adult animals, but recent experiments have shown that it is produced by glial cells and fibroblasts (285–288), is specifically taken up by adrenergic endings and transported in retrograde fashion to sympathetic ganglion cells (289–292), and has at least some effects on neuronal function, morphology, and survival (291, 293–295). It is possible that inadequate transport of NGF following axon interruption might be the basis for the reversible synaptic disjunction and perhaps additional effects that follow axotomy. In any event, it should be possible to test this idea.

A major question that grows out of these studies of peripheral axon interruption is whether nerve cells projecting entirely within the central nervous system undergo similar changes in response to alterations in the nature and extent of their terminal contacts. Cell death appears to be more marked after axon interruption in the central than the peripheral nervous system (see Ref. 229), yet many classes of central neurons do survive, especially if only a portion of their projection is severed (see Ref. 296 for example). The fact that central neurons seem, if anything, more sensitive to axon interruption, suggests that terminal contact may be an important controlling mechanism in the central nervous system as well. It is unlikely that this point will be decided until the techniques of intracellular recording and quantitative electronmicroscopy can be applied in parallel to neurons ramifying entirely within the central nervous system.

SPECIFICITY OF PERIPHERAL RE-INNERVATION

In addition to long-term regulatory influences provided by a cell's synaptic input or by its target organ, those mechanisms that lead to specific connections during development might persist in adult animals and form the basis for another system of long-term control. If this were the case, then particular nerve or muscle cells would maintain their predilection for specific sources of innervation throughout life. Such a mechanism would be especially important if synapses are lost and regained at a low rate in mature animals, as suggested by the observations of Barker and Ip (297). Since re-innervation occurs readily in the vertebrate peripheral nervous system, it might seem an easy matter to settle the question of whether re-innervation is specific or not, and what the underlying mechanisms are. In spite of intensive study of this question, though, many aspects of peripheral re-innervation are not understood.

The term "specificity" has been used in several senses by different authors. In the following discussion specificity is used in a topographical sense: specific re-innervation is taken to mean the return of interrupted axons to the same class of muscle fibers or nerve cells they originally innervated. A related term is "selectivity," used here to refer to the ability of a nerve or muscle cell to select and promote the establishment of correct (native) terminals in the presence of incorrect (foreign) nerve fibers, or, in a more extreme case, after foreign synapses have already become established.

Specific Re-innervation of Muscle Fibers

Clinically, it has long been clear that functional recovery of higher vertebrate skeletal muscle function is at best disappointing following denervation: a patient who suffers a transection of a motor nerve supplying many muscles will generally not regain full coordination of the re-innervated part. This outcome does not reflect the inability of motor nerves to regenerate, or the ability of muscle to become re-innervated, as denervated muscle fibers (or muscles treated with botulinum toxin) avidly reaccept appropriate nerve fibers (94–96, 142) and will induce sprouting from nearby intact terminals after partial denervation (298) (Ref. 299 provides a review of the older literature). However, denervated skeletal muscle fibers of higher vertebrates will also accept inappropriate skeletal motor nerve terminals if they are made available (94–96, 142, 144, 145, 218) (see also the discussion of cross-innervation below). When a motor nerve supplying several different muscles regenerates, this lack of discrimination works to the disadvantage of the animal since inappropriate connections form nearly as readily as appropriate ones (300, 301). A similar lack of discrimination has been observed during the regeneration of motor nerves to mammalian smooth muscles (302). Not only will skeletal muscles accept synapses from motoneurons which originally went to other nearby muscles, but also from completely different types of neurons. Thus, vertebrate skeletal muscle can be re-innervated with parasympathetic preganglionic fibers (98, 303), preganglionic sympathetic fibers (97, 99), and possibly sensory fibers as well (100) (see also Ref. 304). Indeed there appears to be little discrimination even at the level of fast and slow fibers in mammals; slow- and fast-twitch muscle fibers are re-innervated by either "fast" or "slow" nerve fibers, and this situation persists indefinitely (305) (although see Ref. 306). The only example of specific motor nerve regeneration in mammals appears to be the re-innervation of intrafusal fibers (307).

An apparent criterion of acceptability for re-innervating foreign fibers is that the nerve terminal be capable of releasing the normal transmitter; in the case of vertebrate skeletal muscle, acetylcholine. Langley and Anderson (97), for example, failed to observe functional innervation of the tongue by postganglionic sympathetic fibers. Even this limitation, however, has been questioned in the re-innervation of smooth muscle; there is suggestive evidence that cholinergic fibers can, under some conditions, re-innervate the dilator muscle of an iris which has been denervated by extirpation of the superior cervical ganglion (308, 309). In Ceccarelli's experiments (309) an especially interesting finding is that the pharmacology of the newly formed synapses appears to be adrenergic. Additional evidence for such highly abnormal re-innervation has recently been obtained in tissue culture where presumably adrenergic neurons are found to form cholinergic synapses on other ganglion cells (310) (see also Refs. 311 and 312). One possible explanation for some of these results is that synthesis of a novel transmitter is induced by contact with the "foreign" effector. However, a number of more conventional explanations (e.g., some adrenergic fibers running in the preganglionic nerve; some cholinergic neurons in the superior cervical ganglion) are equally likely. Since such bizarre interactions are relevant to the way in which synapses normally form, further results of these investigations are awaited with interest.

In the light of this general lack of specificity of re-innervation, it is surprising that there are at least two instances in which re-innervation of vertebrate muscle appears to be reasonably specific. Both occur in lower vertebrates. (There is also some evidence for specific re-innervation of invertebrate muscle (313).) The first example is the re-innervation of teleost muscle which was studied in a series of experiments carried out by Sperry and Arora (314). It had long been known that, unlike mammals, some lower vertebrates (some amphibians and teleosts for example) are capable of apparently complete functional recovery following denervation. A proposed mechanism of recovery was random peripheral re-innervation and subsequent reorganization within the central nervous system (a process called "myotypic modulation") (315–319). Sperry and Arora (314) re-examined this proposition by studying re-innervation of the four extrinsic ocular muscles supplied by the oculomotor nerve in a cichlid fish. They tested the efficacy of re-innervation by tilting the animal and eliciting ocular reflexes, and found that within two weeks of nerve section, eye movements became entirely normal. They next sectioned individual branches of the oculomotor nerve and implanted them into an inappropriate eye muscle whose own nerve had been resected, and found that the "foreign" nerve re-innervated the muscle reasonably well, but maintained a reflex discharge appropriate to its original connections (that is, the oculomotor reflexes remained abnormal). Moreover, if foreign and native nerves were allowed to re-innervate a muscle simultaneously, the native nerve eventually dominated (i.e., was "selected" by the muscle). This latter phenomenon occurred even if the foreign nerve was allowed to innervate the muscle before the native nerve arrived. In addition to discrediting the idea of myotypic modulation, these results (see also Refs. 320–323) suggested that the fish is capable of a remarkable degree of both specificity and selectivity during re-innervation. Re-innervation of salamander limb muscles also shows specificity (and selectivity) (324, 325) and is further discussed below.

A second example of specific re-innervation of muscle is of a rather different sort, but again seems to be limited to lower vertebrates. Re-innervation of fast- and slow-twitch muscle fibers in mammals is nonspecific (305), but for frog slow muscle fibers (in the iliofibularis or pyriformis) there is some evidence for specificity (and perhaps selectivity) of re-innervation. The muscles contain both true slow-fibers (which are normally of smaller diameter, incapable of generating an action potential, and which have a distributed innervation) (186, 326–328), and twitch fibers which give an all or nothing spike in response to depolarization, and generally have a single synaptic contact along their length. The motor nerve in turn contains large diameter axons which normally contact the fast muscle fibers, and smaller nerve axons which innervate the slow fibers (327). Following denervation of the muscle, the slow muscle fibers acquire the ability to generate action potentials although their contractile properties (which normally lead to a characteristic contracture) (326) do not change much (188, 329, 330). Re-innervation of these slow fibers by axons that normally innervate a fast muscle does not alter the ability of the fiber to give an action potential, but does cause its contractile response to become faster (329, 331). These several findings allow a fairly demanding test of whether or not slow fibers are specifically re-innervated by their original axons when the motor nerve (containing axons going to both fast and slow fibers) is cut: if slow muscle fibers (distinguished throughout by their characteristic passive electrical and morphological properties) are not specifically re-innervated, they should remain capable of action potentials and give a twitch-like contractile response. After initial re-innervation slow muscle fibers remain abnormal (330), but at longer times after re-innervation (several months) they again become incapable of generating action potentials and respond to depolarization by contracture (188, 329, 330). The eventual return to normalcy probably represents re-innervation by native axons, since slow fibers remained abnormal for at least 18 months when re-innervated with fast nerve axons alone (331) (although see also Ref. 330). The fact that initial re-innervation by the normal mixed nerve leaves slow fibers abnormal for some time (330) suggests selectivity as well: one interpretation is that axons to fast fibers also re-innervate slow fibers initially, but are eventually replaced by native "slow" axons (329) (see also Ref. 330). However, in the absence of more definite identification of the types of axon re-innervating particular muscle fibers the evidence for selectivity remains somewhat tentative. Other experiments in which muscle tension was measured in response to nerve stimulation in the toad also support the view that re-innervation of fast and slow muscle fibers is specific (332) (see also Ref. 333).

In summary, re-innervation of muscle fibers in higher vertebrates is largely non-specific (with the possible exception of muscle spindle re-innervation (307)), a finding which is especially disappointing from the vantage of clinical neurology. In some lower vertebrates, however, there is considerable evidence for specificity of muscle re-innervation, and some evidence for the ability of native fibers to displace foreign ones if both are available (selectivity). The absence of selectivity during re-innervation of mammalian skeletal muscle is all the more puzzling since some ability to select appropriate connections exists during de novo synapse formation in development. Muscle fibers are initially innervated by several nerve fibers; however, over the few weeks of life all but one of these endings are rejected (334, 335). Similar findings have recently been reported for re-innervated neonatal mammalian muscle (336).

Because the neuromuscular junction is an easier place to examine competitive interactions than at neuronal synapses, nerve-muscle preparations have been, and continue to be, the objects of intense study in an effort to determine the possible mechanism of specific re-innervation and selectivity (see below).

Specific Re-innervation of Neurons

In spite of the limited ability of higher vertebrates to regenerate neuromuscular connections specifically, there is evidence that the re-innervation of peripheral neurons in the mammalian autonomic nervous system is capable of at least some degree of specificity, and perhaps selectivity as well. This finding is all the more interesting since denervated mammalian autonomic neurons, as denervated muscle fibers, will accept synaptic contacts from a number of foreign sources. Superior cervical ganglion neurons have been re-innervated by preganglionic parasympathetic nerves (146, 337–339), somatic motor nerves (128, 340–342), and visceral afferent fibers (338, 343). Specific re-innervation of autonomic neurons was first suggested by straightforward experiments carried out by J.N. Langley around the turn of the century (302, 344, 345). Langley observed that autonomic neurons could be re-innervated after cutting the cervical sympathetic trunk of cats: after some weeks, electrical stimulation proximal to the lesion produced normal peripheral sympathetic effects (contraction of the nictitating membrane, widening of the pupil and the palpebral fissure, vasoconstriction, and piloerection) (302, 345). In a series of detailed studies, he determined that the normal superior cervical ganglion is innervated in a specific fashion (at least to some extent) (302, 344, 345). Briefly, if the rami communicantes emerging from the upper thoracic spinal segments (T1-3) are stimulated, they generally cause a greater sympathetic response of the smooth muscles of eye than of the ear (stimulation of T1, for example, caused widening of the pupil and fissure, and contraction of the nictitating membrane, but little, if any, vasoconstriction of the ear). Conversely, stimulating T4 constricted the ear vessels but caused little or no perceptible change in the eye. This suggested that axons of preganglionic sympathetic motor neurons emerging from different segmental levels of the spinal cord go preferentially to different groups of cells in the superior cervical ganglion. Langley then tested whether the degree of specificity seen in normal animals is re-established during re-innervation (302, 345). After cutting the cervical trunk and allowing peripheral function to be restored, he again examined the effects of stimulating the thoracic rami. The result was that upon re-innervation particular classes of neurons were generally excited from the same thoracic levels as before, and therefore that re-innervation had been specific. His summary account of these experiments sounds thoroughly modern, although it was written nearly 80 years ago:

> The only feasible explanation appears to me to be that the sympathetic fibers grow out along the peripheral piece of nerve,—as nerve fibers usually are supposed to grow out—spreading amongst the cells of ganglion, and that there is some special chemical relation between each class of nerve fiber and each class of nerve cell, which induces each fiber to grow towards a cell of its own class and there to form its terminal branches. At bottom then the phenomenon would be a chemiotactic one (Ref. 345, p. 284).

This type of experiment has been repeated and extended by Guth and Bernstein (346) who have generally confirmed Langley's results (see also Refs. 135, 338, and 347).

A different experiment along these lines was carried out by Landmesser and Pilar (126) using the ciliary ganglion of pigeons. They took advantage of the fact that there are two types of neurons in the avian ganglion: choroidal cells which are innervated by small preganglionic fibers, and ciliary cells innervated by large fibers. Each cell type has a characteristic latency of ganglionic transmission and a characteristic preganglionic synaptic mechanism (ciliary cells have an electrical component of synaptic transmission, choroidal cells do not) (348–350). Thus during re-innervation it is possible to ask whether the original connections are re-formed, or whether random innervation occurs. The outcome was clear: the normal situation, with smaller fibers innervating choroidal cells solely by a chemical mechanism, and larger fibers innervating ciliary neurons by a dual mode of transmission, was re-established. Thus within the limits of these criteria, re-innervation in the avian ganglion is highly specific. In none of the above experiments is there evidence for the ultimate level of detail at which specific reconnections are made. In a simpler invertebrate system, however, specific re-innervation has been found to occur on the level of individual neurons (351).

There is also evidence for selectivity during the re-innervation of mammalian sympathetic neurons. Murray and Thompson (135) (see also Ref. 346) demonstrated that when 80–90% of the preganglionic fibers reaching the cat superior cervical ganglion are removed (by cutting the rami communicantes of T1-T3), the remaining fibers (T4-7) sprout and re-innervate neurons that they ordinarily would not contact. However, Guth and Bernstein (346) found that if they allowed regeneration of the T1-T3 preganglionic fibers initially interrupted in the thorax, then 6 months after such partial denervation the normal connections were re-established: stimulation of the upper, but not the lower, thoracic rami caused their usual sympathetic effects on the eye. Since one month after T1-T3 interruption stimulation of the lower thoracic rami caused eye effects (as had been found by Murray and Thompson (135)), Guth and Bernstein (346) concluded that "foreign" sprouts which initially re-innervated the deprived neurons became functionally inactive when the correct preganglionic fibers finally arrived. The ability of mammalian peripheral neurons to select between native and foreign fibers has recently been tested in experiments in which vagal (foreign) and sympathetic preganglionic fibers have been allowed to simultaneously re-innervate the superior cervical ganglion (146). Intracellular recordings from ganglion cells showed that initially about equal numbers of neurons were re-innervated by native and foreign fibers, and that about half the neurons recorded from were innervated from both sources. This situation appeared to remain reasonably stable for at least a year, with about half the neurons sampled still receiving dual innervation after 8–12 months. These results conflict with those of Guth and Bernstein (346) which appear to show that ganglionic neurons will reject "foreign" terminals when correct ones become available. Moreoever, they raise the more general question of how it is possible for ganglion cells to distinguish T1 terminals from T4 terminals, when they are unable to distinguish sympathetic from vagal endings. Perhaps the best way to resolve these questions is to repeat Langley's original experiments (302, 344, 345) using intracellular recording methods. Until this is done it is safer to regard specificity and selectivity of mammalian neuronal re-innervation as somewhat uncertain.

Evidence for specific re-innervation is much more impressive in the central nervous system than the peripheral autonomic system. While functional re-innervation seems not to occur to a very significant extent in the adult higher vertebrate central nervous system (see for example Refs. 252–254), it has been known since the experiments of Matthey (352) that a high order of specific re-innervation is possible in the visual system of some amphibians and fish. The initial experiments were extended by Stone (353–358), and more recently by a number of other investigators (see Refs. 359 and 360 for reviews). If the optic nerve of a salamander is severed, the animal recovers vision within a few weeks as judged by behavioral tests (353, 354, 357). If the eye is rotated 180° after optic nerve section, the behavioral response following recovery is permanently abnormal, the animal reacting as if retinal ganglion cells had grown back to the neurons in the tectum which they originally contacted (355, 356) (see also Ref. 358). This latter finding suggests that there is a cell-to-cell specificity in the regeneration of the optic nerve in these animals, or at least a very high degree of specific reconnection. Despite the use of electrophysiological techniques to study tectal re-innervation in lower vertebrates (see Refs. 359 and 360 for extensive reviews of these and related experiments), it is not yet known whether re-innervation is specific at a cell-to-cell level. The recent results of Yoon and others (361–364) have shown that central re-innervation is also capable of considerable flexibility.

In the central nervous system of higher vertebrates the only known examples of regeneration of pathways of any length involve fibers of monoaminergic neurons located in the locus ceruleus and elsewhere (see for example Refs. 365 and 366). Whether re-innervation would be specific if significant central regeneration occurred is therefore moot.

Since both central and peripheral neurons in lower vertebrates and invertebrates can regenerate specifically, these experiments on the specificity of re-innervation raise the question of why mammals have lost this ability in their central nervous system as well as, to a greater or lesser extent, in their peripheral nervous system? While there is no obvious answer at present, understanding the mechanisms of specific regeneration in those instances where it does occur offers some hope for understanding the relative lack of these abilities in mammals.

Possible Mechanisms of Specific Re-innervation in the Peripheral Nervous System

It should be said at the outset that little is known about the mechanism of re-innervation, the specificity and selectivity of the process aside. In studies of muscle re-innervation it is clear that nerve axons do not form synapses with muscle fibers at random points, but tend to return to the region of the end-plate (101, 367, 368). This appears to be the case even when skeletal muscle is re-innervated by a foreign source such as preganglionic autonomic fibers (99, 303). A similar preferential return to vacated synapses sites has been reported for the re-innervation of autonomic ganglion cells (134). That there is something special about the end-plate region vis-à-vis synapse formation is further suggested by the fact that muscle fiber activity will prevent the formation of extrajunctional synapses, but not re-innervation of the original end-plate (144, 145). Similarly, although innervation normally prevents formation of additional synaptic contacts (see above), extrajunctional innervation by an implanted nerve will not initially prevent re-innervation of the original end-plate (143–145). The critical difference between end-plate and extrajunctional membrane remains unknown. There is apparently nothing inherently different about the region where end-plates normally form since end-plates induced by foreign nerves at abnormal sites develop similar stable properties; for example, "ectopic" end-plates produced by foreign nerve implantation in temporarily denervated muscles can be re-innervated in the presence of the original innervation (145). Because synapse formation is associated with high membrane sensitivity to ACh both in development (325, 369) and after denervation (see above), it might be imagined that ACh receptors provide some critical impetus to the establishment of nerve connections. This possibility has been made less likely by the demonstration that de novo synapse formation can occur in the presence of blocking concentrations of curare (370) or *Naja naja* toxin (a neurotoxin which binds very strongly and specifically to ACh) (371), and that re-innervation proceeds normally in rats chronically paralyzed with α-bungarotoxin (372). Nor is presynaptic release of ACh necessary for the development of localized sensitivity at the site of nerve-muscle contact (371). In short, the membrane at a point of innervation appears to be permanently changed even if the innervation is subsequently removed; one aspect of this transformation is that the region is a preferred site for re-innervation. The reasons for such preference remain unclear.

In spite of the fact that so little is known of the factors controlling re-innervation, an attractive hypothesis has been suggested by Mark and his collaborators (373) which might explain, in a rough way, how specificity and selectivity come about during re-innervation. This mechanism, generally referred to as functional synaptic repression, has generated much interest (see, for example, Ref. 374) because, among other things, it can be tested with conventional techniques. The proposal is that outgrowing nerve fibers initially connect with the denervated target in more or less random fashion. However, nerve fibers which happen to have made "correct" synaptic connections have the ability to repress the function of "incorrect" endings on the same cell without causing any change in the ultrastructural appearance of incorrect endings. One controversial aspect of this idea is that non-functional synapses would appear ultrastructurally identical to functioning ones, requiring a revision of the long-held belief that the presence of a chemical synapse seen in the electron microscope is tantamount to a functional contact between cells. A mechanism which could turn synapses "off," while leaving them in place (and presumably capable of being turned back on), is also attractive in the context of a variety of higher nervous system functions. Indeed, Mark (375, 376) has suggested that functional synaptic repression might be a basis for learning and memory.

Evidence for functional synaptic repression comes from two sources: the specific and apparently selective re-innervation of fish eye muscles briefly described above, and re-innervation of partially denervated muscles in salamander limbs. The original experiments carried out by Mark, Marotte, and Mart (323, 377–379) can be summarized as follows. The third cranial nerve innervates the inferior oblique muscle of the fish eye, while the fourth cranial nerve innervates the superior oblique muscle. An assay for normal muscle function (and presumably normal innervation of the extraocular muscles) is the behavioral response to tilting the fish in the vertical axis. A normal fish responds to downward tilting of the head by rotating its eye upward, a movement that depends on the inferior oblique muscle (nerve III); conversely, when the fish is tilted head up, the eye rotates downward, a movement which depends on the superior oblique muscle (nerve IV). This is the same type of behavioral response Sperry and Arora (314) used to determine that re-innervation of eye muscles in the fish is specific and probably selective. Mark and Marotte (323) cross-innervated the superior oblique muscle by resecting the inferior oblique muscle and implanting its nerve in the denervated superior oblique muscle. Initially the behavioral response was found to be inappropriate, as might be expected if successful cross-innervation had occurred: upward tilting of the fish now caused the eye to rotate upward. The critical part of the experiment was a study of the effect of rotation when the original nerve was allowed to regenerate. As Sperry and Arora had found (314), the original nerve grew back after an initial delay of some weeks, and the reflex returned to normal. Moreover, Marotte and Mark (323) reported that takeover by the original nerve occurred rather abruptly, developing over a period of as little as 48 hours. Surprisingly, when the muscle was examined histologically during the period of takeover by the original nerve, no degenerating synapses were seen (377) (however, see Ref. 380). These results were extended by electrophysiological studies which showed that action potentials still occurred in ineffectual foreign nerves (378). When the original nerve was sectioned *after* takeover (379) and the muscle examined electron microscopically 2 days later, degenerating terminals as well as intact synapses were seen. Mark and his co-workers concluded from these findings that the arrival of the native terminals caused a functional repression of the foreign synapses without causing any morphological change.

This conclusion was supported by experiments on salamanders in which the effects of misdirecting spinal nerve roots to the hindlimb were examined (324). In normal animals, stimulation of spinal nerve roots supplying the hindlimb showed a characteristic pattern of innervation. Following interruption and misrouting of the two major nerve trunks running to the hindlimb (carrying axons from four spinal nerves), the normal pattern of innervation was re-established. Because of the grossly abnormal paths which regenerating nerves were forced to take during re-innervation, it seemed likely that numerous incorrect contacts had been formed, but that correct ones had eventually been selected. The nature of this competition was studied by taking advantage of the fact that when a single spinal nerve root is cut, the territories of adjacent roots spread to take over denervated muscle fibers (381, 382), presumably by sprouting. Again the critical observations concerned the sequence of events which occurred when the native nerve (now a single spinal root) was allowed to regenerate. After some weeks the original pattern of innervation was re-established. If, however, the previously severed root was cut a second time, then establishment of functional synapses by the adjacent nerve roots took only 3 days instead of 3 weeks, as after the first interruption. Since this period is brief for functional contacts to form by sprouting of adjacent nerves, the interpretation was again that functional repression had occurred (324).

The difficulty with these experiments is that, in the absence of intracellular recording and electron microscopic study of the same fiber, one cannot exclude a mechanism other than functional repression. Scott (383) recently repeated the experiments on competitive re-innervation of goldfish eye muscles, but arrived at a different conclusion from that of Mark and his co-workers. She found that junctional potentials could be elicited in single re-innervated superior oblique muscle fibers by both native and foreign nerve stimulation more than 100 days after implanting the foreign nerve; in these fibers functional repression of synapses had not occurred. Indeed, tension recordings from re-innervated muscles generally showed good responses to both third and fourth nerve stimulation. She also observed a few animals which showed reversion to the correct behavioral response after re-innervation by the original nerve, as described by Marotte and Mark (323). In these animals, however, significant regeneration and re-innervation of the resected inferior oblique muscle was found. Contraction of this regenerated muscle opposed the response due to cross-innervation, and therefore caused an apparent disappearance of the reversed response without causing any morphological changes of synapses in the superior oblique muscle. Similar results have also been obtained during competitive re-innervation of perch gill muscle, where again persistent innervation of single fibers by both native and foreign nerves has been found (384). This is in agreement with similar experiments on mammalian muscle fibers and autonomic neurons where native and foreign innervation can evidently co-exist indefinitely (145, 146). Thus, in the fish, the existence of functional repression is in some doubt.

Although these experiments (145, 146, 383, 384) were aimed at testing selectivity rather than specificity, the negative results obtained necessarily raise additional doubts about the original observations on re-innervation of the fish muscles (314). If muscle fibers (or neurons) can remain dually innervated indefinitely, it is hard to imagine how full (correct) functional recovery can occur when a nerve supplying several targets (the oculomotor nerve, for example) regenerates. A careful study of oculomotor nerve regeneration in the fish using intracellular recording should be able to settle the issue of how specific such re-innervation really is.

It would also be useful to repeat the original experiments of Cass et al. (324) in the salamander with methods less ambiguous than those used initially. Such a study has recently been undertaken by Yip and Dennis (personal communication) using salamander limb muscle. During competitive re-innervation they find that the quantal content of the response to foreign nerve stimulation declines when native axons re-innervate muscle fibers. This confirms that some selectivity does occur in the re-innervation of salamander muscle, and leaves open the possibility that functional synaptic repression may be the basis for it.

Whichever way the controversy over the existence of selectivity in muscle re-innervation (and functional repression as a mechanism to explain it) is resolved, the question will remain whether these phenomena occur at synapses between neurons.

CONCLUSION

Three areas of current interest in the neurobiology of long-term regulation in the vertebrate peripheral nervous system have been discussed: 1) the effects of denervation and possible underlying mechanisms, 2) the effects of axon interruption and their possible causes, and 3) the degree to which re-innervation is specific and selective in the peripheral nervous system. One way of summarizing the current state of each of these topics is to re-emphasize the major questions which remain unanswered, but which appear capable of solution given current technology.

With regard to denervation-induced changes in muscle, the question remains whether activity level is a sufficient explanation for the physiological regulation of muscle fiber properties. While there is good evidence for trophic factor interactions in some situations, the extent of their role in normal nerve-muscle function is unclear. Which, if either, of these mechanisms normally serves to regulate neuronal properties is not known, but experiments analogous to those carried out at the neuromuscular junction can be done in neuronal preparations. Perhaps the greatest obstacle at the moment is the poor understanding of the neuronal reaction to denervation, due in large part to technical problems which are now being overcome.

The effects of axotomy have recently been better defined at the cellular level. There is growing evidence that some aspects of axonal extension to the periphery influences the degree to which a neuron is innervated, and ultimately whether it survives. Although it seems likely that some factor (or factors) transported by the axon is the basis for this, other mechanisms are possible. Defining the way in which a nerve cell's "peripheral" contact determines its physiological and anatomical properties is another major question which appears solvable.

Finally, the ability of mature nerve and muscle cells to become specifically re-innervated has continued to be a source of controversy. There has been doubt, certainly justified in some cases, concerning the degree to which specific re-innervation and selectivity really occur in the peripheral nervous system. These uncertainties will continue until older studies, using largely behavioral measures of the "correctness" of re-innervation, are repeated using the techniques of intracellular recording and electron microscopy.

The difficulty of carrying out and interpreting these general types of studies in the central nervous system will no doubt make some of the peripheral preparations discussed in this article useful for a long time to come.

ACKNOWLEDGMENTS

I am especially grateful for helpful criticism from M. Dennis, E. Frank, and C.M. Rovainen. This review was completed in August, 1975. Support of my own work by the National Institutes of Health and the Muscular Dystrophy Associations of America is acknowledged.

REFERENCES

1. Katz, B. (1966). Nerve, muscle and synapse. (McGraw-Hill Book Co.: New York.) 193 pp.
2. Katz, B. (1969). The release of neural transmitter substances. (Liverpool University Press: Liverpool.) 60 pp.
3. Harris, A.J. (1974). Inductive functions of the nervous system. Ann. Rev. Physiol. 36: 251.
4. Tower, S.S. (1939). The reaction of muscle to denervation. Physiol. Rev. 19: 1.
5. Cannon, W.B., and Rosenblueth, A. (1949). The supersensitivity of denervated structures: a law of denervation. Macmillan, New York. 243 pp.
6. Langley, J.N. (1907). On the contraction of muscle, chiefly in relation to the presence of "receptive" substances. Part I. J. Physiol. (Lond.) 36:347.
7. Ginetsinskii, A.G., and Shamarina, N.M. (1942). The tonomotor phenomenon in denervated muscle. Department of Scientific and Industrial Research, Translation RTS 1710. Usp. Sovrem. Biol. 15:283.
8. Kuffler, S.W. (1943). Specific excitability of the endplate region in normal and denervated muscle. J. Neurophysiol. 6:99.
9. del Castillo, J., and Katz, B. (1955). On the localization of acetylcholine receptors. J. Physiol. (Lond.) 128:157.
10. Axelsson, J., and Thesleff, S. (1959). A study of supersensitivity in denervated mammalian muscle. J. Physiol. (Lond.) 147:178.
11. Miledi, R. (1960). The acetylcholine sensitivity of frog muscle after complete or partial denervation. J. Physiol. (Lond.) 151:1.
12. Miledi, R. (1960). Junctional and extra-junctional acetylcholine receptors in skeletal muscle fibers. J. Physiol. (Lond.) 151: 24.
13. Miledi, R. (1962). Induction of receptors. *In* J.L. Mongar and A.V.S. de Reuck (eds.), Ciba Foundation Symposium on Enzymes and Drug Action, pp. 220–238. Little, Brown and Company, Boston.
14. McMahan, U.J., Spitzer, N.C., and Peper, K. (1972). Visual identification of nerve terminals in living isolated skeletal muscle. Proc. Roy. Soc. Lond. B 181:421.

15. Peper, K., and McMahan, U.J. (1972). Distribution of acetylcholine receptors in the vicinity of nerve terminals on skeletal muscle of the frog. Proc. Roy. Soc. Lond. B 181:431.
16. Dreyer, F., and Peper, K. (1974). Iontophoretic application of acetylcholine: advantages of high resistance micropipettes in connection with an electronic current pump. Pflügers Arch. 348:263.
17. Dreyer, F., and Peper, K. (1974). The acetylcholine sensitivity in the vicinity of the neuromuscular junction of the frog. Pflügers Arch. 348:273.
18. Kuffler, S.W., and Yoshikami, D. (1975). The distribution of acetylcholine sensitivity at the post-synaptic membrane of vertebrate skeletal twitch muscles: iontophoretic mapping in the micron range. J. Physiol. (Lond.) 244:703.
19. Anderson, M.J., and Cohen, M.W. (1974). Fluorescent staining of acetylcholine receptors in vertebrate skeletal muscle. J. Physiol. (Lond.) 237:385.
20. Daniels, M.P., and Vogel, Z. (1975). Immunoperoxidase staining of α-bungarotoxin binding sites in muscle endplates shows distribution of acetylcholine receptors. Nature 254:339.
21. Harzell, H.C., and Fambrough, D.M. (1972). Acetylcholine receptors. Distribution and extrajunctional density in rat diaphragm after denervation correlated with acetylcholine sensitivity. J. Gen. Physiol. 60:248.
22. Fertuck, H.C., and Salpeter, M.M. (1974). Localization of acetylcholine receptor by ^{125} I-labelled α-bungarotoxin binding of mouse motor endplates. Proc. Nat. Acad. Sci. USA 71:1376.
23. Feltz, A., and Mallart, A. (1971). An analysis of acetylcholine responses of junctional and extrajunctional receptors of frog muscle fibers. With an appendix by R. Kahn and A. le Yaouanc. J. Physiol. (Lond.) 218:85.
24. Feltz, A., and Mallart, A. (1971). Ionic permeability changes induced by some cholinergic agonists on normal and denervated frog muscles. J. Physiol. (Lond.) 218:101.
25. Miledi, R., and Zelena, J. (1966). Sensitivity to acetylcholine in rat slow muscle. Nature 210:855.
26. Albuquerque, E.X., and Thesleff, S. (1968). A comparative study of membrane properties of innervated and chronically denervated fast and slow muscle of the rat. Acta Physiol. Scand. 73:471.
27. Nasledov, G.A., and Thesleff, S. (1974). Denervation changes in frog skeletal muscle. Acta. Physiol. Scand. 90:370.
28. Katz, B., and Miledi, R. (1964). Further observations on the distribution of acetylcholine-reactive sites in skeletal muscle. J. Physiol. (Lond.) 170:379.
29. Chang, C.C., and Huang, M.C. (1975). Turnover of junctional and extrajunctional acetylcholine receptors of the rat diaphragm. Nature 253:643.
30. Berg, D.K., and Hall, Z.W. (1974). Fate of α-bungarotoxin bound to acetylcholine receptors of normal and denervated muscle. Science 184:473.
31. Berg, D.K., and Hall, Z. Loss of α-bungarotoxin from junctional and extrajunctional acetylcholine receptors in rat diaphragm muscle *in vivo* and in organ culture. J. Physiol. (Lond.) In press.
32. Albuquerque, E.X., and McIsaac, R.J. (1970). Fast and slow mammalian muscles after denervation. Exp. Neurol. 26:183.
33. Frank, E., Gatuvik, K., and Sommerschild, H. Cholinergic receptors at denervated mammalian end-plates. Acta Physiol. Scand. In press.
34. Fambrough, D.M. (1970). Acetylcholine sensitivity of muscle fiber membranes: mechanism of regulation by motoneurons. Science 168:372.
35. Grampp, W., Harris, J.B., and Thesleff, S. (1972). Inhibition of denervation changes in skeletal muscle by blockers of protein synthesis. J. Physiol. (Lond.) 221:743.
36. Devreotes, P.N., and Fambrough, D.M. Turnover of ACh receptor in skeletal muscle. Cold Spring Harbor Symp. Quant. Biol. In press.
37. Sakmann, B. (1975). Reappearance of extrajunctional acetylcholine sensitivity in denervated rat muscle after blockage with α-bungarotoxin. Nature 255:415.
38. Brockes, J.P., and Hall, Z.W. (1975). Synthesis of acetylcholine receptor by denervated rat diaphragm muscle. Proc. Natl. Acad. Sci. USA 72:1368.
39. Katz, B., and Miledi, R. (1964). The development of acetylcholine sensitivity in nerve-free segments of skeletal muscle. J. Physiol. (Lond.) 170:389.

40. Lee, C.Y., Tseng, L.F., and Chiu, T.H. (1967). Influence of denervation on localization of neurotoxins from clapid (sic) venoms in rat diaphragm. Nature 215:1177.
41. Miledi, R., and Potter, L.T. (1971). Acetylcholine receptors in muscle fibers. Nature 233:599.
42. Berg, D.K., Kelly, R.B., Sargent, P.B., Williamson, P., and Hall, Z.W. (1972). Binding of α-bungarotoxin to acetylcholine receptors in mammalian muscle. Proc. Natl. Acad. Sci. USA 69:147.
43. Fambrough, D.M. (1974). Acetylcholine receptors. Revised estimate of extrajunctional receptor density in denervated rat diaphragm. J. Gen. Physiol. 64:468.
44. Devreotes, P.N., and Fambrough, D.M. (1975). ACh receptor turnover in membrane of denervated muscle fiber. J. Cell. Biol. 65:335.
45. Beránek, R., and Vyskočil, F. (1967). The action of tubocurarine and atropine on the normal and denervated rat diaphragm. J. Physiol. (Lond.) 188:53.
46. Lapa, A.J., Albuquerque, E.X., and Daly, J. (1974). An electrophysiological study of the effects of d-tubocurarine, atropine, and α-bungarotoxin on the cholinergic receptor in innervated and chronically denervated mammalian skeletal muscles. Exp. Neurol. 43:375.
47. Katz, B., and Miledi, R. (1972). The statistical nature of the acetylcholine potential and its molecular components. J. Physiol. (Lond.) 224:665.
48. Mallart, A., and Trautman, A. (1973). Ionic properties of the neuromuscular junction of the frog: effects of denervation and pH. J. Physiol. (Lond.) 234:553.
49. Brockes, J.P., and Hall, Z.W. (1975). Acetylcholine receptors in normal and denervated rat diaphragm muscle. I. Purification and interaction with ^{125}I-α-bungarotoxin. Biochemistry 14:2092.
50. Brockes, J.P., and Hall, Z.W. (1975). Acetylcholine receptors in normal and denervated rat diaphragm muscle. II. Comparison of junctional and extrajunctional receptors. Biochemistry 14:2100.
51. Trendelenburg, U. (1966). Mechanisms of supersensitivity and subsensitivity to sympathomimetic amines. Pharmacol. Rev. 18:629.
52. Langer, S.Z., Draskólzy, P.R., and Trendelenburg, U. (1967). Time course of development of supersensitivity to various amines in the nictitating membrane of the cat. J. Pharmacol. Exp. Ther. 157:255.
53. Takeuchi, A., and Takeuchi, N. (1964). The effect on crayfish muscle of iontophoretically applied glutamate. J. Physiol. (Lond.) 170:296.
54. Takeuchi, A., and Takeuchi, N. (1965). Localized action of gamma-amino butyric acid on the crayfish muscle. J. Physiol. (Lond.) 177:225.
55. Usherwood, P.N.R., Machili, P., and Leaf, G. (1968). L-glutamate at insect excitatory nerve-muscle synapses. Nature 219:1169.
56. Hoy, R.R., Bittner, G.D., and Kennedy, D. (1967). Regeneration in crustacean motoneurons: evidence for axonal fusion. Science 156:251.
57. Hoy, R.R. (1969). Degeneration and regeneration in abdominal flexor motor neurons in the crayfish. J. Exp. Zool. 172:219.
58. Atwood, H.L., Govind, C.K., and Bittner, G.D. (1973). Ultrastructure of nerve terminals and muscle fibers in denervated crayfish muscle. Z. Zellforsch mikrosk. Anat. 146:155.
59. Bittner, G.D. (1973). Degeneration and regeneration in crustacean neuro-muscular systems. Amer. Zool. 13:379.
60. Frank, E. (1974). The sensitivity to glutamate of denervated muscles of the crayfish. J. Physiol. (Lond.) 242:371.
61. Usherwood, P.N.R. (1969). Glutamate sensitivity of denervated insect muscle fibers. Nature 223:411.
62. Rosenblueth, A., and Luco, J.V. (1937). A study of denervated mammalian skeletal muscle. Amer. J. Physiol. 120:781.
63. Thesleff, S. (1963). Spontaneous electrical activity in denervated rat skeletal muscle. *In* E. Gutmann and P. Hník (eds.), Effect of Use and Disuse on Neuromuscular Functions, pp. 41–51. Czechoslovakian Academy of Sciences, Prague.
64. Belmar, J., and Eyzaguirre, C. (1966). Pacemaker site of fibrillation potentials in denervated mammalian muscle. J. Neurophysiol. 29:425.
65. Harvey, A.M., and Kuffler, S.W. (1943). Synchronization of spontaneous activity in denervated human muscle. Archs. Neurol. Psychiat. (Chicago) 48:495.

66. Hník, P., and Skorpil, V. (1962). Fibrillation activity in denervated muscle. *In* E. Gutmann (ed.), The Denervated Muscle, pp. 135–150. Czechoslovakian Academy of Sciences, Prague.
67. Li, C.L., Shy, G.M., and Wells, J. (1957). Some properties of mammalian skeletal muscle fibers with particular reference to fibrillation potentials. J. Physiol. (Lond.) 135:522.
68. Li, C.L., Engel, K., and Klatzo, I. (1959). Some properties of cultured chick skeletal muscle with particular reference to fibrillation potentials. J. Cell Comp. Physiol. 53:421.
69. Purves, D., and Sakmann, B. (1974). Membrane properties underlying spontaneous activity in denervated mammalian muscle fibers. J. Physiol. (Lond.) 239:125.
70. Thesleff, S., and Ward, M.R. (1975). Studies on the mechanism of fibrillation potentials in denervated muscle. J. Physiol. (Lond.) 244:313.
71. Muchnik, S., Ruarte, A.C., and Kotsias, B.A. (1973). On the mechanism of denervatory electrical activity of single muscle fibers as tested *in vivo* with tetrodotoxin. Acta. Physiol. Lat. Amer. 23:24.
72. Muchnik, S., Ruarte, A.C., and Kotsias, B.A. (1973). Effects of actinomycin D on fibrillation activity in denervated skeletal muscles of the rat. Life Sci. 13:1763.
73. Redfern, P., and Thesleff, S. (1971). Action potential generation in denervated rat skeletal muscle. II. The action of tetrodotoxin. Acta Physiol. Scand. 82:70.
74. Harris, J.B., and Thesleff, S. (1971). Action potential generation in denervated rat skeletal muscle. I. Quantitative aspects. Acta Physiol. Scand. 81:557.
76. Colquhoun, D., Rang, H.P., and Ritchie, J.M. (1974). The bidning of tetrodotoxin and α-bungarotoxin to normal and denervated mammalian muscle. J. Physiol. (Lond.) 240:199.
77. Cullen, M.J., Harris, J.B., Marshall, M.W., and Ward, M.R. (1975). An electrophysiological and morphological study of normal and denervated chicken latissimus dorsi muscles. J. Physiol. (Lond.) 245:371.
78. Nicholls, J.G. (1956). The electrical properties of denervated skeletal muscle. J. Physiol. (Lond.) 131:1.
79. Westgaard, R.H. (1975). Influence of activity on the passive electrical properties of denervated soleus muscle fibers in the rat. J. Physiol. (Lond.) 251:683.
80. Albuquerque, E.X., Schuh, F.T., and Kauffman, F.C. (1971). Early membrane depolarization of the fast mammalian muscle after denervation. Pflügers Arch. 328:36.
81. Hille, B. (1970). Ionic channels in nerve membranes. Prog. Biophys. Mol. Biol. 21:1.
82. Armstrong, C.M. (1975). Ionic pores, gates and gating current. Quart. Rev. Biophys. 7:179.
83. Jenkinson, D.H., and Nicholls, J.G. (1961). Contractures and permeability changes produced by acetylcholine in depolarized denervated muscles. J. Physiol. (Lond.) 159:111.
84. Pellegrino, C., and Franzini, C. (1963). An electron microscope study of denervation atrophy in red and white skeletal muscle fibers. J. Cell Biol. 17:327.
85. Miledi, R., and Slater, C.R. (1969). Electron-microscopic structure of denervated skeletal muscle. Proc. Roy. Soc. Lond. B, 174:253.
86. Gutmann, E. (1962). Metabolic reactivity of the denervated muscle. *In* E. Gutmann (ed.), The Denervated Muscle, pp. 377–426. Czechoslovakian Academy of Sciences, Prague.
87. Buller, A.J., Eccles, J.C., and Eccles, R.M. (1960). Differentiation of fast and slow muscles in the cat hind limb. J. Physiol (Lond.) 150:399.
88. Buller, A.J., Eccles, J.C., and Eccles, R.M. (1960). Interactions between motoneurons and muscles in respect of the characteristic speeds of their response. J. Physiol. (Lond.) 150:417.
89. Eccles, J.C., Eccles, R.M., and Kozak, W. (1962). Further investigations on the influence of motoneurons on the speed of muscle contraction. J. Physiol. (Lond.) 163:324.
90. Close, R.I. (1972). Dynamic properties of mammalian skeletal muscle. Physiol. Rev. 52:129.
91. Buller, A.J. (1975). The physiology of skeletal muscle. *In* C.C. Hunt (ed.), MTP Science Series Vol. III, pp. 279–302. (MTP, Lancaster.)

92. Gauthier, G.F., and Dunn, R.A. (1973). Ultrastructural and cytochemical features of mammalian skeletal muscle fibers following denervation. J. Cell Sci. 12:525.
93. Hogan, E.L., Dawson, D.M., and Romanul, F.C.A. (1965). Enzymatic changes in denervated muscle. II. Biochemical studies. Arch. Neurol. 13:274.
94. Elsberg, C.A. (1917). Experiments on motor nerve regeneration and the direct neurotization of paralysed muscles by their own and by foreign nerves. Science 45:318.
95. Aitken, J.T. (1950). Growth of nerve implants in voluntary muscle. J. Anat. 84:38.
96. Hoffman, H. (1951). A study of the factors influencing innervation of muscles by implanted nerves. Aust. J. Exp. Biol. Med. Sci. 29:289.
97. Langley, J.N., and Anderson, H.K. (1904). The union of different kinds of nerve fibers. J. Physiol. (Lond.) 31:365.
98. Landmesser, L. (1972). Contractile and electrical responses of vagus-innervated frog sartorius muscles. J. Physiol. (Lond.) 213:707.
99. Bennett, M.R., McLachlan, E.M., and Taylor, R.S. (1973). The formation of synapses in mammalian striated muscle reinnervated with autonomic preganglionic nerves. J. Physiol. (Lond.) 233:501.
100. Vera, C.L., and Luco, J.V. (1967). Reinnervation of smooth and striated muscle by sensory nerve fibers. J. Neurophysiol. 30:620.
101. Miledi, R. (1960). Properties of regenerating neuromuscular synapses in the frog. J. Physiol. (Lond.) 154:190.
102. Lüllman-Rauch, R. (1971). The regeneration of neuromuscular junction during spontaneous reinnervation of the rat diaphragm. Z. Zellforsch. Mikrosk. Anat. 121:593.
103. McArdle, J.J., and Albuquerque, E.X. (1973). A study of the reinnervation of fast and slow mammalian muscles. J. Gen. Physiol. 61:1.
104. Cannon, W.B., and Rosenblueth, A. (1936). The sensitization of a sympathetic ganglion by preganglionic denervation. Amer. J. Physiol. 116:408.
105. Simeone, F.A., Cannon, W.B., and Rosenblueth, A. (1938). The sensitization of the superior cervical ganglion to nerve impulses by partial denervation. Amer. J. Physiol. 122:94.
106. Rosenblueth, A., and Cannon, W.B. (1939). The effects of preganglionic denervation on the superior cervical ganglion. Amer. J. Physiol. 125:276.
107. Chien, S. (1960). Supersensitivity of denervated superior cervical ganglion to acetylcholine. Amer. J. Physiol. 198:949.
108. McMahan, U.J., and Kuffler, S.W. (1971). Visual identification of synaptic boutons on living ganglion cells and of varicosities in postganglionic axons in the heart of the frog. Proc. Roy. Soc. Lond. B 177:485.
109. Dennis, M.J., Harris, A.J., and Kuffler, S.W. (1971). Synaptic transmission and its duplication by focally applied acetylcholine in parsympathetic neurons in the heart of the frog. Proc. Roy. Soc. Lond. B 177:509.
110. Harris, A.J., Kuffler, S.W., and Dennis, M.J. (1971). Differential chemosensitivity of synaptic and extrasynaptic areas on the neuronal surface membrane in parasympathetic neurons of the frog, tested by microapplication of acetylcholine. Proc. Roy. Soc. Lond. B 177:541.
111. Kuffler, S.W., Dennis, M.J., and Harris, A.J. (1971). The development of chemosensitivity in extrasynaptic areas of the neuronal surface after denervation of parasympathetic ganglion cells in the heart of the frog. Proc. Roy. Soc. Lond. B 177:555.
112. Diamond, J. (1963). Variation in the sensitivity of gamma-amino butyric acid of different regions of the Mauthner neurone. Nature 199:773.
113. Lang, W. (1967). Nomarski differential interference-contrast microscopy. Photographic Applications in Science and Industry, p. 114.
114. McMahan, U.J., and Purves, D. (1976). Visual identification of two kinds of nerve cells and their synaptic contacts in a living autonomic ganglion of the mudpuppy (*Necturus maculosus*). J. Physiol. (Lond.) 254:405.
115. Federov, B.G. (1935). Essai de l'étude intravitale des cellules nerveuses et des connexions interneuronales dans les système nerveux autonome. Trav. Laborat. Recherche. Biolog. Madrid 30:403.
116. Majorov, V.N. (1969). Morphology of reactive changes in vegetative interneuronal synapse (vital experimental investigation). I.P. Pavlov Institute of Physiology Academy of Sciences of the USSR, Leningrad. (In Russian, English summary.)

117. Roper, S. (1976). The acetylcholine sensitivity of the surface membrane of multiply innervated parasympathetic ganglion cells before and after partial denervation. J. Physiol. (Lond.) 254:427.
118. Roper, S. (1976). An electrophysiological study of synapses on neurons in the parasympathetic ganglion of the heart of the mudpuppy (*Necturus maculosus*). J. Physiol. (Lond.) 254:455.
119. Szentagothai, J. (1957). Zum Elementaren bau der interneuronalen Synapse. Acta. Anat. 30:827.
120. Govaerts, J. (1935). Apparition d'un tonus cardio-accélérateur dans le ganglion stellaire déconnecté centralement. C.R. Soc. Biol. 119:1181.
121. Govaerts, J. (1936). Etude oscillographique de l'activité éléctrique du ganglion stellaire déconnecté du névraxe. C.R. Soc. Biol. 121:854.
122. Govaerts, J. (1939). Nouvelles recherches sur l'activité spontaneé des ganglions sympathiques déconnectés du névraxe. Arch. Int. Physiol. 44:426.
123. Takeshige, C., and Volle, R.L. (1963). Cholinoceptive sites in denervated sympathetic ganglia. J. Pharmacol. 141:206.
124. McLennan, H., and Pascoe, J.E. (1954). The origin of certain non-medullated fibers which form synapses in the inferior mesenteric ganglion of the rabbit. J. Physiol. (Lond.) 124:145.
125. Hunt, C.C., and Nelson, P.G. (1965). Structural and functional changes in the frog sympathetic ganglion following cutting of the presynaptic nerve fibers. J. Physiol. (Lond.) 177:1.
126. Landmesser, L., and Pilar, G. (1970). Selective re-innervation of two cell populations in the adult pigeon ciliary ganglion. J. Physiol. (Lond.) 211:203.
127. Perri, V., Sacchi, O., and Casella, C. (1970). Synaptically mediated potentials elicited by the stimulation of post-ganglionic trunks in the guinea-pig superior cervical ganglion. Pflügers Arch. 314:55.
128. McLachlan, E. (1974). The formation of synapses in mammalian sympathetic ganglia re-innervated with preganglionic or somatic nerves. J. Physiol. (Lond.) 237:217.
129. Precht, W., Shimeizu, H., and Markham, C.H. (1966). A mechanism of central compensation of vestibular function following hemilabyrinthectomy. J. Neurophysiol. 29:996.
130. Hamlyn, L.H. (1954). The effect of preganglionic section on the neurons of the superior cervical ganglion in rabbits. J. Anat. 88:184.
131. Taxi, J. (1965). Contribution à l'étude des connexions des neurones moteurs du système nerveux autonome. Ann. Sci. Nat. Zool. (12) 7:413.
132. Sotelo, C. (1968). Permanence of postsynaptic specialization in the frog sympathetic ganglion cells after denervation. Exp. Brain Res. 6:294.
133. Hamori, J., Lang, E., and Simon, L. (1968). Experimental degeneration of preganglionic fibers in the superior cervical ganglion of the cat. Z. Zell. Mikrosk. Anat. 90:37.
134. Raisman, G., Field, P.M., Ostberg, A.J.C., Iversen, L.L., and Zigmond, R.E. (1974). A quantitative ultrastructural and biochemical analysis of the process of re-innervation of the superior cervical ganglion of the adult rat. Brain Res. 71:1.
135. Murray, J.G., and Thompson, J.W. (1957). The occurrence and function of collateral sprouting in the sympathetic nervous system of the cat. J. Physiol. (Lond.) 135:133.
136. Williams, T.H., and Jew, J. (1970). Collateral nerve sprouts produced experimentally. Nature 228:862.
137. Cowan, W.M. (1970). Anterograde and retrograde transneuronal degeneration in the central and peripheral nervous system. *In* S.O.E. Ebbesson and W.J.H. Nauta (eds.), Contemporary Research Methods in Neuroanatomy, pp. 217–251. Springer-Verlag, New York.
138. Cook, W.H., Walker, J.H., and Barr, M.L. (1951). A cytological study of transneuronal atrophy in the cat and rabbit. J. Comp. Neurol. 94:267.
139. Matthews, M.R., Cowan, W.M., and Powell, T.P.S. (1960). Transneuronal cell degeneration in the lateral geniculate nucleus of the macaque monkey. J. Anat. 94:145.
140. Wiesel, T.N., and Hubel, D.H. (1963). Effects of visual deprivation on morphology and physiology of cells in the cat's lateral geniculate body. J. Neurophysiol. 26:978.
141. Wiesel, T.N., and Hubel, D.H. (1965). Extent of recovery from the effects of visual deprivation in kittens. J. Neurophysiol. 28:1060.

142. Fex, S., Sonesson, B., Thesleff, S., and Zelená, J. (1966). Nerve implants in botulinum poisoned mammalian muscle. J. Physiol. (Lond.) 184:872.
143. Gutmann, E., and Hanzlikovâ, V. (1967). Effects of accessory nerve supply to muscle achieved by implantation into muscle during regeneration of its nerve. Physiol. Bohemoslov 16:244.
144. Frank, E., Jansen, J.K.S., Lømo, T., and Westgaard, R. (1974). Maintained function of foreign synapses on hyperinnervated skeletal muscle fibers of the rat. Nature 247:375.
145. Frank, E., Jansen, J.K.S., Lømo, T., and Westgaard, R.H. (1975). The interaction between foreign and original motor nerves innervating the soleus muscle of rats. J. Physiol. (Lond.) 247:725.
146. Purves, D. (1975). Persistent innervation of mammalian sympathetic neurones by native and foreign fibers. Nature 256:589.
147. Stein, D.G., Rosen, J.J., and Butters, N. (eds.) (1974). Plasticity and the recovery of function in the central nervous system. Academic Press, New York. 516 p.
148. Bernstein, J.J., and Goodman, D.C. (eds.) (1973). Neuromorphological Plasticity (S. Karger, Basel.) 164 pp.
149. Black, I.B., Hendry, I.A., and Iversen, L.L. (1971). Trans-synaptic regulation of growth and development of adrenergic neurons in a mouse sympathetic ganglion. Brain Res. 34:229.
150. Black, I.B., Hendry, I.A., and Iversen, L.L. (1971). Effects of surgical decentralization and nerve growth factor on the maturation of adrenergic neurons in a mouse sympathetic ganglion. J. Neurochem. 19:1367.
151. Black, I.B., and Geen, S. (1973). Transynaptic regulation of adrenergic neuron development: inhibition by ganglionic blockade. Brain Res. 63:291.
152. Hendry, I.A., Iversen, L.L., and Black, I.B. (1973). A comparison of the neural regulation of tyrosine hydroxylase activity in sympathetic ganglia of adult rats and mice. J. Neurochem. 20:1683.
153. Guth, L. (1968). Trophic influences of nerve on muscle. Physiol. Rev. 48:645.
154. Miledi, R. (1963). An influence of nerve not mediated by impulses. *In* E. Gutmann and R. Hník (eds.), The Effect of Use and Disuse of Neuromuscular Functions. pp. 35–40. Elsevier, Amsterdam.
155. Vrbová, G. (1967). Induction of an extrajunctional chemosensitive area in intact innervated muscle fibers. J. Physiol. (Lond.) 191:20.
156. Jones, R., and Vyskočil, F. (1975). An electrophysiological examination of the changes in skeletal muscle fibers in response to degenerating nerve tissue. Brain Res. 88:309.
157. Lømo, T., and Westgaard, R.H. Control of ACh sensitivity in rat muscle fibers. Cold Spring Harbor Symp. Quant. Biol. In press.
158. Birks, R., Katz, B., and Miledi, R. (1960). Physiological and structural changes at the frog myoneural junction, in the course of nerve degeneration. J. Physiol. (Lond.) 150:145.
159. Miledi, R., and Slater, C.R. (1968). Electrophysiology and electron-microscopy of rat neuromuscular junctions after nerve degeneration. Proc. Roy. Soc. Lond. B 169:289.
160. Dennis, M.J., and Miledi, R. (1974). Electrically induced release of acetylcholine from denervated Schwann cells. J. Physiol. (Lond.) 237:432.
161. Dennis, M.J., and Miledi, R. (1974). Non-transmitting neuromuscular junctions during an early stage of end-plate reinnervation. J. Physiol. (Lond.) 239:553.
162. Dennis, M.J., and Miledi, R. (1974). Characteristics of transmitter release at regenerating frog neuromuscular junctions. J. Physiol. (Lond.) 239:571.
163. Bennett, M.R., Pettigrew, A.G., and Taylor, R.S. (1973). The formation of synapses in re-innervated and cross re-innervated adult avian muscle. J. Physiol. (Lond.) 230:331.
164. Koening, J., and Pecot-Dechavassine, M. (1971). Relations entre l'apparition des potentials miniatures spontanes et l'ultrastructure des plaques motrices en voie de ré-innervation et de neoformation chez le rat. Brain Res. 27:43.
165. Tonge, D.A. (1974). Physiological characteristics of re-innervation of skeletal muscle in the mouse. J. Physiol. (Lond.) 241:141.
166. Luco, J.V., and Eyzaquirre, C. (1955). Fibrillation and hypersensitivity to ACh in denervated muscle: effect of length of degenerating nerve fibers.

167. Emmelin, N., and Malm, L. (1965). Development of supersensitivity as dependent on the length of degenerating nerve fibers. Quart. J. Exp. Physiol. 50:142.
168. Miledi, R., and Slater, C.R. (1970). On the degeneration of rat neuromuscular junctions after nerve section. J. Physiol. (Lond.) 207:507.
169. Harris, J.B., and Thesleff, S. (1972). Nerve stump length and membrane changes in denervated skeletal muscle. Nature New Biol. 236:60.
170. Ambache, N. (1974). Peripheral action of botulinum toxin. Nature 161:482.
171. Burgen, A.S.V., Dickens, F., and Zatman, L.J. (1949). The action of botulinum toxin on the neuro-muscular junction. J. Physiol. (Lond.) 109:10.
172. Thesleff, S. (1960). Supersensitivity of skeletal muscle produced by botulinum toxin. J. Physiol. (Lond.) 151:598.
173. Brooks, V.B. (1956). An intracellular study of the action of repetitive nerve volleys and of botulinum toxin on miniature end-plate potentials. J. Physiol. (Lond.) 134: 264.
174. Duchen, L.W., and Strich, S.J. (1968). The effects of botulinum toxin on the pattern of innervation of skeletal muscle in the mouse. Quart. J. Exp. Physiol. 33:84.
175. Duchen, L.W. (1971). An electron microscopic study of the changes induced by botulinum toxin in the motor end-plates of slow and fast skeletal muscle fibers of the mouse. J. Neurol. Sci. 14:47.
176. Tonge, D.A. (1974). Chronic effects of botulinum toxin on neuromuscular transmission and sensitivity to acetylcholine in slow and fast skeletal muscles of the mouse. J. Physiol. (Lond.) 241:127.
177. Drachman, D.B. (1974). The role of acetylcholine as a neurotrophic transmitter. Ann. N.Y. Acad. Sci. 228:161.
178. Lømo, T., and Rosenthal, J. (1972). Control of ACh-sensitivity by muscle activity in the rat. J. Physiol. (Lond.) 221:493.
179. Duchen, L.W., and Stefani, E. (1971). Electrophysiological studies of neuro-muscular transmission in "hereditary motor end-plate disease" in the mouse. J. Physiol. (Lond.) 212:535.
180. Albuquerque, E.X., Warnick, J.E., Tasse, J.R., and Sansone, F.M. (1972). Effects of vinblastine and colchicine on neural regulation of the fast and slow skeletal muscle fibers of the rat. Exp. Neurol. 37:607.
181. Hoffman, W.W., and Thesleff, S. (1972). Studies on the trophic influence of nerve on muscle. Eur. J. Pharmacol. 20:256.
182. Bray, J.J., and Harris, A.J. (1975). Dissociation between nerve-muscle transmission and nerve trophic effects on rat diaphragm using type-D botulinum toxin. J. Physiol. (Lond.) 253:53.
183. Harris, A.J., and Miledi, R. (1971). The effect of type D botulinum toxin on frog neurosmuscular junctions. J. Physiol. (Lond.) 217:497.
184. Spitzer, N.C. (1972). Miniature end-plate potentials at mammalian neuromuscular junctions poisoned by botulinum toxin. Nature New Biol. 237:26.
185. Miledi, R., and Spitzer, N.C. (1974). Absence of action potentials in frog slow muscle fibers paralyzed by botulinum toxin. J. Physiol. (Lond.) 241:183.
186. Kuffler, S.W., and Vaughan-Williams, E.M. (1953). Small nerve junctional potentials. The distribution of small motor nerves to frog skeletal muscle, and the membrane characteristics of the fibers they innervate. J. Physiol. (Lond.) 121:289.
187. Burke, W., and Ginsborg, B.L. (1956). The electrical properties of the slow muscle fiber membrane. J. Physiol. (Lond.) 132:587.
188. Miledi, R., Stefani, E., and Steinbach, A.B. (1971). Induction of the action potential mechanism in slow muscle fibers of the frog. J. Physiol. (Lond.) 217:737.
189. Dahlström, A. (1968). Effect of colchicine on transport of amine storage granules in sympathetic nerves of rat. Eur. J. Pharmacol. 5:111.
190. Dahlström, A. (1971). Axoplasmic transport (with particular reference to adrenergic neurons). Phil. Trans. Roy. Soc. Lond. B 261:325.
191. Kreutzberg, G.W. (1969). Neuronal dynamics and axonal flow. IV. Blockage of intra-axonal enzyme transport with colchicine. Proc. Natl. Acad. Sci. USA 62:722.
192. Cangiano, A. (1973). Acetylcholine supersensitivity: the role of neurotrophic factors. Brain Res. 58:255.

193. Kaufman, F.C., Warnick, J.E., and Albuquerque, E.X. (1974). Uptake of [^{3}H]colchicine from silastic implants by mammalian nerve and muscle. Exp. Neurol. 44:404.
194. Lømo, T. (1974). Neurotrophic control of colchicine effects on muscle? Nature (Lond.) 249:473.
195. Cangiano, A., and Fried, J.A. (1974). Neurotrophic control of skeletal muscle of the rat. J. Physiol. (Lond.) 239:31 P.
196. Jones, R., and Vrbová, G. (1970). Effect of muscle activity on denervation hypersensitivity. J. Physiol. (Lond.) 210:144 P.
197. Jones, R., and Vrbová, G. (1971). Can denervation hypersensitivity be prevented? J. Physiol. (Lond.) 217:69 P.
198. Salmons, S., and Vrbová, G. (1969). The influence of activity on some contractile characteristics of mammalian fast and slow muscles. J. Physiol. (Lond.) 201:535.
199. Vyskočil, F., Moravec, J., and Jansky, L. (1971). Resting state of the myoneural junction in a hibernator. Brain Res. 34:381.
200. Solandt, D.Y., Partridge, R.C., and Hunter, J. (1943). The effect of skeletal fixation on skeletal muscle. J. Neurophysiol. 6:17.
201. Fischbach, G.D., and Robbins, N. (1971). Effect of chronic disuse of rat soleus neuromuscular junctions on post-synaptic membrane. J. Neurophysiol. 34:562.
202. Solandt, D.Y., and Magladery, J.W. (1942). A comparison of effects of upper and lower motor neuron lesions on skeletal muscle. J. Neurophysiol. 5:373.
203. Johns, T.R., and Thesleff, S. (1961). Effects of motor inactivation on the chemical sensitivity of skeletal muscle. Acta Physiol. Scand. 51:136.
204. Sokoll, M.D., Sonesson, B., and Thesleff, S. (1968). Denervation changes produced in an innervated skeletal muscle by long-continued treatment with a local anesthetic. Eur. J. Pharmacol. 4:179.
205. Libelius, R., Sonesson, B., Stamenović, B.A., and Thesleff, S. (1970). Denervation-like changes in skeletal muscle after treatment with a local anaesthetic (marcaine). J. Anat. 106:297.
206. Robert, E.D., and Oester, Y.T. (1970). Absence of supersensitivity to acetylcholine in innervated muscle subject to a prolonged pharmacological nerve block. J. Pharmacol. Exp. Ther. 174:133.
207. Drachman, D.B., and Witzke, F. (1972). Trophic regulation of acetylcholine sensitivity in muscle: effect of electrical stimulation. Science 176:514.
208. Purves, D., and Sakmann, B. (1974). The effect of contractile activity on fibrillation and extrajunctional acetylcholine-sensitivity of rat muscle maintained in organ culture. J. Physiol. (Lond.) 237:157.
209. Cohen, S.A., and Fischbach, G.D. (1973). Regulation of acetylcholine sensitivity by muscle activity in cell culture. Science 181:76.
210. Berg, D.K., and Hall, Z.W. (1975). Increased extrajunctional acetylcholine sensitivity produced by chronic post-synaptic neuromuscular blockade. J. Physiol. (Lond.) 244:659.
211. Auerbach, A., and Betz, W. (1971). Does curare affect transmitter release? J. Physiol. (Lond.) 213:691.
212. Lømo, T., and Westgaard, R.H. (1975). Further studies on the control of ACh sensitivity by muscle activity in the rat. J. Physiol. (Lond.) 252:603.
213. Jansen, J.K.S., Lømo, T., Nicolaysen, K., and Westgaard, R. (1973). Hyperinnervation of skeletal muscle fibers: dependence on muscle activity. Science 181:559.
214. Lømo, T., Westgaard, R.H., and Dahl, H.A. (1974). Contractile properties of muscle: control by pattern of muscle activity in the rat. Proc. Roy. Soc. Lond. B 187:99.
215. Brown, M.D. (1973). Role of activity in the differentiation of slow and fast muscles. Nature 244:178.
216. Jones, R., and Vrbová, G. (1974). Two factors responsible for the development of denervation hypersensitivity. J. Physiol. (Lond.) 236:517.
217. Guth, L., and Zalewski, A.A. (1963). Disposition of cholinesterase following implantation of nerve into innervated and denervated muscle. Exp. Neurol. 7:316.
218. Schwarzacher, H.G. (1960). Untersuchungen über die Skeletmuskel-sehnenverbindung. Acta Anat. 40:59.

219. Harris, A.J., Heinemann, S., Schubert, D., and Tarakis, H. (1971). Trophic interaction between cloned tissue culture lines of nerve and muscle. Nature 231:296.
220. Lentz, T.L. (1972). Development of the neuromuscular junction. III. Degeneration of motor end-plates after denervation and maintenance *in vitro* by nerve explants. J. Cell Biol. 55:93.
221. Zalewski, A. (1974). Neuronal and tissue specifications involved in taste bud formation. Ann. N.Y. Acad. Sci. 228:344.
222. Aguilar, C.E., Bisby, M.A., Cooper, E., and Diamond, J. (1973). Evidence that axoplasmic transport of trophic factors is involved in the regulation of peripheral nerve fields in salamanders. J. Physiol. (Lond.) 234:449.
223. Hamburger, V., and Levi-Montalcini, R. (1949). Proliferation, differentiation and degeneration in the spinal ganglia of the chick embryo under normal and experimental conditions. J. Exp. Zool. 111:457.
224. Hughes, A.F.W. (1968). Aspects of Neural Ontogeny. Academic Press, New York. 249 p.
225. Landmesser, L., and Pilar, G. (1974). Synaptic formation during embryogenesis on ganglion cells lacking a periphery. J. Physiol. (Lond.) 241:715.
226. Landmesser, L., and Pilar, G. (1974). Synaptic transmission and cell death during normal ganglionic development. J. Physiol. (Lond.) 241:737.
227. Prestige, M.C. (1974). Axon and cell numbers in the developing nervous system. Brit. Med. Bull. 30:107.
228. Hamburger, V. (1975). Changing concepts in developmental neurobiology. Perspect. Biol. Med. 18:162.
229. Lieberman, A.R. (1971). The axon reaction: a review of the principal features of perikaryal responses to axon injury. Int. Rev. Neurobiology 14:49.
230. Matthews, M.R., and Raisman, G. (1972). A light and electron microscopic study of the cellular response to axonal injury in the superior cervical ganglion of the rat. Proc. Roy. Soc. Lond. B 181:43.
231. Matthews, M.R. (1973). An ultrastructural study of axonal changes following constriction of postganglionic branches of the superior cervical ganglion in the rat. Phil. Trans. Roy. Soc. B 264:479.
232. Cerf, J.A., and Chacko, L.W. (1958). Retrograde reaction in motoneuron dendrites following ventral root section in the frog. J. Comp. Neurol. 109:205.
233. Sumner, B.E.H., and Watson, W.E. (1971). Retraction and expansion of the dendritic tree of motor neurons of adult rats induced *in vivo*. Nature 233:273.
234. Purves, D. (1975). Functional and structural changes of mammalian sympathetic neurons following interruption of their axons. J. Physiol. (Lond.) 252:429.
235. Tweedle, C.D., Pitman, R.M., and Cohen, M.J. (1973). Dendritic stability of insect central neurons subjected to axotomy and deafferentation. Brain Res. 60:471.
236. Watson, W.E. (1965). An autoradiographic study of the incorporation of nucleic-acid precursors by neurons and glia during nerve regeneration. J. Physiol. (Lond.) 180: 741.
237. Watson, W.E. (1968). Observations on the nucleolar and total cell body nucleic acid of injured nerve cells. J. Physiol. (Lond.) 196:655.
238. Watson, W.E. (1969). The response of motor neurons to intramuscular injection of botulinum toxin. J. Physiol. (Lond.) 202:611.
239. Watson, W.E. (1970). Some metabolic responses of axotomized neurons to contact between their axons and denervated muscle. J. Physiol. (Lond.) 210:321.
240. Watson, W.E. (1972). Some quantitative observations upon the responses of neuroglial cells which follow axotomy of adjacent neurons. J. Physiol. (Lond.) 225:415.
241. Brattgård, S.-O., Edström, J.E., and Hydén, H. (1957). The chemical changes in regenerating neurons. J. Neurochem. 1:316.
242. Blinzinger, K., and Kreutzberg, G. (1968). Displacement of synaptic terminals from regenerating motoneurons by microglial cells. Z. Zellforsch. Mikrosk. Anat. 85:145.
243. Torvik, A., and Skjörten, F. (1971). Electron microscopic observations on nerve cell regeneration and degeneration after axon lesions. II. Changes in the glial cells. Acta Neuropath. 17:265.
244. Sjöstrand, J. (1971). Neuroglial proliferation in the hypoglossal nucleus after nerve injury. Exp. Neurol. 30:178.

245. Kerns, J.M., and Hinsman, E.J. (1973). Neuroglial response to sciatic neurectomy. I. Light microscopy and autography. J. Comp. Neurol. 151:237.
246. Cowan, W.M. (1973). Neuronal death as a regulative mechanism in the control of cell number in the nervous system. *In* M. Rockstein and M.L. Sussman (eds.), Development and Ageing in the Nervous System, pp. 19–41. Academic Press, New York.
247. Levinsohn, G. (1903). Uber das Verhalter des Ganglion cervicale supremum nach Durschschneidung seiner prae-bezw. postcellularen fasern. Arch. Anat. Physiol. Lpz.:438.
248. Acheson, G.H., and Schwarzacher, H.G. (1956). Correlations between the physiological changes and the morphological changes resulting from axotomy in the inferior mesenteric ganglion of the cat. J. Comp. Neurol. 106:247.
249. Torvik, A., and Skjörten, F. (1971). Electron microscopic observations on nerve cell regeneration and degeneration after axon lesions. I. Changes in the nerve cell cytoplasm. Acta Neuropath. 17:248.
250. Romanes, G.J. (1946). Motor localization and the effects of nerve injury on ventral horn cells in the spinal cord. J. Anat. 80:117.
251. Price, D.L. (1974). Influence of the periphery on spinal motor neurons. Ann. N.Y. Acad. Sci. 228:355.
252. Windle, W.F. (1956). Regeneration of axons in the vertebrate central nervous system. Physiol. Rev. 36:427.
253. Guth, L., and Windle, W.F. (1970). The enigma of central nervous regeneration. Exp. Neurol. Suppl. 5:1.
254. Clemente, C.O. (1964). Regeneration in the vertebrate central nervous system. Int. Rev. Neurobiol. 6:257.
255. Acheson, G.H., Lee, E.S., and Morison, R.S. (1942). A deficiency in the phrenic respiratory discharges parallel to retrograde degneration. J. Neurophysiol. 5:269.
256. Brown, G.L., and Pascoe, J.E. (1954). The effect of degenerative section of ganglionic axons on transmission through the ganglion. J. Physiol. (Lond.) 123:565.
257. Acheson, G.H., and Remolina, J. (1955). The temporal course of the effects of postganglionic axotomy on the inferior mesenteric ganglion of the cat. J. Physiol. (Lond.) 127:603.
258. Pilar, G., and Landmesser, L. (1972). Axotomy mimicked by localized colchicine application. Science 177:1116.
259. Hunt, C.C., and Riker, W.K. (1966). Properties of frog sympathetic neurons in normal ganglia and after axon section. J. Neurophysiol. 29:1096.
260. Matthews, M.R., and Nelson, V. (1975). Detachment of structurally intact nerve endings from chromatolytic neurons of the rat superior cervical ganglion during depression of synaptic transmission induced by post-ganglionic axotomy. J. Physiol. (Lond.) 245:91.
261. Campbell, B., Mark, V.H., and Gasteiger, E.L. (1949). Alteration of neuron excitability by retrograde degeneration. Amer. J. Physiol. 158:457.
262. Downman, C.B.B., Eccles, J.C., and McIntyre, A.K. (1953). Functional changes in chromatolysed motoneurons. J. Comp. Neurol. 98:9.
263. Eccles, J.C., Libet, B., and Young, R.R. (1958). The behavior of chromatolysed motoneurons studied by intracellular recording. J. Physiol. (Lond.) 143:11.
264. McIntyre, A.K., Bradley, and Brock, L.G. (1959). Responses of motoneurons undergoing chromatolysis. J. Gen. Physiol. 42:931.
265. Kuno, M., and Llinás, R. (1970). Alterations of synaptic action in chromatolysed motoneurons of the cat. J. Physiol. (Lond.) 210:823.
266. Kuno, M., and Llinás, R. (1970). Enhancement of synaptic transmission by dendritic potentials in chromatolysed motoneurons of the cat. J. Physiol. (Lond.) 210:807.
267. Pitman, R.M., Tweedle, C.D., and Cohen, M.J. (1972). Electrical responses of insect central neurons: augmentation by nerve section or colchicine. Science 178:507.
268. Pitman, R.M. (1975). The ionic dependence of action potentials induced by colchicine in an insect motoneuron cell body. J. Physiol. (Lond.) 247:511.
269. Hamberger, A., Hansson, H-A, and Sjöstrand, J. (1970). Surface structure of isolated neurons. Detachment of nerve terminals during axon regeneration. J. Cell Biol. 47:319.
270. Sumner, B.E.H., and Sutherland, F.I. (1973). Quantitative electron microscopy on the injured hypoglossal nucleus in the rat. J. Neurocytol. 2:315.

271. Cull, R.E. (1974). Role of nerve-muscle contact in maintaining synaptic connections. Exp. Brain Res. 20:307.
272. Sumner, B.E.H. (1975). A quantitative analysis of the response of presynaptic boutons to postsynaptic motor neuron axotomy. Exp. Neurol. 46:605.
273. Cragg, B.G. (1970). What is the signal for chromatolysis? Brain Res. 23:1.
274. Purves, D. Functional and structural changes in mammalian sympathetic neurones following colchicine application to postganglionic nerves. J. Physiol. (Lond.) In press.
275. Perísić, M., and Cuénod, M. (1972). Synaptic transmission depressed by colchicine blockage of axoplasmic flow. Science. 175:1140.
276. Cuénod, M., Sandri, C., and Akert, K. (1972). Enlarged synaptic vesicles in optic nerve terminals induced by intraocular injection of colchicine. Brain Res. 39:285.
277. Kuno, M., Miyata, Y., and Muńoz-Martinez, E.J. (1974). Differential reaction of fast and slow α-motoneurons to axotomy. J. Physiol. (Lond.) 240:725.
278. Kuno, M., Miyata, Y., and Muñoz-Martinez, E.J. (1974). Properties of fast and slow alpha motoneurons following motor re-innervation. J. Physiol. (Lond.) 242:273.
279. Frank, E. (1973). Matching of facilitation at the neuromuscular junction of the lobster: a possible case for the influence of muscle on nerve. J. Physiol. (Lond.) 233:635.
280. Watson, W.E. (1974). Cellular responses to axotomy and to related procedures. Brit. Med. Bull. 30:112.
281. Thoenen, H., and Tranzer, J.P. (1968). Chemical sympathectomy by selective destruction of adrenergic nerve endings with 6-hydroxydopamine. Naunyn-Schmiedebergs Arch. Pharak. u. exp. Path. 261:271.
282. Thoenen, H., and Tranzer, J.P. (1973). The pharmacology of 6-hydroxydopamine. Pharm. Rev. 13:169.
283. Levi-Montalcini, R. (1964). Growth control of nerve cells by a protein factor and its antiserum. Science 143:105.
284. Levi-Montalcini, R., and Cohen, S. (1960). Effects of the extract of the mouse submaxillary glands on the sympathetic system of mammals. Ann. N.Y. Acad. Sci. 85:324.
285. Hendry, I.A., and Iversen, L.L. (1973). Reduction in the concentration of nerve growth factor in mice after sialectomy and castration. Nature 243:500.
286. Longo, A.M., and Penhoet, E.E. (1974). Nerve growth factor in rat glioma cells. Proc. Natl. Acad. Sci. USA 71:2347.
287. Oger, J., Arnason, B.G.W., Pantazis, N., Lehrich, J., and Young, M. (1974). Synthesis of nerve growth factor by L and 3T3 cells in culture. Proc. Natl. Acad. Sci. USA 71:1554.
288. Young, M., Oger, J., Blanchard, M.H., Asdourain, H., Amos, H., and Arnason, B.G.W. (1975). Secretion of nerve growth factor by primary chick fibroblast culture. Science 187:361.
289. Stöckel, K., Paravicini, U., and Thoenen, H. (1974). Specificity of the retrograde axonal transport of nerve growth factor. Brain Res. 75:413.
290. Henry, I.A., Stöckel, K., Thoenen, H., and Iversen, L.L. (1974). The retrograde axonal transport of nerve growth factor. Brain Res. 68:103.
291. Paravicini, U., Stöckel, K., and Thoenen, H. (1975). Biological importance of retrograde axonal transport of nerve growth factor in adrenergic neurons. Brain Res. 84:279.
292. Iversen, L.L., Stöckel, K., and Thoenen, H. (1975). Autoradiographic studies of the retrograde axonal transport of nerve growth factor in mouse sympathetic neurons. Brain Res. 88:37.
293. Angeletti, P.U., Levi-Montalcini, R., and Caramia, F. (1971). Ultrastructural changes in sympathetic neurons of newborn and adult mice treated with nerve growth factor. J. Ultrastruct. Res. 36:24.
294. Levi-Montalcini, R., Aloe, L., Mugnaini, E., Oesch, F., and Thoenen, H. (1965). Nerve growth factor induces volume increase and enhances tyrosine hydroxylase synthesis in chemically axotomized sympathetic ganglia of newborn rats. Proc. Natl. Acad. Sci. USA 72:595.
295. Banks, B.E.C., Charlwood, K.A., Edwards, D.C., Vernon, C.A., and Walter, J.J. (1975). Effects of nerve growth factors from mouse salivary glands and snake venom on sympathetic ganglia of neonatal and developing mice. J. Physiol. (Lond.) 247:289.

296. Fry, F.J., and Cowan, W.M. (1972). A study of retrograde cell degeneration in the lateral mammillary nucleus of the cat, with special reference to the role of axonal branching in the preservation of the cell. J. Comp. Neurol. 144:1.
297. Barker, D., and Ip, M.C. (1966). Sprouting and degeneration of mammalian motor axons in normal and deafferented skeletal muscle. Proc. Roy. Soc. Lond. B 163:538.
298. Edds, M.V. (1953). Collateral nerve regeneration. Quart. Rev. Biol. 28:260.
299. Guth, L. (1956). Regeneration in the mammalian peripheral nervous system. Physiol. Rev. 36:441.
300. Weiss, P., and Hoag, A. (1946). Competitive re-innervation of rat muscles by their own and foreign nerves. J. Neurophysiol. 9: 413.
301. Bernstein, J.J., and Guth, L. (1961). Non-selectivity in establishment of neuromuscular connections following nerve regeneration in the rat. Exp. Neurol. 4:262.
302. Langley, J.N. (1897). On the regeneration of pre-ganglionic and of post-ganglionic visceral nerve fibers. J. Physiol. (Lond.) 22:215.
303. Landmesser, L. (1972). Pharmacological properties, cholinesterase activity and anatomy of nerve-muscle junctions in vagus innervated frog sartorius. J. Physiol. (Lond.) 220:243.
304. Weiss, P., and Edds, M.V. (1945). Sensory-motor nerve cross in the rat. J. Neurophysiol. 8:173.
305. Miledi, R., and Stefani, E. (1969). Non-selective re-innervation of slow and fast muscle fibers in the rat. Nature 222:569.
306. Feng, T.P., Wu, W.Y., and Yang, F.Y. (1965). Selective re-innervation of "slow" or "fast" muscle by its original motor supply during re-innervation of a mixed nerve. Scientia Sin. 14:1717.
307. Bessou, P., Laporte, Y., and Pages, B. (1965). Observations sur la re-innervation de fuseaux neuro-musculaires de chat. C.R. Soc. Biol. 160:408.
308. Vera, C.L., Vial, J.D., and Luco, J.V. (1957). Re-innervation of the nictitating membrane of the cat by cholinergic fibers. J. Neurophysiol. 20:363.
309. Ceccarelli, B., Clementi, F., and Mantegazza, P. (1972). Adrenergic re-innervation of smooth muscle of nictitating membrane by preganglionic sympathetic fibers. J. Physiol. (Lond.) 220:211.
310. O'Lague, P.H., Obata, K., Claude, P., Furshpan, E.J., and Potter, D.D. (1974). Evidence for cholinergic synapses between dissociated rat sympathetic neurons in cell culture. Proc. Natl. Acad. Sci. USA 71:3602.
311. Rees, R., and Bunge, R.P. (1974). Morphological and cytochemical studies of synapses formed in culture between isolated rat superior cervical ganglion neurons. J. Comp. Neurol. 157:1.
312. Nurse, C., and O'Lague, P. (1975). Formation of cholinergic synapses between dissociated sympathetic neurons and skeletal myotubes of the rat in cell culture. Proc. Natl. Acad. Sci. USA 72:1955.
313. Bittner, G.D., and Johnson, A.L. (1974). Degeneration and regeneration in crustacean peripheral nerves. J. Comp. Physiol. 89:1.
314. Sperry, R.W., and Arora, H.L. (1965). Selectivity in regeneration of the oculomotor nerve in the cichlid fish, *Astronotus ocellatus*. J. Embryol. Exp. Morphol. 14:307.
315. Weiss, P. (1937). Further experimental investigations on the phenomenon of homologous response in transplanted amphibian limbs. I. Functional observations. J. Comp. Neurol. 66:181.
316. Weiss, P. (1937). Further experimental investigations on the phenomenon of homologous response in transplanted amphibian limbs. II. Nerve regeneration and the innervation of transplanted limbs. J. Comp. Neurol. 66:481.
317. Weiss, P. (1937). Further experimental investigations on the phenomenon of homologous response in transplanted amphibian limbs. III. Homologous response in the absence of sensory innervation. J. Comp. Neurol. 66:537.
318. Weiss, P. (1937). Further experimental investigations on the phenomenon of homologous response in transplanted amphibian limbs. IV. Reverse locomotion after the interchange of right and left limbs. J. Comp. Neurol. 66:269.
319. Weiss, P. (1956). Special vertebrate organogenesis: Nervous system. *In* B.H. Willier, P.A. Weiss, and V. Hamburger (eds.), pp. 346–401. W.B. Saunders, Philadelphia.
320. Sperry, R.W. (1941). The effect of crossing nerves to antagonistic muscles in the hind limb of the rat. J. Comp. Neurol. 45:1.

321. Grimm, L. (1971). An evaluation of myotypic respecification in Axolotls. J. Exp. Zool. 178:479.
322. Mark, R.F. (1965). Fin movement after regeneration of neuromuscular connections: an investigation of myotypic specificity. Exp. Neurol. 12:292.
323. Marotte, L.R., and Mark, R.F. (1970). The mechanism of selective re-innervation of fish eye muscles. I. Evidence from muscle function during recovery. Brain Res. 19:41.
324. Cass, D.T., Sutton, T.J., and Mark, R.F. (1973). Competition between nerves for functional connexions with axolotl muscles. Nature 243:201.
325. Dennis, M.J., and Ort, C.A. Physiological properties of nerve-muscle junctions developing *in vivo*. Cold Spring Harbor Symp. Quant. Biol. In press.
326. Kuffler, S.W., and Vaughan-Williams, E.M. (1953). Properties of the "slow" skeletal muscle fibers of the frog. J. Physiol. (Lond.) 121:318.
327. Orkand, R.K. (1963). A further study of electrical responses in slow and twitch muscle fibers of the frog. J. Physiol. (Lond.) 167:181.
328. Stefani, E., and Steinbach, A.B. (1969). Resting potential and electrical properties of frog slow muscle fibers. Effect of different external solutions. J. Physiol. (Lond.) 203:383.
329. Elul, R., Miledi, R., and Stefani, E. (1970). Neural control of contracture in slow muscle fibers of the frog. Acta Physiol. Lat. Amer. 20:194.
330. Stefani, E., and Schmidt, H. (1972). Early stages of re-innervation of frog slow muscle fibers. Pflügers Arch. 336:271.
331. Miledi, R., and Orkand, P. (1966). Effect of a "fast" nerve on "slow" muscle fibers in the frog. Nature 209:717.
332. Hoh, J.F.Y. (1971). Selective re-innervation of fast-twitch and slow-graded muscle fibers in the toad. Exp. Neurol. 30:263.
333. Guth, L. (1963). The problem of nerve and end organ following nerve regeneration. *In* E. Gutmann and R. Hník (eds.), The Effect of Use and Disuse on Neuromuscular Functions, pp. 135–142. Publishing House of the Czechoslovakian Academy of Science, Prague.
334. Redfern, P.A. (1970). Neuromuscular transmission in new-born rats. J. Physiol. (Lond.) 209: 701.
335. Bagust, J., Lewis, D.M., and Westerman, R.A. (1973). Polyneuronal innervation of kitten skeletal muscle. J. Physiol. (Lond.) 229:241.
336. Jansen, J.K.S., Brown, M.C., and Van Essen, D. Formation and elimination of synapses in rat skeletal muscle. Cold Spring Harbor Symp. Quant. Biol. In press.
337. Langley, J.N. (1898). On the union of cranial autonomic (visceral) fibers with the nerve cells of the superior cervical ganglion. J. Physiol. (Lond.) 23:240.
338. De Castro, F. (1951). Aspects anatomique de la transmission synaptique ganglionnaire chez les mammifères. Arch. Int. Physiol. 59:479.
339. Guth, L. (1956). Functional recovery following vagosympathetic anastomosis in the cat. Amer. J. Physiol. 185:205.
340. Langley, J.N., and Anderson, H.K. (1904). On the union of the fifth cervical nerve with the superior cervical ganglion. J. Physiol. (Lond.) 30:439.
341. Cannon, W.B., Binger, C.A.L., and Fitz, R. (1914). Experimental hypertension. Am. J. Physiol. 36:363.
342. De Castro, F. (1935). Note sur la régénération fonctionelle hètèrogèné figure dans les anastomoses des nerfs pneumogastrique et hypoglosse avec le sympathique cervical. Trav. Lab. recherches Biol. Univ. Madrid. 30:397.
343. Matsumura, M., and Koelle, G.B. (1961). The nature of synaptic transmission in the superior cervical ganglion following re-innervation by the afferent vagus. J. Pharmacol. Exp. Ther. 134:28.
344. Langley, J.N. (1892). On the origin from the spinal cord of the cervical and upper thoracic sympathetic fibers, with some observations on white and grey rami communicantes. Phil. Trans. 183B:85.
345. Langley, J.N. (1895). Note on regeneration of pre-ganglionic fibers of the sympathetic. J. Physiol. (Lond.) 18:280.
346. Guth, L., and Bernstein, J.J. (1961). Selectivity in the re-establishment of synapses in the superior cervical sympathetic ganglion of the cat. Exp. Neurol. 4:59.
347. Eccles, J.C. (1935). The action potential of the superior cervical ganglion. J. Physiol. (Lond.) 85:179.

348. Martin, A.R., and Pilar, G. (1963). Dual mode of synaptic transmission in the avian ciliary ganglion. J. Physiol. (Lond.) 168:443.
349. Marwitt, R., Pilar, G., and Weakly, J.N. (1971). Characterization of two ganglion cell populations in avian ciliary ganglia. Brain Res. 25:317.
350. Hess, A., Pilar, G., and Weakly, J.N. (1969). Correlation between transmission and structure in avian ciliary ganglion synapses. J. Physiol. (Lond.) 202:339.
351. Jansen, J.K.S., and Nicholls, J.G. (1972). Regeneration and changes in synaptic connections between individual nerve cells in the central nervous system of the leech. Proc. Natl. Acad. Sci. USA 69:636.
352. Matthey, R. (1926). Récuperation de la vue après résection des nerfs optiques chez le triton. C.r. Séanc. Soc. Biol. 93:904.
353. Stone, L.S., and Zaur, I.S. (1940). Reimplantation and transplantation of adult eyes in the salamander (*triturus viridescens*) with return of vision. J. Exp. Zool. 85:243.
354. Stone, L.S., and Chace, R.R. (1941). Experimental studies on the regenerating lens and the eye in adult *Triturus viridescens*. Anat. Rec. 79:333.
355. Stone, L.S. (1944). Functional polarization in the retinal development and its re-establishment in regenerating retinae of rotated grafted eyes. Proc. Soc. Exp. Biol. Med. 57:13.
356. Sperry, R.W. (1943). The effect of 180-degree rotation of the retinal field on visuomotor co-ordination. J. Exp. Zool. 92:263.
357. Sperry, R.W. (1944). Optic nerve regeneration with return of vision in anurans. J. Neurophysiol. 7:57.
358. Attardi, D.G., and Sperry, R.W. (1963). Preferential selection of central pathways by regenerating optic fibers. Exp. Neurol 7:46.
359. Gaze, R.M. (1970). The Formation of Nerve Connections. Academic Press, London. 288 pp.
360. Jacobson, M. (1970). Developmental Neurobiology. Holt, Rinehart and Winston, New York. 465 p.
361. Gaze, R.M., and Sharma, S.C. (1970). Axial differences in the re-innervation of the goldfish optic tectum by regenerating optic fibers. Exp. Brain Res. 10:171.
362. Yoon, M. (1971). Reorganization of retinotectal projection following surgical operations on the optic tectum in goldfish. Exp. Neurol. 33:395.
363. Chung, S.H., Keating, M.J., and Bliss, T.V.P. (1974). Functional synaptic relations during the development of the retino-tectal projection in amphibians. Proc. Roy. Soc. B 187:449.
364. Yoon, M.G. (1975). Effects of post-operative visual environments on reorganization of retinotectal projection in goldfish. J. Physiol. (Lond.) 246:673.
365. Stenevi, U., Bjðrklund, A., and Moore, R.Y. (1973). Morphological plasticity of central adrenergic neurons. *In* J.J. Bernstein and D.C. Goodman (eds.), Neuromorphological Plasticity, pp. 110–134. S. Karger, Basel.
366. Bjðrkland, A., Johansson, B., Steneni, U., and Svengaard, N-A. (1975). Re-establishment of functional connections by regenerating central adrenergic and cholinergic axons. Nature 253:446.
367. Saito, A., and Zacks, S.I. (1969). Fine structure of the neuromuscular junction after nerve section and implantation of nerve in denervated muscle. Exp. Molec. Pathol. 10:256.
368. Bennett, M.R., McLachlan, E.M., and Taylor, R.S. (1973). The formation of synapses in re-innervated mammalian striated muscle. J. Physiol. (Lond.) 233:481.
369. Diamond, J., and Miledi, R. (1962). A study of fetal and newborn rat muscle fibers. J. Physiol. (Lond.) 162:393.
370. Cohen, M.W. (1972). The development of neuro-muscular connexions in the presence of D-tubocurarine. Brain Res. 41:457.
371. Steinbach, J.H., Harris, A.J. Patrick, J., Schubert, D., and Heinemann, S. (1973). Nerve-muscle interaction *in vitro:* role of acetylcholine. J. Gen. Physiol. 62:255.
372. Van Essen, D., and Jansen, J.K.S. (1974). Re-innervation of the rat diaphragm during perfusion with α-bungarotoxin. Acta Physiol. Scand. 91:571.
373. Mark, R.F. (1974). Selective innervation of muscle. Brit. Med. Bull. 30:122.
374. Nature (1973). 243:185.
375. Mark, R.F. (1970). Chemospecific synaptic repression as a possible memory store. Nature 225:178.

376. Mark, R.F. (1974). Memory and Nerve Cell Connections. Clarendon Press, Oxford. 156 p.
377. Marotte, L.R., and Mark, R.F. (1970). The mechanism of selective re-innervation of fish eye muscles. II. Evidence from electron microscopy of nerve endings. Brain Res. 19:53.
378. Marotte, L.R., and Mark, R.F. (1972). The mechanism of selective re-innervation of fish eye muscles. III. Functional electrophysiological and anatomical analysis of recovery from section of the IIIrd and IVth nerves. Brain Res. 46:131.
379. Mark, R.F., Marotte, L.R., and Mart, P.E. (1972). Mechanism of selective re-innervation of fish eye muscles. IV. Identification of repressed synapses. Brain Res. 46:149.
380. Fangboner, R.F., and Vanable, J.W. (1974). Formation and regression of inappropriate nerve sprouts during trochlear nerve regeneration in *Zenopus laevis*. J. Comp. Neurol. 157:391.
381. Stirling, R.V. (1970). Central adaptation in the salamander spinal cord. J. Physiol. (Lond.) 210:184P.
382. Stirling, R.V. (1973). The effect of increasing the innervation field sizes of nerves on their reflex response time in salamanders. J. Physiol. (Lond.) 229:657.
383. Scott. S.A. (1975). Persistence of foreign innervation of re-innervated goldfish extraocular muscles. Science. 189:644.
384. Frank, E., and Jansen, J.K.S. (1976). Interaction between foreign and original nerves innervating gill muscles in fish. J. Neurophysiol. 39:84.

International Review of Physiology
Neurophysiology II, Volume 10
Edited by Robert Porter
 University Park Press Baltimore

4
Neurophysiology of Nociception

M. ZIMMERMANN
II. Physiologisches Institut, Universität Heidelberg, D-6900 Heidelberg, Neuenheimer Feld 326, Deutschland (Germany)

Pain has always been a most important factor in the behavior of man. Presumably, of all the sensory modalities, it is the perception of and the reaction to pain that have undergone the fewest changes during the evolution of sensory experience. In contrast to this outstanding significance our knowledge of the neurophysiological basis of pain is rather fragmentary. Many of the fundamental statements and theories on the physiological mechanisms have been derived indirectly from psychophysical experiments and clinical observations.

BASIC PROBLEMS IN THE APPROACH TO PAIN PHYSIOLOGY

There are two main reasons, one semantic, the other technical, for the lack of a coherent physiological picture. The semantic problem inherent in any discussion on pain is that the same name is used for a variety of meanings (13), e.g., pain as a sensation produced by experimental stimuli, pain as a chronic suffering in patients, pain as an emotional state (opposite to pleasure), pain in animals. One should not expect all of these different phenomena to share the same physiological mechanisms. Hence, to avoid confusion the topic of a study on pain should be defined as explicitly as possible.

The technical problem refers to neurophysiological experiments, in man and animals: it is difficult to establish a "pain stimulus," in the sense of a quantifiable adequate stimulus. This is a prerequisite for any experimental studies on the neurophysiology of sensory events.

Nociception versus Pain

The emphasis of this article will be on the neuronal representation of noxious stimuli, in the peripheral and central nervous system. In order to avoid semantic confusion the term "nociception" will be used to denote this particular experimental approach in man and animals, as has been proposed by Sherrington (1).

This concept implies that animals respond with motor and autonomic reflexes (e.g., flexor reflex, pressor reflex, tachypnea), escape and pseudoaffective behavior (e.g., vocalization), which are identical or similar to responses seen in man upon stimulation which is felt as painful. Thus nociception is defined as the processing of stimuli which eventually leads to one or several of these responses. These in turn have been termed as nocifensive responses.

The study of the neurophysiological basis of nociception using animal experiments may yield models for the physiological mechanisms underlying the sensation of pain in man, evoked by these stimuli. However, generalizations which try to explain the many types of pain in man in terms of these models must be considered as highly speculative.

A Test for a Stimulus To Be Nociceptive in Animals

A stimulus used in neurophysiological experiments on nociception must be established to really be nociceptive. This could be done by applying the quantified stimulus to human subjects, and measuring the threshold intensity for pain sensation. This has been done with a variety of setups, sometimes called algesimeters. However, it is known that cutaneous sensitivity shows considerable local variation over the body, and also interspecies differences. Therefore the best approach would be to measure, in behavioral experiments in animals, the nocifensive responses to a stimulus used for neurophysiological investigations on nociception, in the same species.

Heating the skin has been used frequently in studies on the psychophysics and the neurophysiology of nociception. Is this a suitable stimulus for animal experiments? This question has been investigated using quantitative measurements of escape withdrawal in cats. As is seen in Figure 1, the unrestrained animal started to lift one of its hind feet when the supporting platform was heated. The threshold for the response was about 45° C, which is liminal for reports of pain in human experiments. A quantitative measure of the length of time that one or other of the cat's feet was withdrawn during heating was obtained by recording the force exerted by the weight of the animal on the foot plates (Figure 1*B*). The percentage of time for which the foot was lifted is plotted against the temperature of heating in Figures 1, *C* and *D*. A practically linear relationship was obtained between the temperature and the withdrawal response. This graded behavioral response is conspicuously reminiscent of a psychometric approach to heat pain in man, by which the dol-scale of subjective heat intensity was established (2).

Two items in the results of Figure 1 deserve consideration: 1) to explain the increase in response with increasing temperatures an afferent line is required to carry information about the intensity of noxious heat; 2) no avoidance response to subnoxious stimuli could be conditioned in the animal by repeating the experiments. This indicates that warming the hind foot to temperatures below those producing a nocifensive escape response is probably not perceived by the cat. An explanation for this is the virtual absence of sensitive warm receptors in the skin of the hind foot, as has been established by a systematic search for thermosensitive receptors in an electrophysiological analysis of single afferent fibers (3). This is probably different from the situation in the forefoot and in the

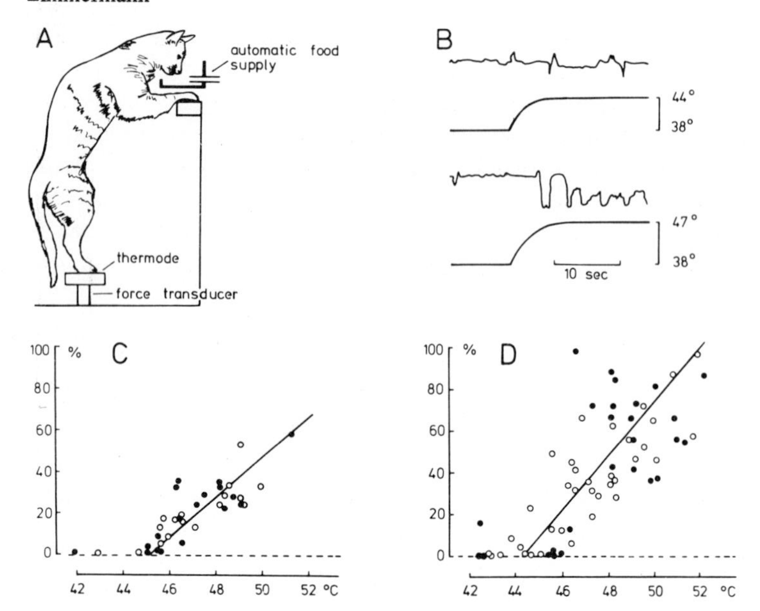

Figure 1. Nocifensive withdrawal reaction to heating the sole of the cat's hind foot. *A*, experimental arrangement of unrestrained cat standing on two thermodes, one for each hind foot. The temperatures of the thermodes could be changed independently. The forces exerted on the thermodes by the weight of the animal were recorded by strain gauges. *B*, two simultaneous records of temperature and force (lower and upper trace of each pair, respectively) for one of the thermodes, during heating to 44°C and 47°C. *C* and *D*, the ordinate plots the proportion of time of withdrawal in a 10-s analysis interval, during heating to various levels (*abscissa*). Data from several experiments are superimposed on the graphs; each graph is from one animal. Measurements are from right (○) and left (●) foot, respectively. Regression lines fitted by eye (From Rossmann, Sassen, and Zimmermann, unpublished.)

face, with which the cat might be able to detect small local gradients in temperature (4, 5).

Therefore, the results of this behavioral approach reveal that heating the cat's hind foot produces a virtually selective nociceptive input, which provides an excellent opportunity for studying central nervous system activity related to such peripheral noxious events (see section "Central Nervous System").

PAIN THEORIES

Peripheral Nervous System

In the past 100 years three basic concepts have been formulated to explain the peripheral encoding of painful stimuli. The keywords to label each of these concepts are: intensity, pattern, and specificity.

The hypothesis of intensity coding (6) claimed that low threshold receptors, normally excited, e.g., by touch, would respond with excessively high impulse frequency to strong stimuli. The subject would interpret this excessive activation as pain.

The basis of pattern theory (7, 8) was the assumption that cutaneous and visceral nerve endings generally respond to a variety of stimuli, but that each individual afferent fiber has a characteristic spectrum or profile of effective stimuli. A population of fibers could therefore cover the qualitative and quantitative continua of relevant cutaneous stimulations. Information about the different aspects of painful and non-painful stimuli impinging upon the skin was thought to be encoded in the spatio-temporal pattern of a set of afferent fibers.

The specificity hypothesis (9) postulated distinct types of receptors, responding exclusively to stimuli of sufficiently high intensity, whose activity leads directly to the sensation of pain. Evidence from neurophysiological experiments is in favor of the specificity concept, i.e., the existence of specialized nociceptors. However, excitation of receptors other than nociceptors will normally contribute to the sensation of a noxious stimulus. This has led to the proposal of a modified pattern concept (10).

Central Nervous System

Hypotheses on the central nervous mechanism involved in pain may be subdivided according to two basic assumptions. The idea of specificity, inherent in the more anatomical and neurosurgical ways of thinking, claims that specific pain tracts and centers exist in the spinal cord and the brain. This therefore implies a simple extension of peripheral specificity. Finally, when the cortex was parceled into areas of distinct physical and psychical capabilities (11), pain perception was attributed to a small strip of the postcentral gyrus (area 3a).

Alternative concepts are classified in this review as theories of interaction. They trace back to the postulate, conceived by Head (52), as a result of self-observation of altered skin sensitivity during nerve regeneration, of two parts of the somatosensory system, the epicritic (fast conducting afferents, recent in phylogeny) and the protopathic (slowly conducting, old in phylogeny). The original idea had subsequently been developed to a more general psycho-physiological dichotomy of the nervous system (12). According to this view the epicritic component is correlated with the informational and discriminative, the protopathic component with the motivational and affective aspects of sensation and behavior; all sensations share both aspects to different degrees.

These two components of sensations are thought to interact in various ways, e.g., by mutual masking, inhibition, or facilitation. In particular the suppression of pain observed during stimulation of the epicritic components has long been known as an example of such interactions (14).

The Gate Control Hypothesis (15)

This was a neurophysiological model formulated as an explanation for the possible interaction between the two systems. The basic mechanism was assumed to be presynaptic inhibition of primary afferent fibers in the spinal cord (16). The details of the gate control hypothesis are depicted in Figure 2.

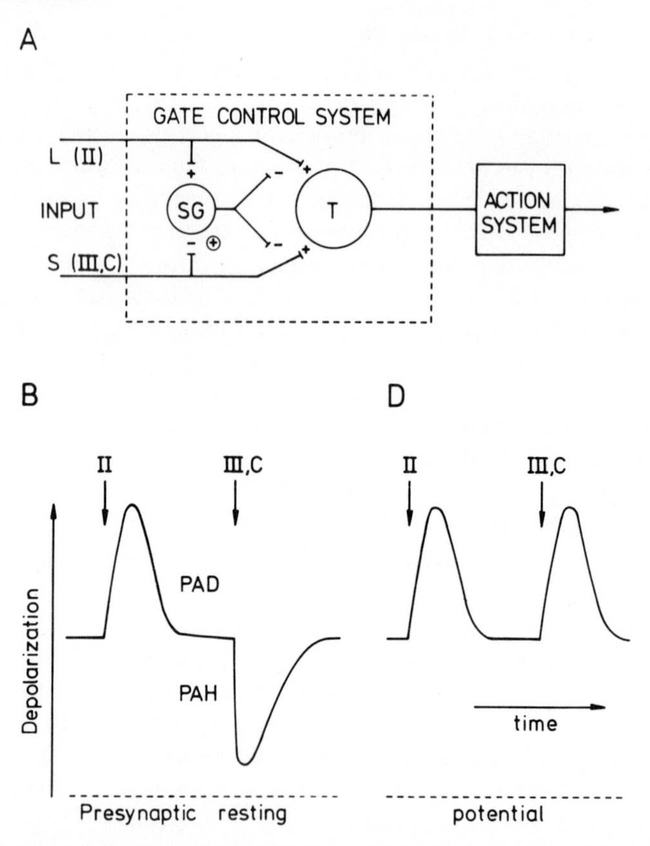

Figure 2. The gate control hypothesis. *A*, large (L: group II) and small (S: group III and C) fibers converge on spinal transmission cells T. Both L and S are inhibited presynaptically by interneuron SG, the SG neuron being excited by L, and inhibited by S afferents. As indicated by ⊕, many findings indicate that S afferents may also excite the SG cells, hence producing presynaptic inhibition rather than disinhibition or facilitation. *B*, diagrammatic drawing of presynaptic slow potential changes upon large (II) and small (III, C) fiber input, as postulated in the gate control hypothesis. PAD and PAH indicate primary afferent depolarization, and hyperpolarization, respectively. *D*, same as *B*, indicating, however, the finding of PAD produced by small fiber input, as reported by many authors. (*A* according to Melzack and Wall (15).)

Large (L) and small (S) afferent fibers converge onto dorsal horn cells, which have an axon projecting to the brain. Interneurons (SG) located in the substantia gelatinosa, a cyto-architectonic layer of the dorsal horn, exert presynaptic depolarization and hence presynaptic inhibition onto both L and S afferents.

These inhibitory interneurons (SG) are excited by the large (L) and inhibited by the small (S) afferents. Thus, a predominance of peripheral excitation of L afferents will increase the level of presynaptic inhibition; when, however, the excitation in the periphery is shifted to a predominant activation of S afferents, the presynaptic inhibition is reduced. Hence stimulation of S afferents should produce a relative hyperpolarization of all the afferents, i.e., a presynaptic facilitation. This presynaptic hyperpolarization upon stimulation of small afferents, i.e., of Aδ (group III) and C (group IV) fibers, should be reflected in a dorsal root potential (DRP) of positive polarity (17).

There is now no doubt that presynaptic hyperpolarization may occur whenever a tonic depolarizing effect is diminished. However, it is found that hyperpolarization is not specifically produced by excitation of the thin afferent fibers (III and C) of skin nerves (22); instead, these afferents predominantly produce a depolarization (18–21, 23, 24).

The idea of presynaptic facilitation produced by small fiber input, a major and basically new constituent of the gate control hypothesis, therefore has to be discarded. On the contrary, the observations of presynaptic depolarization following small fiber stimuli suggest that these produce inhibition in large fibers; thus most, if not all, types of noxious and non-noxious input participate in a system of mutual inhibition, which has been established to show a specialized organization (23, 25).

Although some of the most interesting new ideas of the spinal gate control hypothesis must now be rejected, it had a very stimulating effect: never before have so many laboratories been involved in a scientific approach to the pain problem!

NEUROPHYSIOLOGY OF NOCICEPTORS

Electrical Stimulation of Cutaneous Nerves

Electrical stimulation sufficient to recruit small myelinated (III, Aδ) and non-myelinated (IV, C) fibers of cutaneous nerves of animals evokes profound autonomic reflexes. This nocifensive response is mainly produced by an increased sympathetic activity (26). In conscious human subjects electrical stimulation of nerve branches dissected from the skin has been reported to evoke painful sensations, whenever the stimulus excited the III and C fibers (27–29). Stimulation of Group II (Aβ) afferents alone never produced such sensations.

By using graded stimulus intensities, and differential block of myelinated versus non-myelinated fibers, it was established that activation of the III fibers or C fibers produced the so-called fast and slow pain, respectively. Fast pain (III fibers) was described to be a short lasting and well localized sensation of pin-prick quality, whereas the slow pain (C fibers) was of relatively long duration, not well localized, and dull and burning in character.

Electrophysiological recording from single cutaneous nerve fibers in the cat and the monkey revealed that nociceptive afferents exist in both fiber classes (see section "Cutaneous Nociceptors"). However, various types of low threshold receptors have also been found among III or C afferents, i.e., mechanoreceptors (e.g., from hair follicles), warm receptors, cold receptors (40, 49). Single fiber activity recorded in man has likewise revealed that not only nociceptors exist in these fiber groups (30, 31), but also sensitive warm and cold receptors (32, 33). Therefore the identity assumed previously from the experiments with electrical nerve stimulation in man, that all the small diameter fibers are pain fibers, is no longer tenable.

Cutaneous Nociceptors

Neurophysiological analysis of sensory receptors started in 1926, when Adrian and Zotterman (34) made the first recordings from identified single afferent fibers. In his pioneering studies Zotterman (35, 36) applied this approach to the neurophysiological elucidation of "pain." With an admirable intuition he recognized single Aδ and C fiber impulses in his multi-unit records, which could be attributed to noxious stimuli. His early findings, which he interpreted as the experimental proof of the existence of the specific nociceptors ("pain receptors"), have subsequently been confirmed and extended by various authors, using different techniques for recording from single nerve fibers.

According to the adequate stimuli to excite the nociceptors found thus far in the skin they might be partitioned into the following groups: mechanosensitive nociceptors, heat nociceptors, and polymodal nociceptors. This classification must be regarded as preliminary, since in the range of high stimulus intensities the question of the adequate stimulus for a particular nerve ending often cannot be settled unambiguously, in contrast to other cases, e.g., of low threshold mechanoreceptors (see section "Criteria for the Specificity of Nociceptors").

Mechanosensitive Nociceptors Receptors responding to mechanical stimuli, which were designated as noxious when applied to man, have been found in cat and monkey among both III and C afferent fibers (3, 38–41). Their respective fields consist either of single spots (C fibers), or of multiple spots (III fibers), distributed over an area of 1 to 8 cm^2 (Figure 3, *D* to *F*). These receptors respond to stimuli such as pinching the skin with serrated clips (Figure 3*C*) or pin prick (Figure 3*B*), but not if the same force which was suprathreshold when applied with a pin was exerted by a blunt probe, hence producing moderate pressure (Figure 3*A*).

A clear-cut demarcation between medium threshold mechanoreceptors and nociceptors could often not be achieved on the basis of threshold measurements, particularly in the group III fibers. Additional criteria are necessary, as is pointed out below.

Thermosensitive Nociceptors Receptors responding to noxious heat (Figure 4) have been reported to occur frequently in the C fiber group (3, 42–45). They have small receptive fields (less than 3 mm^2). Their thresholds are at skin surface temperatures between 40°C and 45°C, the temperature levels at which heat sensations are known to be evoked in man (46), and nocifensive behavior in animals (Figure 1).

Long-lasting stimuli (e.g., 5 min) evoke sustained discharges (3). With increasing skin temperature the discharge rate goes up; in the example of Figure 4 the correlation was linear. The relatively low variability, reflected by the high correlation coefficient (greater than 0.9), implies that these receptors might be good for signaling information on the intensity of the noxious heat. Analysis of Figures 1 and 4 suggests that the excitation of such heat nociceptors underlies the graded withdrawal responses of the conscious animal.

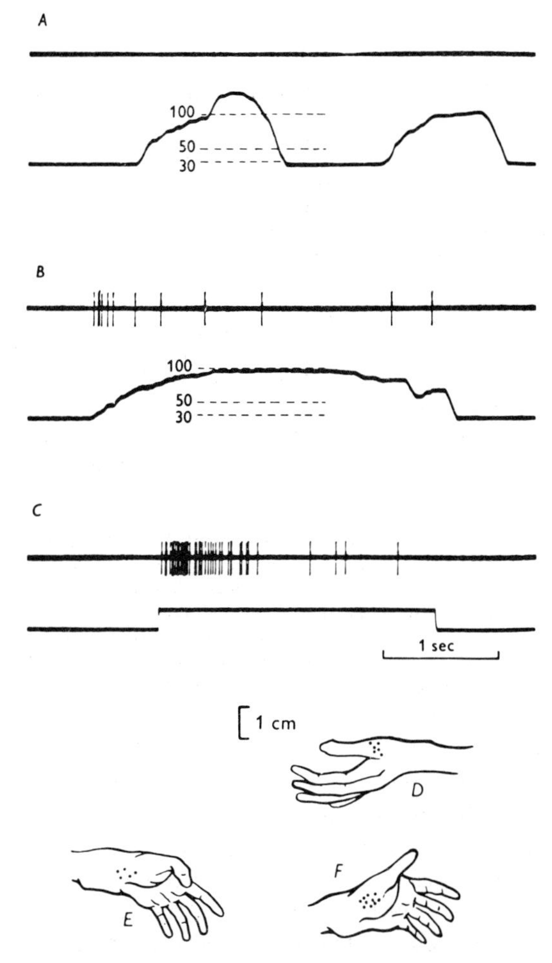

Figure 3. Nociceptors with group III afferents. In *A* to *C* upper traces show discharges of the fiber with a conduction velocity of 22 m/s. *A*, pressure with 2.2 mm diameter probe; output from strain gauge showing the force on lower trace, calibration in grams. *B*, pressure with needle; baseline of lower trace indicating force (strain gauge) of 20 g. *C*, forcible pinch with serrated forceps; lower trace is stimulus marker. *D* to *F*, multiple receptive fields of nociceptor afferents in the monkey's hand. (From Perl (41).)

Polymodal Nociceptors Nociceptors which respond to both mechanical and thermal stimulation are termed polymodal nociceptors. They have been reported to occur commonly among C fibers (3, 38), and less frequently among III fibers (3, 43). The nociceptors with C fibers so far recorded in man (30, 47) have probably all been of the polymodal type.

The proportion of the nociceptive fibers is different in the various cutaneous nerves. Results vary between 10% and 20% for the III fibers, and between 20% and 50% for the C fibers (3, 38, 39).

Sensitization of Nociceptors Frequent stimulation with noxious skin temperatures, or heat stimuli of long duration, produces sensitization of C heat

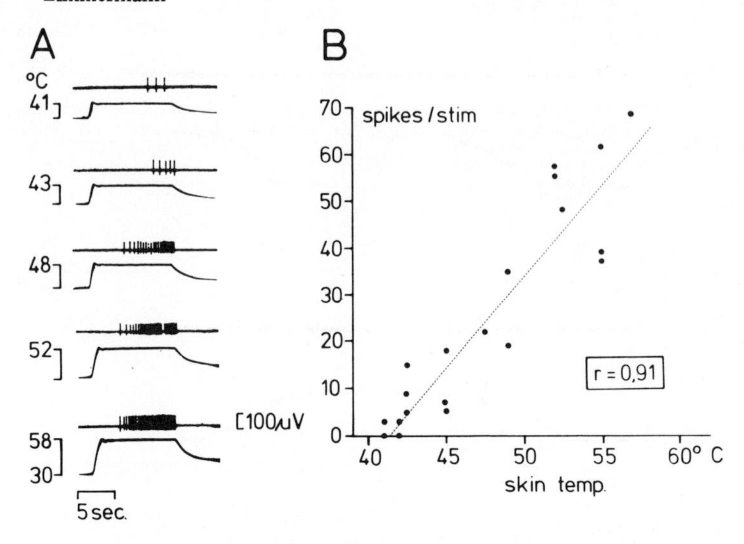

Figure 4. Nociceptor with C afferent during radiant heat stimulation of the skin. *A*, specimen records at various degrees of heating. Time course of skin surface temperature displayed below. *B*, relationship between the surface temperature of the skin and the total number of spikes/stimulus. Calculated regression line, correlation coefficient 0.91. Stimuli of different intensities were presented in random order with interstimulus intervals of 3 min. Receptive field of unit was in the plantar hairy skin of the cat's second toe. Conduction velocity 0.63 m/s. (From Beck, Handwerker, and Zimmermann (3).)

nociceptors (3), and of polymodal nociceptors (38), i.e., a fall in threshold, and an increase in discharge frequency at a certain temperature (3). These phenomena might be caused by chemical substances, released from damaged tissue (see section "Nociceptors and Algogenic Substances").

Following regeneration after experimental lesions of the cat's plantar nerves, thermosensitive C fibers have been found; their thresholds, however, were considerably lower than those of the heat nociceptors normally found in these nerves (51). On the assumption that these regenerated fibers are in fact redeveloped heat nociceptors, one must infer that a persistent sensitization is found in these new nociceptive afferents.

Both cases of receptor sensitization may be related to the hyperpathia and hyperalgesia observed in man under comparable conditions (52, 53).

Criteria for the Specificity of Nociceptors

The most obvious criterion for a nociceptor is its threshold to the adequate stimulus. However, as has been mentioned above, precise determination of the adequate stimulus of a nociceptor is often difficult. For example, about half of the receptors excited by noxious heat in the cat's hind foot (Figure 4) also respond to strong mechanical stimulation of the skin (3), therefore they have been termed polymodal nociceptors. In a study of the hairy skin of the cat's leg all C fibers responding to heat could also be excited by strong mechanical stimuli (38). For this reason it has been claimed that virtually all thermosensitive nociceptors are polymodal, some having very high thresholds to mechanical stimulation. To clarify this question further, it will be necessary to apply well defined and reproducible mechanical stimuli, to yield, for example, the relationship between the discharge frequency of the receptors and the intensity of the stimulus, and the time course of the discharge during long-lasting stimuli.

By experiments of this type it should be possible to establish the ability of the receptors to encode information on the intensity and the duration of mechanical stimuli, compared with the ability to carry information about noxious thermal stimuli (Figure 4). Such comparisons are expected eventually to yield consistent criteria for the classification of nociceptors in terms of stimulus specificity. Not only would it be possible to decide whether or not a nociceptor is polymodal, but it would also be possible to decide whether or not a receptor is in fact a nociceptor at all!

Visceral Nociceptors

Many types of mechanosensitive and chemosensitive receptors have been described in the viscera (54). Most of them respond to physiological changes of the milieu interne, and therefore may be involved in homeostasis. Normally the firing of these receptors is not consciously perceived.

In contrast, conscious pain sensation can arise from all the inner organs, sometimes reaching excruciating intensity. Stimuli which evoke visceral pain in man, and nocifensive reactions in animals, include excessive distension of hollow organs, chemical agents and ischemia, but not the thermal stimuli which are so effective on the skin.

Direct electrophysiological recording from visceral afferent nerves has yielded some hints of the existence of nociceptors; however, knowledge of their physiological characteristics is far less advanced than for the cutaneous nociceptors. A major problem in the identification of a deep receptor (e.g., visceral or muscle) as a nociceptor is the investigator's lack of direct experience with the parameters of stimuli to these inner organs, which are felt as painful. This is a situation completely different from that of the skin, where many familiar situations make us aware of the objects and the events producing pain.

To establish quantitatively the precise stimuli which produce pain, it will be necessary to study the threshold, and the intensity functions of the nocifensive responses in conscious animals (cf. section "A Test for a Stimulus To Be Nociceptive in Animals"). The stimulus thus defined could then serve as a standard for neurophysiological analysis of nociceptors and central neurons. So far, this comparative approach to visceral nociception has only been done in a few investigations, e.g., on cardiac ischemia (section "Cardiac Nociceptors").

Mechanosensitive Visceral Nociceptors Mechanoreceptors with predominantly afferent C fibers have been found in the smooth muscle wall of *hollow organs*, such as the intestines, the urethra, and the urinary bladder (50, 55, 56). Some respond with moderate spike frequencies to either passive distension or active contraction of the smooth muscles. Their response to bradykinin is due to smooth muscle contractions (50). A dramatic increase in discharge frequently has been observed in such receptors when contractions of the smooth muscle occurred under isometric conditions, i.e., when the exit of the organ (e.g., the urinary bladder) was obstructed. This is the situation in spasmodic colic pain; therefore it has been concluded that impulses in these afferents will induce such pain states.

So far it is not clear which of the intramural distension receptors can be regarded as nociceptors. This difficulty in classification is because of large fluctuations in the threshold of individual units, probably produced by uncontrollable changes in tone of the in series smooth muscle. Thus a major criterion for the identification of a receptor as a nociceptor, i.e., its threshold, could not be established unambiguously (57). With a better understanding of the function of the myenteric plexuses (60), which are involved in the variations of intestinal smooth muscle tone, it is hoped that control of the experimental conditions for studying these mechanoreceptors will be possible.

In the *lung*, two types of receptors have been described which probably have nociceptive functions: the "type J receptors" with C afferents, and the "lung irritant receptors" with afferents in the Aδ range, all running in the vagus nerve (61, 62). The type J receptors (juxta-capillary receptors) are located in the interstitial space close to the capillaries, the lung irritant receptors in the epithelia of the lung and its airways. They are activated by a variety of stimuli such as pulmonary congestion, microembolism, atelectasis, and pneumothorax, all of which produce mechanical distortions within the lung. In addition they are excited by irritant gases (chlorine gas, ammonia vapor), and, particularly the lung irritant receptors, by aerosols and dust particles. Their activity is claimed to give rise to reflexes such as bradycardia, apnea or tachypnea, and probably to sensations of dyspnea and pain in man. All these effects are abolished by blocking the vagus nerves, both in animals and in man.

Nociceptors in Skeletal Muscle

There have been few approaches to identify nociceptors in muscle. As in cutaneous and visceral nerves, the III and C fibers were regarded as highly likely to contain such nociceptive afferents.

In fact, the majority of the muscle III fibers investigated so far in single fiber studies were found to be mechanosensitive, having high thresholds (63–65). They responded to strong localized pressure on the muscle, and very irregularly, if at all, to large stretches of the muscle. Most of the receptors with III fibers had rapidly adapting discharges, and were not excited by ischemia, which is a very effective stimulus for muscle C fibers.

C fibers from muscle usually reveal high threshold mechanical excitability, which is comparable to that of the III fibers (66, 67). In addition, these fibers are excited by ischemia combined with contraction of the muscle (66, 68), which produces muscle pain in man. Neither ischemia nor tetanic contraction alone elicited activity in these fibers.

It has been suggested that one or more of the chemical substances bradykinin, 5-hydroxytryptamine, and histamine (cf. section "Nociceptors and Algogenic Substances") play a role in these responses to ischemia and contraction.

Therefore the effects of these agents have been tested on muscle C fibers. In a systematic study 75 of 153 single C units of the gastrocnemius-soleus muscle have been found to be excited by one or more of these substances, when administered by close arterial injection (67). Bradykinin was the most potent stimulant. However, some of the C fibers identified in this study may not have been nociceptors, but chemoceptors involved in respiratory regulation during muscular exercise (69).

In the muscles studied so far (predominantly the gastrocnemius-soleus) experimental access to the fiber endings is difficult, because most of them are located deep within the bulk of the muscle. This is a major reason why a satisfactory elucidation of the adequate stimuli has not been possible. Preparations should be used which provide better access to the nerve endings (e.g., the cat tenuissimus muscle).

Cardiac Nociceptors

A frequent source of pain from the inner organs is the heart. Angina pectoris, the most common form of cardiac pain, is a symptom of recurrent reversible ischemia of ventricular muscle. In the conscious dog ischemia induced by occlusion of a coronary artery evoked pseudoaffective behavior. This sign of discomfort in animals, and cardiac pain in man, are both totally abolished by neurosurgical transection of T_1 to T_4 dorsal roots, or of the corresponding sympathetic ganglia (71). Thus it was concluded that the nociceptive afferents run via these structures, and not via the vagus nerve.

Recordings have been made of single units in the T_2 and T_3 rami communicantes (72). A considerable number of Aδ and C fibers were found which started to discharge during ischemia by coronary occlusion. It was suggested that the Aδ fibers were stimulated by the unphysiological motions of the ischemic heart, since their discharges were correlated with the cardiac cycle. The C fibers exhibited irregular discharges not related to the cardiac cycle.

The authors assumed that the C fibers respond to chemical agents (72, 78) such as K^+, lactic acid, bradykinin, and hyperosmolarity, which all occur during cardiac ischemia, whereas the Aδ fibers are thought to be sensitized by these substances and thus become mechanosensitive. These findings are in contrast to previous work in which it was shown that cardiac Aδ fibers excited by ischemia neither responded to most of these chemical substances, nor to mechanical stimuli (73).

More work needs to be done in this important field in order to resolve the question of the adequate stimulus of these probable nociceptors, and to separate them from the other types of cardiac receptors (e.g., mechanoreceptors) frequently encountered in these investigations. Moreover, it should be clarified whether the receptors excited by ischemia of the heart and of skeletal muscle are of the same functional type.

Nociceptors and Algogenic Substances

A large number of chemical substances evoke pain in man, when applied to the base of an experimental cantharidin blister (74), and pseudoaffective reactions in animals when applied intra-arterially to visceral organs (70, 75). Some of these substances, e.g., bradykinin, serotonin, and histamine (70, 74, 75), or hydrogen ions (76), are liberated as a consequence of tissue damage. It has been suggested that they are necessary for the excitation of any nociceptor. This hypothesis implies that nociceptors are chemoreceptors.

Direct tests using electrophysiological recording from single units in cutaneous and muscle nerves (37, 67, 77) have revealed that many nociceptors can be excited by one or more of these algogenic chemicals. This is certainly the basis for the pain sensations and nocifensive reactions produced by these substances.

Assuming that some of these chemical agents are necessary steps in the transducer process of nociceptors, one might postulate that nociceptors of one distinct type should all respond consistently to the same chemical or set of chemicals, when tested directly. This has not been found when investigating the homogeneous group of thermosensitive nociceptors of the cat's foot (37): some of these C heat receptors could be excited by a close arterial injection of bradykinin, but did not respond to serotonin, some behaved in the opposite way, and others could be excited by both substances.

A search for visceral receptors excited by intra-arterial injections of bradykinin (50) revealed that only visceral mechanoreceptors responded to this substance; most probably this excitation was mediated by contractions of the surrounding smooth muscle.

Thus, the evidence is against the existence of specific chemical transduction in nociceptors, and the powerful chemical activation of nociceptors must be considered as an unspecific excitatory effect. This interpretation is supported by the observation that many of the slowly adapting low threshold mechanoreceptors of skin and muscle, which have large myelinated afferent fibers, also respond to one or more of these algogenic substances (37, 77).

However, these chemical agents might produce a sensitization, thus enhancing the response of a receptor to its adequate stimulus. In Figure 5 the effect of a subliminal dose of bradykinin is shown on the responsiveness of a C heat nociceptor (37). This experiment demonstrates sensitization, i.e., fall in temperature threshold, and increase of the discharge rate to a standard heat stimulus.

Such a mechanism could account for the sensitization found upon frequent application of noxious stimuli to the skin (cf. section "Cutaneous Nociceptors"): tissue damage probably occurs, which might produce some of the algogenic substances. They are also released during inflammation (79).

These cases of tissue injury are known to be accompanied in man by primary hyperalgesia or hyperpathia: in these conditions a noxious stimulus is subjectively perceived as stronger than usual, or a normally non-noxious stimulus is felt as painful, respectively.

NOCICEPTION AND THE CENTRAL NERVOUS SYSTEM

The Dichotomy of the Somatosensory System

Anatomical and embryological considerations suggest a division of the somatosensory system into two main components, which are often referred to as the

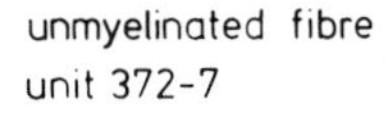

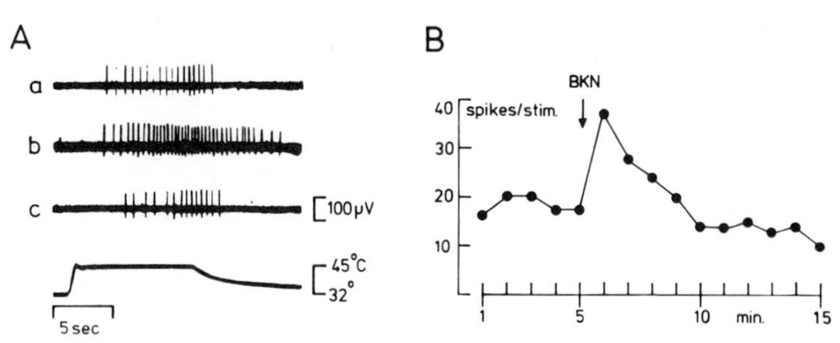

Figure 5. Sensitization of C nociceptor by bradykinin injection. *A*, discharges during identical heat stimuli (45°C) applied to the receptive field before (*a*), 1 min after (*b*), and 6 min after (*c*) close arterial injection of 10 g of bradykinin. *B*, number of spikes per heat stimulus (same series as in *A*), applied at 1-min intervals. Arrow indicates bradykinin (*BKN*) injection. The fiber had a conduction velocity of 0.39 m/s, the receptive field was on the plantar surface of the cat's hind leg. (From Beck and Handwerker (37), and unpublished observations.)

"lemniscal" and the "extra-lemniscal" (synonyms: non-lemniscal, anterolateral) systems, respectively (49, 80–82). The *lemniscal system*, phylogenetically recent, is the somatotopically ordered projection of the cutaneous sensory surface onto the postcentral gyrus of the cortex, mainly involving the large myelinated (group II) afferents, which have specialized mechanoreceptors. This connection runs via the dorsal columns of the spinal cord, dorsal column nuclei of the medulla, medial lemniscal tract, ventrobasal nuclei of the contralateral thalamus, and reaches the two somatosensory projection areas of the cortex, SI and SII. There is a corresponding pathway for the face area, innervated by the trigeminal nerve (84).

Lemniscal neurons, found, for example, in the dorsal column nuclei and in the ventrobasal nuclei of the thalamus, have the following characteristics: they are exclusively excited by one type of group II mechanoreceptors, somatotopically arranged, have restricted receptive fields, high security of synaptic transmission (i.e., they are not susceptible to temporal summation, or facilitation) and lack of afterdischarge.

Information about a tactile stimulus is transmitted to the cortex by a pathway usually consisting of three lemniscal neurons in series. This ensures a fast transfer which preserves information about the stimulus specificity and its configuration in space and time.

Lesions in the lemniscal system impair the ability for spatial discrimination of tactile stimuli. It is therefore agreed that the lemniscal system is the anatomical basis of the *epicritic* component of the cutaneous senses (cf. section "Pain Theories"). In addition, the lemniscal route provides feedback for the control of movement (83).

The pathway of the *extralemniscal system* has been located by most authors to the dorsal horn of the spinal cord, the ascending anterolateral tract, the brainstem reticular formation, the medial, intralaminar and posterior group of nuclei of the thalamus, and the SII cortex.

The nonspecific activating system, which plays an essential part in the control of consciousness, apparently shares the same brainstem and thalamic regions as the extralemniscal system. Knowledge of the neuronal systems involved in these different functions, however, is very fragmentary.

In contrast to the lemniscal system, the components of the extralemniscal system are phylogenetically old, and show no somatotopic arrangement. The ascending pathway involves a large number of synapses in series, and also extensive divergence and convergence of connections. This means that the spatio-temporal information about the stimulus is lost. The input is claimed to consist predominantly of the Aδ (III) and C afferents from skin and viscera. There seems to be an intimate relationship between the extralemniscal system and the autonomic nervous system, and the limbic system.

The sensations and perceptions mediated by the extralemniscal system are thought to be affective-emotional, and motivational in character, lacking all the well determined properties of those related to the lemniscal system. These sensations are called *protopathic* (cf. section "Central Nervous System"); pain is one of the protopathic sensations.

Apparently, the concept of a dichotomy of the somatosensory system permits a straightforward classification of sensory phenomena, and their correlation with brain pathways and nuclei. However, some recent neurophysiological findings do not reconcile with this concept; therefore it should not be used for a thorough and rigorous description of central nervous system function in nociception.

Spinal Neurons Involved in Nociception

Cutaneous and visceral afferents enter the spinal cord via the dorsal roots and synapse onto dorsal horn neurons. This is true for all types of fibers, regardless of fiber diameter. A large proportion of the thick myelinated (group II, Aβ afferents also have collaterals, which ascend directly in the dorsal columns to the medulla oblongata. They form the main pathway of the lemniscal system.

The dorsal horn is a layered structure, the laminae (I to VI) being defined according to cytoarchitectonic criteria (85). The anatomy of the afferent fibers of the dorsal horn neurons, and of their synaptic relations, has been described (86–88).

Neurophysiological analysis has revealed that practically all types of low threshold mechanoreceptors found in the skin (48) may excite dorsal horn neurons by way of group II afferents. Often these neurons have convergent input from more than one type of these low threshold receptors, which clearly differentiate them from lemniscal neurons (cf. section "The Dichotomy of the Somatosensory System").

As regards excitatory input from nociceptors, two basic types of dorsal horn neurons exist, one being exclusively nociceptive (section "Specifically Nociceptive Neurons"), the other having convergent input from low threshold mechanoreceptors and from nociceptors (section "Polymodal Neurons"). They project into ascending tracts (section "Spinal Ascending Tracts"), and have segmental and descending inhibitory influences (section "Inhibition of Dorsal Horn Neurons").

Specifically Nociceptive Neurons The most detailed description of such neurons was obtained using the sacral and coccygeal segments of the cat's spinal cord (89). Electrical stimulation of the dorsal roots excites these cells only when III (Aδ) fibers are recruited (Figure 6, *A* to *C*), the latency of the discharge being appropriate for the conduction velocity in these afferent fibers (i.e., less than 30 m/s).

These neurons could be excited by noxious mechanical stimulation of the skin, either by itself or together with noxious thermal stimulation. This suggests input from mechanosensitive nociceptors with III afferents to all of the neurons, and convergence of these afferents with C fibers from heat nociceptors in others. However, not all of the cells excited via group III afferents had input from nociceptors: some responded to mild warming or cooling of the skin, suggesting afferent connections with sensitive thermoreceptors.

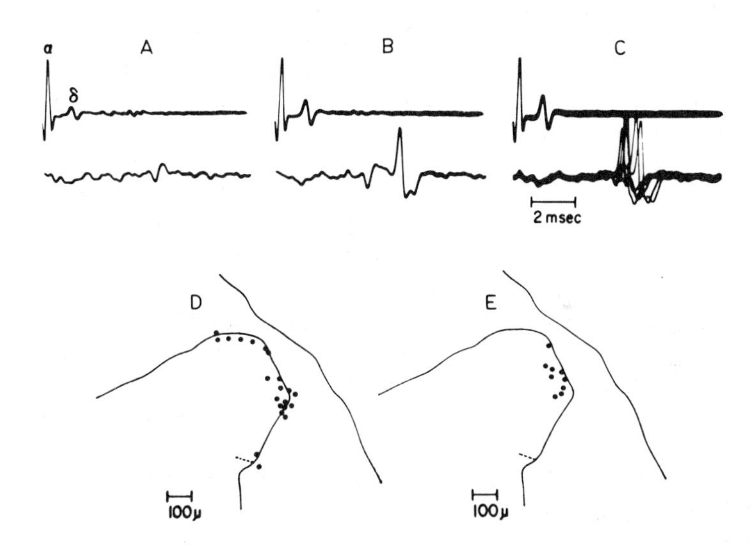

Figure 6. Characteristics of nociceptive dorsal horn neurons in the cat. In *A* to *C* unitary responses to afferent volleys are plotted. Upper trace of each pair is the compound action potential recorded from the coccygeal dorsal root and the lower trace the microelectrode recording from the dorsal horn neuron. *A*, electrical shock sufficient to excite nearly all α-fibers (over 40 m/s; presumably including groups I and II afferents), and some δ-fibers (under 30 m/s). *B*, shock intensity increased slightly above that in *A*. *C*, shock intensity maximal for all myelinated fibers (5 sweeps superimposed). *D*, location of dye marks at sites where neurons specifically activated by slowly conducting afferent fibers were found. Position of each mark is indicated relative to major landmarks on an outline drawing of the dorsal horn in the coccygeal spinal segment. Units excited only by strong mechanical stimulation of the skin. *E*, units excited by both noxious heat and by strong mechanical stimulation of the skin. (From Christensen and Perl (89).)

The locations of the cells were identified by dye ejections from the recording microelectrode. They were found to be accumulated in the marginal layer (lamina I (85)), suggesting that these nociceptive neurons are the Wedeyer cells described previously (90): these are single large cells interspersed in lamina I. From histological sections it is estimated that they comprise less than 1% of all the large neurons of the dorsal horn, most of which are located in laminae IV to VI.

The same functional type of neurons have been described in the monkey lumbosacral cord (91–93), and in the spinal trigeminal nucleus (94). Some of these neurons have an ascending axon in the anterolateral tract (cf. section "The Anterolateral Tract").

Polymodal Neurons It has been frequently suggested that there are neurons in the dorsal horn which have a convergent input from low threshold mechanosensitive afferents, and from nociceptive afferents (95–97). It is now well established that these neurons have input from both cutaneous group II fibers and from C fibers (99–101). Indirect evidence suggests that group III afferents also contribute to their excitation (20, 102). Therefore they will be called polymodal neurons in this review, though the meaning of the term "polymodal" as used here should not be confused with its meaning as used in "polymodal nociceptors" (cf. section "Polymodal Nociceptors"). There is evidence that these neurons, which are much more numerous than the specifically nociceptive neurons (section "Specifically Nociceptive Neurons"), are of importance in the perception of painful stimuli in man (cf. section "The Anterolateral Tract").

The nociceptive input to these neurons cannot be investigated using mechanical stimuli, since the responses to such stimuli would be concealed beneath the concomitant excitation of the cells by their low threshold mechanosensitive input. Therefore, two methods have been used to produce selective activation of C fibers: 1) electrical stimulation of afferent A and C fibers, a selective C input then being produced by differentially blocking the A fibers (Figure 7, *A* to *C*); 2) radiant heat (Figure 7*D*), which has been shown to selectively excite the C heat nociceptors (3). The cat's hind foot is particularly favorable for the use of the latter method, since low threshold warm receptors are virtually absent in this region of skin (3).

These studies (99–101, 103–106) revealed that between 50% and 70% of the dorsal horn neurons encountered in systematic microelectrode trackings have convergent group II and C fiber inputs, i.e., they were polymodal units as defined above. The intensity of a noxious stimulus such as radiant heat is encoded in their discharge frequency (Fig. 7*E*); this observation supports the hypothesis that the polymodal dorsal horn cells contribute to the central processing of nociceptive messages. The remaining units (apart from those described in section "Specifically Nociceptive Neurons") have only afferents from sensitive cutaneous mechanoreceptors. They will not be considered in this review. It was found that C fiber input was particularly likely onto those units which were monosynaptically driven by II fibers, though the probability of a neuron having a C input is independent of the type of the low threshold mechanoreceptors that produces the group II input, and independent of its location in the dorsal horn. It is not known whether C fibers have monosynaptic access to these neurons, or whether there are neurons intercalated in this pathway (see below). Some of these neurons have an axon projecting into the crossed anterolateral tract (section "The Anterolateral Tract"), or into the spinocervical tract (section "Other Spinal Ascending Pathways").

An outstanding feature of the discharges of these neurons is that the firing frequency of a unit to a heat stimulus is much higher (e.g., by a factor of 10) than that of a C heat nociceptor under the same conditions (3, 103); this finding suggests that there is a lot of convergence of C input. In contrast, when driven by mechanoreceptor group II afferents the discharge rates of the neurons are in a range comparable to those of single receptors.

Often the response of the polymodal dorsal horn neurons is augmenting in the course of repetitive electrical C fiber stimulation at a rate of about 1 Hz (wind up) (101). Also during radiant heat stimulation the firing frequency of these neurons sometimes builds up progressively, after a delay of several seconds (Figures 7*D* and 9*A*). Both phenomena indicate that considerable temporal summation of the C input is required to discharge the neurons.

Another phenomenon regularly encountered in heat-evoked activity is an afterdischarge (Figures 7*D* and 9, *A* and *B*), which often outlasts the heat stimulus by 20 s or more (103), whereas the activity in the C heat nociceptors ceases within 2 s after the end of the heat stimulus (3, 103).

Neither temporal summation nor afterdischarge is seen when the same dorsal horn units are activated via the group II afferent route, by either electrical or adequate stimulation. Thus, the deviations from "high fidelity" in the time

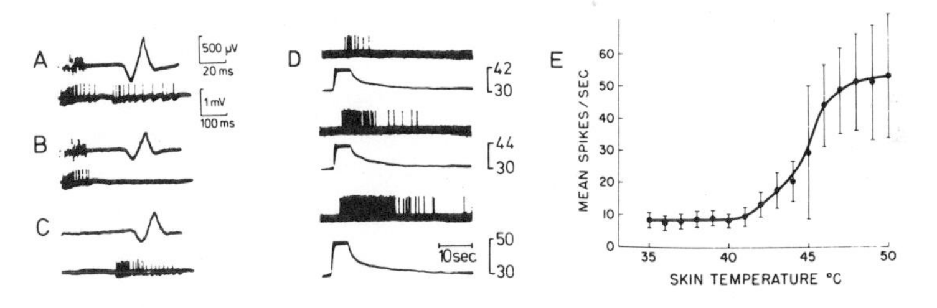

Figure 7. Characteristics of polymodal dorsal horn neurons of the cat and monkey with C fiber input. *A* to *C* show effect of differential blocking of the sural A fibers on the activity of a dorsal horn neuron. Each pair of records shows the spike discharge of the cell (*lower trace*), and the simultaneously recorded nerve compound action potential (C fibers), following electrical stimulation. Note different time scales in the upper and lower traces. The stimulus was a square pulse of 15 V, 0.5 ms in duration in *A* and *C*, 4 V in *B*. In *C* all A fibers were blocked by a polarizing current of 80 μA, applied to the nerve between stimulating and recording sites. *D*, response of a dorsal horn neuron to heating the skin of the foot pad by radiation to 42°C, 44°C, and 50°C, respectively; the skin surface temperatures, measured with a thermocouple, are shown below each record. *E*, skin temperature-mean firing frequency relationship of 11 units responding to noxious skin temperatures found in the monkey's dorsal horn. Vertical bars are standard error of the mean. (*A* to *C* from Gregor and Zimmermann (99); *D* from Handwerker, Iggo, and Zimmermann (103); *E* from Price and Mayer (106).)

domain must be considered to be characteristic for the processing of noxious stimuli by these neurons. These phenomena have counterparts in psychophysics: temporal summation and aftersensation are well known effects in the perception of noxious stimuli.

The neuronal mechanisms underlying these special features in the responses of polymodal neurons to noxious stimuli are unknown. Two possibilities should be investigated: 1) that there is a special neuronal circuit between the C afferent and the polymodal neurons, e.g., in the meshwork of small neurons in the substantia gelatinosa (these neurons have so far never been recorded from (107)); 2) that the C fiber input has special synaptic characteristics. Mechanisms involved could be, for example, low synaptic security of the single synapse, coupled with a large number of convergent C fibers onto one neuron; or a transmitter substance with particular kinetics of release, postsynaptic action, and enzymatic removal.

Apart from the excitatory convergence of A and C fibers there is also an inhibitory action exerted by A volleys on the response of the neurons to the C fiber input (Figure 7, *A* to *C*). This segmental inhibition, as well as the influences originating in the supraspinal CNS, is referred to in separate paragraphs (sections "Inhibitory Interactions at the Segmental Level" and "Supraspinal Inhibition of Dorsal Horn Neurons").

Somatovisceral Convergence and Interactions To explain the occurrence of "referred pain," i.e., the false localization of pain of visceral origin to the skin, several neurophysiological mechanisms have been postulated (58, 108, 109). One of them was the hypothesis that afferents from the viscera and the skin converge onto the same neurons in the spinal cord, and that the ascending axons from these cells are shared for signaling noxious events in either organ.

There is now a body of evidence from neurophysiological investigations to substantiate the hypothesis of somatovisceral convergence: both in the thoracic (111) and lumbar (109, 110, 112, 113) cord, neurons have been described which are excited by cutaneous $A\delta$ and visceral $A\delta$ fibers. So far no information is available as to whether visceral C fibers also excite these neurons. From the spatial distribution of intraspinal potential fields it was concluded that cutaneous $A\beta$ afferents also converge onto the same neurons (109). Some of these units with somatovisceral convergence have an axon ascending in the anterolateral tract (113).

Experiments using adequate stimuli (113) revealed that distension of visceral hollow organs was effective in exciting these neurons, though other visceral stimuli have not been tried. The converging afferents from the skin to these neurons contained both nociceptive and low threshold mechanoreceptive fibers. Therefore these units might be regarded as belonging to the population of polymodal neurons described above (section "Polymodal Neurons"). Some of them have, in addition, group III input from muscle (111), which can be activated by noxious stimuli (cf. section "Nociceptors in Skeletal Muscle"). With the hypothesis not proven so far that the visceral afferents to these cells contain some which come from nociceptors, the convergence of cutaneous and visceral fibers onto common neurons in the dorsal horn might not only serve as an explanation for referred pain. Subliminal activation of the neurons via the visceral afferents will lower the threshold to cutaneous stimuli, and this might then explain the origin of some states of hyperpathic skin tenderness, and secondary hyperalgesia. On the other hand, the common neuron is supposed to be a link in the pathway of vasomotor reflexes, controlling the local circulation: thus, skin stimulation such as warming might produce a reflex hyperemia not only of that skin area, but also of the corresponding viscus (59); this procedure is known to be of therapeutic value in certain visceral diseases.

Mutual inhibition (111, 114, 115) is exerted by the convergent cutaneous and visceral inputs; it has been suggested that this is the neurophysiological basis for the well known effect of "counter-irritation" in man, that is, the suppression of visceral pain by strong cutaneous stimulation (see section "Inhibitory Interactions at the Segmental Level").

So far no neurons have been found which have an exclusive Aδ input, either from viscera alone, or from both skin and viscera. Search for such units, responding specifically to nociceptive visceral afferents, should be intensified, along with the elucidation of the visceral afferent system in terms of its adequate stimuli (cf. section "Visceral Nociceptors").

Inhibition of Dorsal Horn Neurons

A variety of inhibitory factors act at the spinal level to modulate the afferent information. In this paragraph the emphasis is on findings which are probably related to nociception.

Inhibitory Interactions at the Segmental Level A basic observation appears in Figure 7, *A* to *C*: the discharge of a polymodal dorsal horn neuron to an afferent C volley is inhibited by a preceding discharge evoked by an A volley. The neuronal inhibition of messages about noxious events has recently received a lot of attention: electrical stimulation of the dorsal columns and of peripheral nerves is now used in man for the treatment of chronic pain (116, 126), and it has been suggested that the abolition of pain by this method might be due to inhibition of spinal neurons.

In Figure 8 the effect of dorsal column stimulation in the cat is shown on the response of a polymodal neuron to noxious heating of the skin. Although these neurons are all excited by the dorsal column stimuli, via the antidromically conducted action potentials reaching the collaterals which are presynaptic to the neurons, an inhibitory effect often prevails, particularly at high rates of electrical stimulation. This inhibitory effect has been observed also on discharges evoked by noxious mechanical stimulation (95). Inhibition may also be produced by electrical stimulation of the afferent cutaneous nerve from the skin area where the noxious heat stimulus was applied; group II afferent impulses produce this inhibition, judging from the low stimulus strength used (103).

We do not yet have sufficient knowledge of the functional organization of the inhibitory interactions of A and C fibers at the dorsal horn level. Therefore it is not yet clear whether the suppression in Figures 7*A* and 8 is a specific

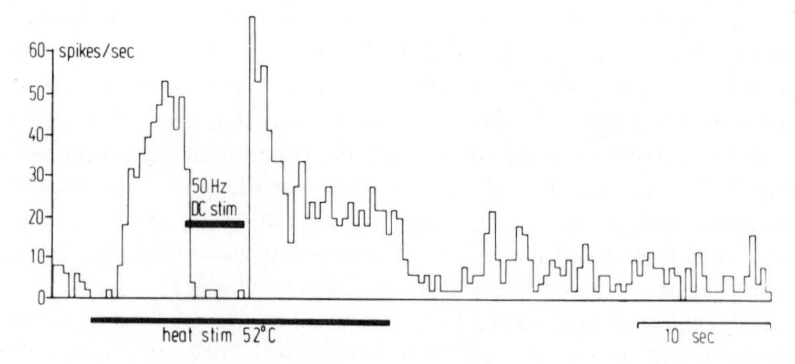

Figure 8. Effect of dorsal column stimulation on heat response of polymodal dorsal neuron in the cat. Histogram of discharge upon radiant heat stimulation; the bar below indicates the duration of the 52°C heat stimulus to skin. As indicated by the bar in the histogram the dorsal columns were stimulated electrically at L_2 with 4 times threshold intensity, 50 Hz. (From Handwerker, Iggo, and Zimmermann (103).)

inhibitory effect of A fiber input onto C fiber evoked discharges, or whether it is a more general inhibition of the neuron which follows excitation by any input, e.g., produced by recurrent inhibition.

As regards the specifically nociceptive lamina I cells, afferent inhibition of the surround type was observed occasionally, being produced by high threshold afferents (89). However, no investigation of the inhibitory actions exerted by electrical stimulation of nerves or dorsal columns has so far been reported. These neurons are ideally suited for the study of such inhibitions, since they do not have excitatory group II input which would interfere with the inhibition. It would be highly desirable to investigate this question, to help to clarify both the functional significance of these neurons, and the mechanism of pain relief by electrical nerve and dorsal column stimulation.

Supraspinal Inhibition of Dorsal Horn Neurons Descending pathways originating in various regions of the brain exert control over the motor, as well as the sensory, functions of the spinal cord. This control may appear as inhibition or facilitation of spinal neurons. The capacity of this descending control to modulate sensory information is demonstrated by its ability to change, for example, 1) the size of the receptive fields of spinal neurons (95, 117), and 2) the modalities of the afferents which excite a particular neuron (118, 119). Presumably these effects are mediated, at least partly, by the systems of presynaptic inhibition of spinal mechanoreceptive afferents (16, 120).

There have been a few reports of descending influences on the responsiveness of spinal neurons to noxious stimuli, or to electrical activation of C fibers. The existence of a tonic descending inhibition was demonstrated by cooling the cord at about L_1 level (95, 103,122). This blocks all descending pathways, and consequently alters the response to noxious stimuli, such as heating the skin(103) or pinching (95).

In Figure 9, *A* and *B,* the effect of a cold block of the cord is shown on the discharge of a polymodal dorsal horn neuron. Both the spontaneous activity and the response to the heat stimulus are enhanced. A corresponding change in the responsiveness of these cells to electrical C fiber stimulation has been observed.

The descending inhibition also affects the *coding of the intensity* of noxious stimuli. Figure 9*C* shows a shift of the intensity function to higher discharge frequencies on the release from descending inhibition. In addition there is a slight increase in the slope, which may be interpreted as a change in the gain of the dorsal horn transmission of information about noxious events. In some extreme cases neurons were found which, though they normally gave practically no response to C fiber stimulation, responded quite well to this type of stimulation during cold block of the descending fibers (103, 122).

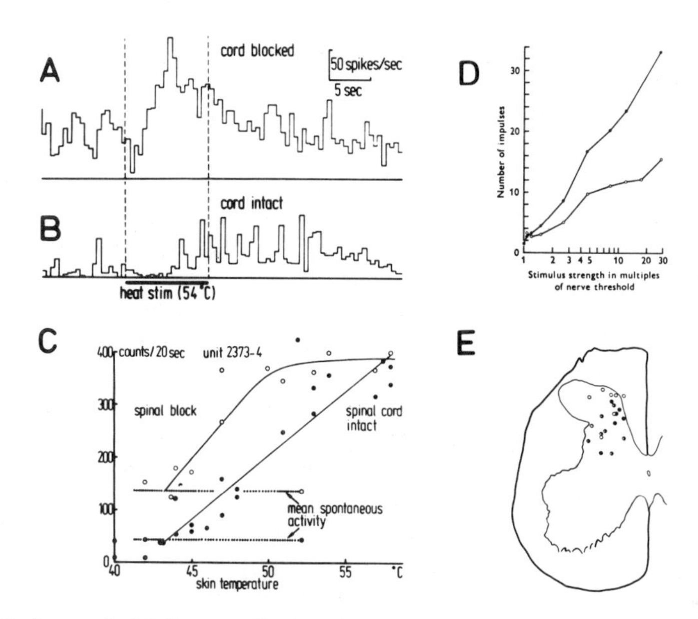

Figure 9. Supraspinal influence affecting the responses of dorsal neurons. In *A* and *D* the effect of a reversible cold block of the spinal cord around L_1 level is shown. *A*, time course of discharge of the polymodal dorsal horn neuron upon radiant heat stimulation (54°C, indicated by the bar below *B*) of the receptive field on the foot, when the cord was blocked. *B*, response of the same unit as in *A* when the cord was intact. *C*, intensity function of a polymodal dorsal horn neuron. The total number of spikes discharged in a period of 20 s during and after radiant heat stimuli to the foot pad (10 s in duration) is plotted against the intensity of the stimulus. Measurements performed when the cord was blocked reversibly at L_1 by cooling (○), and before and after the cold block (●). The dotted lines indicate the spontaneous activity averaged from 10-s periods before each stimulus. Regression lines fitted by eye. *D*, intensity function of a dorsal horn neuron, having an axon in the SCT, which could be excited by cutaneous A and C fibers. The number of impulses evoked in the neuron by electrical stimulation of the medial plantar nerve is plotted against the strength of stimulation in multiples of threshold (T) for the whole nerve. Measurements made when the spinal cord was blocked (●), and intact (○), respectively. *E*, effects of stimulation of the sensorimotor cortex on dorsal horn neurons; the location of the neurons, which had an axon in the anterolateral tract, from different experiments is shown. The cortical stimulation produced either inhibition (●) on the activity evoked in the neurons by low threshold cutaneous input, or a sequence of excitation/inhibition (○), or had no effect at all (○). *A* to *D* were from cats, *E* from monkeys. (*A* to *C* from Handwerker, Iggo, and Zimmermann (103); *D* from Brown (122), *E* from Coulter, Maunz, and Willis (124).)

A clear *change of gain* by block of descending influences has been demonstrated in polymodal neurons (122), as is shown in Figure 9*D*. Single electrical stimuli of a cutaneous nerve were used to excite the neuron, which has an axon in the spinocervical tract (section "Other Spinal Ascending Pathways"). The number of spikes per stimulus increases linearly with the stimulus strength. The slope of this relationship, i.e., the gain of transmission, is much greater in the spinal animal. The range of stimulus intensities used in Figure 9*D* was sufficient to recruit the afferent groups II and III fibers, but not the C fibers. However, there are indications that the C fiber response of the same neuron was influenced in the same way (122).

Focal electrical brain stimulation has been used to locate the sites of origin of the inhibition. Such stimulation has been performed on the cortex, and in the mesencephalic central gray, to study effects possibly relevant in nociception on dorsal horn neurons.

It is well known that the *sensorimotor cortex* exerts a modulatory influence on the spinal cord (120), which is mediated by the pyramidal tract (123). Stimulation of the leg area of either the pre- or postcentral gyrus in the primate influences dorsal horn neurons of the lumbar spinal cord (124); the locations of the cells investigated are shown in Figure 9*E*. The units were either inhibited (●), or showed a sequence of excitation-inhibition (◐), or were not influenced at all (○). The neurons which showed inhibition or excitation-inhibition all had low threshold cutaneous mechanoreceptor input. Some of them responded also to noxious stimuli; these are therefore polymodal neurons (section "Polymodal Neurons"). In these units cortical stimulation inhibited the low threshold input, but had little or no effect on discharges evoked in these cells by noxious stimuli. This was interpreted by the authors as the switching of the neurons from detectors of weak stimuli to detectors of intense stimuli.

Four cells located in the marginal zone of the dorsal horn, and also some at deeper sites (○ in Figure 9*E*), were not affected by cortical stimulation. Most of these neurons had high thresholds to mechanical stimuli in their cutaneous receptive field. They were probably of the specifically nociceptive type (section "Specifically Nociceptive Neurons").

These findings therefore suggest that the cortex is able to produce a highly specific modification of the low threshold afferent input onto the polymodal neurons, possibly operating at the presynaptic level.

The *periaqueductal gray* in the mesencephalon has recently attracted much interest, after it was reported that powerful analgesia could be produced in rats and cats by electrical stimulation of this area (125, 127). In order to test whether inhibition of spinal neurons might contribute to this analgesia, the effect of stimulation of the mesencephalic gray on polymodal dorsal horn neurons has been studied (127). It was found that the neuronal discharges to noxious stimuli, e.g., pinch applied to a skin fold, were strongly inhibited. In contrast, in about half of these units, the discharge to light tactile stimuli was not affected by midbrain stimulation.

The differential effect is obviously complementary to that of inhibition originating in the cortex, as has been reported above in the context of Figure 9*E*.

When acting on the same neuron, both influences would be able to cooperate in shifting the optimum responsiveness of the neuron between stimuli of either low or high intensity.

Spinal Ascending Tracts

The Anterolateral Tract The relief from chronic pain produced by transection (cordotomy) of the spinal anterolateral tract (ALT) was the major argument for attributing a specific function in nociception to this pathway (128, 129). Clinical evidence suggested that the majority of the ALT originates from the contralateral spinal gray, though some axons are of ipsilateral origin.

There has been, and still is, much debate among anatomists about the brain targets of the ALT (135, 141, 181, 190), which will not be reviewed here. It is, however, generally accepted that some of the fibers project to the brainstem reticular formation, others to the thalamus.

The first neurophysiological studies had not been performed until quite recently (91, 113, 130–133, 157). These investigations showed that neurons projecting into the ALT have the same characteristics as other cells in the dorsal horn with cutaneous input. Therefore there are basically two types of neurons in the ALT that transmit information about noxious events, described as specifically nociceptive (section "Specifically Nociceptive Neurons") and polymodal (section "Polymodal Neurons"), respectively; there also appears to be a lot of somatovisceral convergence (113) (section "Somatovisceral Convergence and Interactions"). In addition, axons have been found from neurons responding only to low threshold skin mechanoreceptors, such as hair follicle receptors, and still others with input from muscle and joint afferents. These are obviously not concerned with nociception.

Location of Cells of Origin of the ALT There appear to be species differences in the location of cells projecting to the ALT (130–133). In the monkey they have been found mainly in the dorsal horn, whereas in the cat they occur predominantly in more ventro-medial parts of the gray matter, but some have been found also in the marginal layers of the dorsal horn (92). In the monkey some of the units which respond exclusively to high intensity mechanical and thermal skin stimuli have been located in the uppermost layers of the dorsal horn, but not so distinctively in the border zone of the gray matter as had been reported previously (89); others were found in deeper laminae. It is not clear whether these two separate locations indicate the existence of two types of high threshold units.

Conduction Velocities The conduction velocities revealed by antidromic stimulation were in the range 10–100 m/s, the mean values established by various authors being between 40 and 50 m/s. Thus, the ALT is a fast conducting tract. Histological evaluations in the primate of the fiber diameter spectrum yielded values from 2 to 7 μm, with a peak at 4 μm (135). To explain the fast conducting velocities, a much larger Hursh-factor must be assumed for these central axons, compared with that established for peripheral nerves. It is important to point out that no unmyelinated fibers have been found in the ALT

(136). Thus, transmission in the ALT of nociceptive input from peripheral C fibers is by myelinated axons exclusively.

Rostral Course of ALT The rostral course of ALT axons has been studied using the antidromic stimulation method, and has been found to involve the brainstem reticular formation, the medial and intralaminar thalamic nuclei, and the region of the ventrobasal and the posterior thalamic nuclei (91, 130–133). Here again a species difference appeared: in the monkey, but not in the cat, a large proportion of the ALT axons reach the postero-lateral thalamus. This is in agreement with degeneration studies, suggesting the evolution of a neospinothalamic tract in primates (181). From lesion experiments the neospinothalamic tract has been claimed to be part of the lemniscal system, i.e., to contribute to the discriminative aspects of tactile sensations.

These loci of termination of the ALT have all been postulated to be involved in the processing of noxious stimuli (137–142). One would, however, expect to find different properties in ALT neurons terminating in regions presumed to be associated with the lemniscal system from those terminations in extralemniscal brain regions. So far only one investigation has been published on this aspect, reporting that many ALT neurons projecting into the brainstem reticular formation had large, often ipsilateral or even bilateral receptive fields. This suggests a predominance of extralemniscal features (132). However, the authors admitted that it was premature to draw conclusions from this first study.

Polymodal Neurons and Pain in Man Recently an important contribution to the understanding of the role of the two basic types of nociceptive neurons came from a comparative study on the ALT in man and monkey (106, 144). The thresholds for sensation of pain produced by focal electrical stimulation of the ALT have been measured in conscious patients, undergoing percutaneous cordotomy. Both double pulse stimuli with varying time separations, and single pulse stimuli, have been used. The measurements of threshold for pain sensations under these conditions yielded estimates for the refractory period of the axons whose stimulation was sufficient to evoke these liminal painful sensations.

A study of the thresholds and refractory periods of single ALT units in monkeys (106) revealed that the values of these characteristics for polymodal units were approximately the same as the estimated values for the units producing liminal pain sensations in man. The specific nociceptive units found in the monkey had higher thresholds and a longer refractory period. This evidence therefore suggests that the excitation of polymodal units is sufficient to evoke pain in man; though it does not exclude the possibility that specific nociceptive cells contribute to the pain produced by stronger electrical stimulation of the ALT. The different properties of the two types of unit are due to the differences in their axon diameters. Axons of specific nociceptive cells in the ALT have in fact been found to have lower conduction velocities than those of polymodal cells (91, 106).

Other Spinal Ascending Pathways

Spinocervical Tract (SCT) This tract, which runs in the dorsolateral funiculus of the spinal cord, is found in cats and primates to contain axons from lumbar dorsal horn cells (118, 145–148). The SCT has a synaptic relay in the lateral cervical nucleus, which in turn projects into various contralateral thalamic nuclei.

The SCT neurons are activated by group II afferents from low threshold cutaneous mechanoreceptors. In general more than one type of afferent converges upon the cells, the relative contribution of each type of input being controlled by influences descending from the brain (118) (section "Supraspinal Inhibition of Dorsal Horn Neurons").

About 60% of the SCT neurons have the same characteristics as polymodal dorsal horn neurons (section "Polymodal Neurons"), as judged from activation by electrical C fiber stimuli (104, 122), and radiant heat applied to the skin (118). Apparently, these polymodal SCT neurons do not differ from the non-projecting dorsal horn cells (section "Polymodal Neurons"), neither in the frequency of occurrence, nor in the types of excitatory input.

The specifically nociceptive neurons of the marginal zone of the dorsal horn (section "Specifically Nociceptive Neurons") do not project into the SCT. However, a few units located deep in the dorsal horn have been found which required "more vigorous" manipulations in the receptive field to be excited. The exact identity of these units was, however, not fully established, and it is possible that they did not only have a nociceptive cutaneous input. They could have had a predominant input from muscle (being for example spino-cerebellar tract neurons), or even have been descending fibers excited via a supraspinal loop, being silent in the spinalized animal (118, 149).

The conclusion from studies of the effects of SCT lesions on the behavior of cats and dogs was that this pathway is involved in tactile discrimination (155, 156). However, the large proportion of SCT cells of the polymodal type, i.e., responding to C fibers and heat nociceptors, suggests that this tract may also be involved in nociception. There is an occasional observation of a decreased behavioral responsiveness to harmful stimuli in cats, after the dorsolateral quadrant of the spinal cord has been transected (150, 151).

Systematic work should be done, particularly in primates, to evaluate the significance of this pathway for nociception.

Propriospinal Pathways As early as 1854 Schiff (152) had suggested that axons in the gray matter of the spinal cord might transmit painful stimuli. This idea was later used to explain the observation in man that chronic pain often reappears several months after a cordotomy (even bilateral) has been performed.

Support for this hypothesis came from behavioral experiments on cats and rats, in which two spinal hemisections were made in either half of the cord, separated by various distances (153, 154). When the separation of the hemisections was more than four spinal segments, nocifensive reactions (e.g., vocalizations) occurred on noxious stimulation of the dermatomes of segments caudal to the lesions. This indicates a multisynaptic ascending pathway via short propriospinal axons, establishing multiple connections which cross back and forth between the two spinal cord halves.

Most probably fibers in the white matter contribute to this diffuse pathway; degeneration studies have shown that the bulk of axons in the spinal white matter are short propriospinal axons (158, 218).

The tract of Lissauer, located in the dorsolateral quadrant immediately lateral to the dorsal root entry, consists mainly of nonmyelinated fibers (88). Some of them are primary afferent fibers, the others are axons of the small neurons in the substantia gelatinosa. From lesion experiments it was concluded that this tract carries tonic inhibitory and excitatory influences from one segment to another at distances of up to five spinal segments. These influences affect cutaneous input. This suggestion has not yet been substantiated by neurophysiological experiments.

Supraspinal Neurons

It has been postulated that the extralemniscal, or anterolateral, parts of the somatosensory system in the brain are involved in nociception and pain (80, 81, 137–139, 160). However, the reports on the characteristics of neurons, the so-called extralemniscal neurons, found in these brain areas have produced conflicting conclusions about their role in nociception. The present state of knowledge about the driving of supraspinal cells by noxious stimuli, and about the interactions of these responses with other inputs, is much less than it is for dorsal horn cells (sections "Spinal Neurons Involved in Nociception" and "Inhibition of Dorsal Horn Neurons"). Below some reports on neuron responses in various regions of the brain upon peripheral noxious stimulation are reviewed.

Brainstem Reticular Formation and Nonspecific Thalamus Noxious stimulation and/or electrical stimulation of nerves at Aδ or C strength excite neurons in the reticular formation (178) of the medulla (161–167, 169, 171) and of the midbrain (169, 170, 177), and in the intralaminar and medial nuclei of the thalamus (172–176). They receive input from parts of the ALT, the spinoreticular, and the palaeo-spinothalamic tracts. The long latencies of the responses after electrical nerve stimulation (168) suggest that there might be other pathways (e.g., propriospinal (162), cf. section "Other Spinal Ascending Pathways") which contribute to their excitation.

The available information about their response characteristics is not sufficient for a further functional subdivision of these units. Usually, they have large receptive fields, often bilateral, even comprising the whole body surface. This suggests a much greater spatial convergence than is found in the spinal cord. No somatotopic organization has been found. Most of the units are spontaneously active. Some can be excited by auditory and/or visual stimuli, in addition to their somatic input via the spinal cord or the trigeminal nerve; these have therefore been described as polysensory cells.

Neurons have been reported which are excited by noxious stimuli only, and others which respond to low intensity mechanical stimuli, but which require noxious stimulation for maximum response. The relative proportions of these two classes of neurons as given by the various authors are, however, very different: for example, the purely nociceptive group in the brainstem reticular formation has been reported to be 0% (168), 8% (171), 29% (167), 66% (164), and 70% (163) of the total population.

As a result, different conclusions have been drawn by the various authors. Some interpret the existence of the exclusively nociceptive neurons as evidence for central specificity (163), in the sense of von Frey's definition. Others hypothesize that a noxious stimulus is signaled by the polymodal neurons in the reticular formation when a certain level of activity is exceeded (166, 169, 170, 173), this being an adoption of Goldscheider's intensity hypothesis (cf. section "Pain Theories"). These, and other discrepancies, might be explained by differences in species (rat, cat, monkey), in the methods used for unit sampling and for testing the type of the afferent excitatory input, and in the anesthesia.

Investigations in Unanesthetized Animals Anesthesia may produce specifically nociceptive units, as have been shown by experiments on the conscious monkey (173). Most, if not all, of the units in the medial and intralaminar thalamic nuclei were found to respond to light tactile stimuli. However, many of these units were converted into purely nociceptive neurons by a sedative dose of barbiturate. This finding is of crucial importance in all studies on nociception, particularly at the supraspinal level.

Therefore another approach should be pursued (171, 179, 194): recording with implanted microelectrodes from the unanesthetized unrestrained animal. In cats neuronal activity in the medullary reticular formation was recorded during electrical stimulation at gradually increasing intensities, eventually leading to an escape response of the animal (Figure 10). It was found that the activity in many neurons was correlated in a complex operational manner to the escape response. For example, when the electrical stimulus reached an intensity sufficient to elicit escape, the responses of the units were greatly prolonged, reminiscent of the focusing of attention. The units did not respond to either active or passive movements of the limbs in the absence of electrical stimulation. Electrical stimulation through the recording microelectrode also elicited escape (179).

It would be desirable if this method could be used in the future in various regions of the brain, in combination with well controlled adequate noxious stimuli (e.g., as in Figure 1), instead of the electrical nerve shocks. Hopefully this will greatly enhance the understanding of the interplay of neurons, or neuronal populations, as the basis for nocifensive behavior.

According to guiding principles compiled recently by Wall (180), suffering of the animals must be prevented by the design of the experiments.

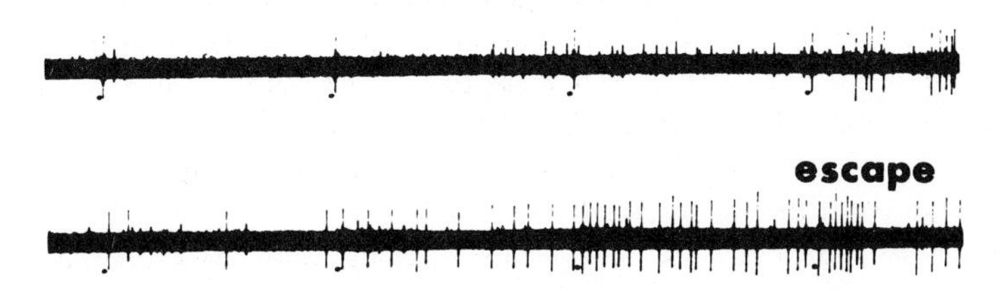

Figure 10. Neurons in the medullary reticular formation of the awake cat. The traces (*continuous recording*) show the discharge of a neuron (*large spike*) to electrical stimuli (*dots*) delivered at 1-s intervals to the superficial radial nerve. The intensity was constant at a level eliciting escape, which occurred as indicated in the lower record. Initially, there was one spike per stimulus (*upper trace*). Unit responded to natural noxious stimuli. (From Casey (171).)

The Posterior Nuclear Group Comparative anatomical studies indicate that the lateral part of the anterolateral tract (ALT, cf. section "The Anterolateral Tract") has undergone a large increase in size during evolution (181). This neospinothalamic tract projects predominantly onto the posterior nuclear group (PO), a region situated between the ventrobasal complex (VB) and the geniculate bodies.

Because of the well known analgesic effect of ALT transection (cordotomy) there have been speculations that the PO is a thalamic center for pain. Since the PO in turn projects to the second somatosensory cortex (S II), it has sometimes been assumed that this area is responsible for the cortical representation of pain (cf. section "Cortex Involvement in Nociception").

In an early neurophysiological investigation (182) the PO neurons in the cat were found to have properties distinctly different from those in the ventrobasal complex (VB) (183). These properties of the PO neurons were then defined as anterolateral, in contrast to the lemniscal properties of the VB neurons. The finding that 60% of the PO units could be excited only by noxious stimuli supported the original hypothesis of the PO being a "pain center."

Recent neurophysiological (184–187) and anatomical (188–192) investigations have however led to a re-evaluation of the extent and types of somatosensory input to the PO. Only the medial division of the PO, the PO_m, was found to be a discrete somatosensory nucleus. A most unexpected finding was that the predominant spinal input to PO_m neurons was from the dorsal columns (186, 187), and from the dorsolateral funiculi (i.e., the SCTs). The excitation produced by ALT input was estimated to be about 12% of that produced by the dorsal columns; the corresponding values for the SCTs were 40% (contralateral) and 20% (ipsilateral), respectively (186). Thus, both PO_m and VB receive their major input from the dorsal columns and the SCT(s), a finding which does not agree with the duality of ascending projections suggested in previous work (182, 183).

There is less disagreement about the general exteroceptive characteristics of the PO_m neurons: large receptive fields, often bilateral; no somatotopic arrangement; most units excited by hair movement; some by convergent auditory input; lower synaptic security than the lemniscal neurons.

However, in contrast to the investigation mentioned above (182), only 5 of 165 units (3%) required noxious stimulation for excitation, when the search was restricted to PO_m (184). In one of these units it was shown, by tract transections, that the major nociceptive input was via the contralateral SCT. These findings led the authors to state that the somatic part of the PO cannot be considered to constitute a "pain center."

The Ventrobasal Complex The notion that the ventrobasal complex (VB) contains only lemniscal neurons (183) has been challenged repeatedly. Neurons have been found which have properties similar to those described in the PO_m (187, 193, 194). The major input to this type of VB units was via the dorsal column nuclei. It has been suggested (187) that this discrepancy is due to a sampling bias, the larger microelectrode tips used in the early studies (182) having selected the presumably large neurons with lemniscal properties. According to this suggestion the nonlemniscal neurons found in the VB would be small neurons, many of which presumably do not project onto the cortex (195).

Reports of a nociceptive input to VB neurons are exceptional. In a study on cats, VB units were classified (196) according to the minimum strength of electrical stimulation of the saphenous nerve required for excitation, as "$\alpha\beta\gamma$" or "δ." 8% in the $\alpha\beta\gamma$, and 50% in the δ class of VB neurons could only be excited with noxious skin stimuli (pinch, pinprick). Unfortunately it is not clear whether all these units were actually in the VB.

After chronic partial transection of the spinal cord in monkeys and cats, which spared the ALT and probably the cervico-thalamic pathway (i.e., the continuation of the contralateral SCT), 6 out of 75 VB units were found to require noxious stimuli (pinch, heat) to well circumscribed sites on the distal extremities (174). It would be interesting to know whether these responses would have been concealed by dorsal column input, when present. This question could best be answered by performing a reversible block of the dorsal columns (e.g., by local cooling (103)).

A population of neurons has been reported to exist in the rat's VB which responded exclusively to heating the tail (197). The mean threshold temperature of 16 such units was 42.6°C (± 0.6°C), the discharge rates increase with temperature up to a maximum at 45.3°C (± 0.4°C). All the units also responded to noxious mechanical stimulation of the tail, but not to light mechanical stimulation. No data on the localization of these neurons are available so far. It is important to know whether they are from the VB, or from other thalamic regions.

So, despite the fact that many neurons exist in the VB which have properties which are not lemniscal, the few reports of nociceptive neurons need to be confirmed.

Cortex Involvement in Nociception Electrical stimulation predominantly of the somatosensory cortex in conscious patients (198–200) evoked paresthesias or, at liminal stimulus intensities, sensations with a natural quality (200). Only in rare cases were weak painful sensations reported (i.e., in 11 out of 426 stimulations (199)). In conscious monkeys cortical stimulation never produced nocifensive behavior, which could be elicited from many subcortical regions and also by peripheral stimulation (201). A variety of effects have been reported to follow ablations in the somatosensory fields SI and SII, and these have been reviewed extensively (138, 139, 198, 202). SI lesions either have no effect on the perception of noxious stimuli, and on chronic pain, or they may even result in a hyperpathia; the latter suggests release from descending inhibition, either in the thalamus or in the spinal cord (cf. context of Figure 9*E*). There is some dispute about whether or not relief from chronic pain may be produced when lesions include SII.

Nociceptive Neurons in SII The anatomical connections (both ascending and descending) of SII with the PO of the thalamus (191), and hence with the neospinothalamic tract, have been used as an argument to suggest that SII plays a part in nociception. Single neuron analysis (185) has revealed that 18% of the PO_m neurons in the cat project to SII, and that 92% were influenced trans-synaptically by stimulation in SI and/or SII (inhibitory or excitatory, or both).

The majority of SII cells, in the anesthetized cat (203) and in the conscious monkey (204), responded to light tactile skin stimulation. Despite the fact that they are both arranged in topically organized columns there are considerable differences between the response characteristics of SII and SI units (205): SII neurons are less reliable at encoding the spatial and temporal aspects of stimuli. It has been proposed that these characteristics result from a concatenation of lemniscal and anterolateral features (204). As in the PO_m, the units have a major input from the dorsal columns (206) and the VB (207). The extent of ALT input has not yet been determined, which is certainly an important question for the nociceptive units.

A small number of SII neurons were found which did not respond unless the stimulation reached a noxious level: only 12 out of 448 units (2.7%) in the cat required noxious stimuli to the skin or periosteum for excitation (203); the proportion in the monkey (204) was not given. They were found particularly near the zone of transition into the second auditory area, a region in which there are many polysensory units, receiving somatic, auditory, and even visual input.

In another study on neurons in the SII forepaw projection of the cat (208) one of 457 units required noxious skin stimuli for activation, although some units could not be excited at all by adequate stimulation.

A response of a nociceptive SII neuron to intense radiant heat is shown in Figure 11. It is evident that the time course of the response (delayed onset of discharge, prolonged afterdischarge) is similar to that seen in polymodal dorsal horn cells (Figures 7*D*, 8, 9, *A* and *B*). The authors (203) stated that both the peak frequency and the duration of the afterdischarge increased as the intensity and duration of the stimulus were increased. Unfortunately, in this documentation of a noxious input, which is exceptionally rare in studies on supraspinal neurons, the skin temperature was not measured, which prevented any quantitative study. The same neuron, which was studied for many hours, also responded to pinching in all regions of the body.

Nociceptive Neurons in Other Areas of the Cortex It is well established that the SI area provides a highly ordered, place and modality specific image of the body surface, whose input comes almost exclusively from low threshold mechanoreceptors. Nociceptive neurons have very rarely been observed. In the monkey, 12 out of 593 (2%) units with cutaneous input responded exclusively to noxious stimulation (209). After transection of the dorsal half of the spinal cord, which interrupts the SCTs and the dorsal columns, many SI neurons still respond to low threshold cutaneous stimuli (210); however, a greater percentage now required squeezing and pinching. These findings are concordant with those obtained in the VB under similar conditions (174).

Electrical stimulation of afferent Aδ fibers from the tooth pulp, which is claimed to provide a purely nociceptive input, gave rise to evoked potentials in the region of overlapping SI and SII facial and lingual projection (211). The

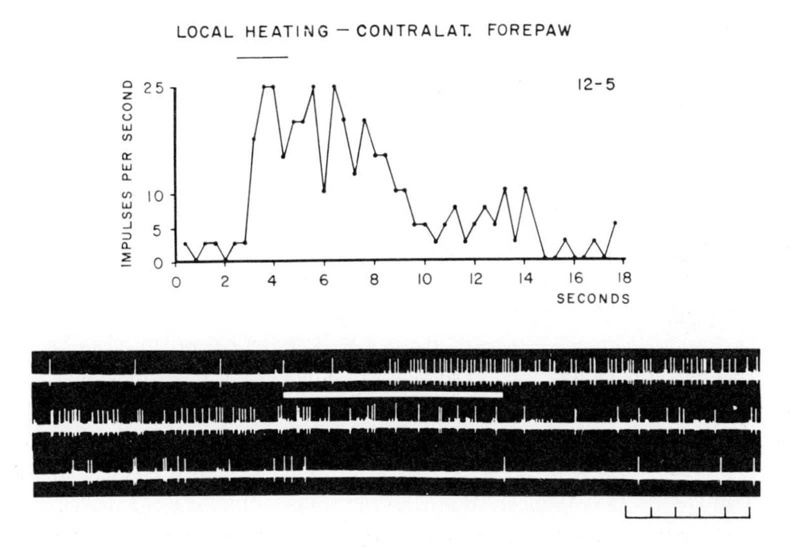

Figure 11. Nociceptive neuron of the cat SII cortex. The effect of radiant heat to the contralateral forefoot is shown. *Upper,* graph of the discharge frequency plotted against time. *Lower,* continuous oscillographic record of the discharge. Horizontal lines indicate duration of the stimulus. The time marker is 0.2 s between each vertical line. The unit was activated by pinching from all body regions, but not by non-noxious stimuli. (From Carreras and Anderson (203).)

evoked activity extended into the region around the sulcus orbitalis. This is part of the orbital cortex, which has been claimed to be non-primary cortex with polysensory convergence (80). In this context it may be stated that localization in neurophysiological work on cat's cortex is often rather poor. Reference should be made to a comprehensive cytoarchitectonic map (212), for localization and nomenclature.

From a few units (211) it became apparent that the tooth pulp stimulation produced activity in cells which could also be excited by stimulation of low threshold fibers of the infraorbital nerve. The convergence of these two inputs occurred in the trigeminal nucleus of the brainstem (211). Therefore the cortical units do not respond specifically to the tooth pulp afferents. Furthermore, in the conscious cat stimulation of these afferents does not elicit signs of discomfort (211); therefore it may be questioned whether this and other neurophysiological work using tooth pulp stimulation is related to nociception at all.

Corticofugal Control SI and SII cortex exert descending influences, both excitatory and inhibitory, on most synaptic stations of afferent activity, including the spinal cord, the dorsal column nuclei, the medial thalamus as well as VB and PO (120–124, 185, 191). The example given (section "Supraspinal Inhibition of Dorsal Horn Neurons") of the differential influence on polymodal dorsal horn cells might illustrate the functional capabilities of such corticofugal controls for the processing of noxious stimuli. However, no investigations have been done so far which contribute to our understanding of the corticothalamic interactions in nociception.

CONCLUSIONS

So where is the neuronal substrate for nociception, the "pain system"? This review made an encouraging start which supported the ideas of specificity, in the realm of cutaneous receptors. It is quite certain that several types of specific receptors exist in the skin, and others are going to be revealed in muscles and visceral organs by careful and quantitative search.

However, most of the specificity in the periphery is not used in the central nervous system. Despite the existence of some central neurons which respond, more or less exclusively, to noxious stimuli, it is obvious that information about these stimuli appears in neurons which also have non-noxious input. These constitute the bulk of cells projecting to the spinal ascending tracts, and evidence is accumulating that their activation by noxious stimuli is not a marginal effect. It seems to be a general rule that the proportion of selectively nociceptive cells gets smaller, the higher one goes in the somatosensory system.

Why not ask the question about the neurophysiological basis of nociception the other way around? There are myriads of neurons: we should study with an unbiased mind what the image of the sensory surface is in this neuronal space. This is not searching for central specificity, neither in terms of receptors, nor of psychophysics. The approach is to apply a battery of standard stimuli, noxious and non-noxious, and to reconstruct the profiles of their neuronal representation. Conceptually, this approach will be similar to that used by Karl Lashley in his hunt for the engram.

After years of "pinching" and "squeezing" neurophysiologists involved in nociception have now started to use quantified stimuli. The regions of the brain where the neuronal image of nociceptive stimuli should be studied have been established. Furthermore, mechanisms of interaction have been disclosed, e.g., those of forward and backward inhibition; they might operate as gain control, differentially affecting the various inputs of a neuron population. First steps towards a correlation of neurophysiological findings with the behavioral responses of the animal have been made, by recording from neurons in conscious animals. More investigations of this type, carefully designed and preferably in unrestrained animals, are needed.

So far we have been concerned with nociception. But what do physiologists contribute to the understanding of chronic pain? A few neurophysiological models are going to be developed. Experimental neuromas (213) in animals might help to elucidate the mechanisms of those pain states which arise in the periphery. Hopefully the understanding of pain of central origin will profit greatly from the recent work on developmental neurobiology, synaptic plasticity, and regeneration in the central nervous system (214, 215), which brought a renaissance of Cannon's and Rosenblueth's idea on denervation supersensitivity (216, 217). The application of these approaches may provide one of the keys necessary for the transition from neurophysiology of nociception to pain physiology.

ACKNOWLEDGMENTS

The author is greatly indebted to Mrs. Ursula Nothoff for the bibliographic work and the typing of the manuscript, to Mrs. Almut Manisali for the graphic work, and to Mr. Simon Fleminger for improving the English. The author's work has been supported by grants from the Deutsche Forschungsgemeinschaft.

REFERENCES

1. Sherrington, C.S. (1906). The Integrative Action of the Nervous System. Yale University Press, New Haven.
2. Hardy, I.D., Wolff, H.G., and Goodell, H. (1947). Studies on pain: discrimination of differences in intensity of a pain stimulus as a basis of a scale of pain intensity. J. Clin. Invest. 26:1152.
3. Beck, P.W., Handwerker, H.O., and Zimmermann, M. (1974). Nervous outflow from the cat's foot during noxious radiant heat stimulation. Brain Res. 67:373.
4. Finger, S., and Norrsell, U. (1974). Temperature sensitivity of the paw of the cat: a behavioural study. J. Physiol. (Lond.) 239:631.
5. Kenshalo, D.R., Duncan, D.G., and Weymark, C. (1967). Thresholds for thermal stimulation of the inner thigh, footpad, and face of cats. J. Comp. Physiol. Psychol. 63:133.
6. Goldscheider, A. (1920). Das Schmerzproblem. Springer, Berlin.
7. Sinclair, D.C. (1955). Cutaneous sensation and the doctrine of specific energy. Brain 78:584.
8. Weddell, G. (1955). Somesthesis and the chemical senses. Annu. Rev. Psychol. 6:119.
9. von Frey, M. (1869). Untersuchungen über die Sinnesfunctionen der menschlichen Haut. I. Druckempfindung und Schmerz. Abh. der Kgl. sächs. Ges. d. Wissensch. 23:175.
10. Burgess, P.R. (1974). Patterns of discharge evoked in cutaneous nerves and their significance for sensation. *In* J.J. Bonica (ed.), Advances in Neurology, Vol. 4, pp. 11–18. Raven Press, New York.
11. Kleist, K. (1934). Gehirnpathologie. Barth, Leipzig.
12. Hebb, D.O. (1955). Drives and the C.N.S. (conceptual nervous system). Psychol. Rev. 62:243.
13. Sternbach, R.A. (1968). Pain. A Psychological Analysis. Academic Press, New York and London.
14. Noordenbos, W. (1959). Pain. Elsevier, Amsterdam.
15. Melzack, R., and Wall, P.D. (1965). Pain mechanisms: a new theory. Science 150:971.
16. Schmidt, R.F. (1971). Presynaptic inhibition in the vertebrate central nervous system. Ergebn. Physiol. 63:21.
17. Mendell, L.M., and Wall, P.D. (1964). Presynaptic hyperpolarization: a role for fine afferent fibres. J. Physiol. (Lond.) 172:274.
18. Burke, R.E., Rudomin, P., Vyklicky, L., and Zajac III, F.E. (1971). Primary afferent depolarization and flexion reflexes produced by radiant heat stimulation of the skin. J. Physiol. (Lond.) 213:185.
19. Franz, D.N., and Iggo, A. (1968). Dorsal root potentials and ventral root reflexes evoked by nonmyelinated fibers. Science 162:1140.
20. Gregor, M., and Zimmermann, M. (1973). Dorsal root potentials produced by afferent volleys in cutaneous Group III fibres. J. Physiol. (Lond.) 232:413.
21. Jänig, W., and Zimmermann, M. (1971). Presynaptic depolarization of myelinated afferent fibres evoked by stimulation of cutaneous C fibers. J. Physiol. (London) 214:29.

22. Mendell, L. (1972). Properties and distribution of peripherally evoked presynaptic hyperpolarization in cat lumbar spinal cord. J. Physiol. (Lond.) 226:769.
23. Whitehorn, D., and Burgess, P.R. (1973). Changes in polarization of central branches of myelinated mechanoreceptor and nociceptor fibers during noxious and innocuous stimulation of the skin. J. Neurophysiol. 36:226.
24. Zimmermann, M. (1968). Dorsal root potentitals after C fiber stimulation. Science 160:896.
25. Jänig, W., Schmidt, R.F., and Zimmermann, M. (1968). Two specific feedback pathways to the central afferent terminals of phasic and tonic mechanoreceptors. Exp. Brain Res. 6:116.
26. Schmidt, R.F., and Weller, E. (1970). Reflex activity in the cervical and lumbar sympathetic trunk induced by unmyelinated somatic afferents. Brain Res. 24:207.
27. Collins, W.F., Nulsen, F.E., and Randt, C.T. (1960). Relation of peripheral nerve fiber size and sensation in man. Arch. Neurol. 3:381.
28. Pattle, R.E., and Weddell, G. (1948). Observations on electrical stimulation of pain fibres in an exposed human sensory nerve. J. Neurophysiol. 11:93.
29. Torebjørk, H.E., and Hallin, R.G. (1973). Perceptual changes accompanying controlled preferential blocking of A and C fibre responses in intact human skin nerves. Exp. Brain Res. 16:321.
30. van Hees, J., and Gybels, J.M. (1972). Pain related to single afferent C fibers from human skin. Brain Res. 48:397.
31. Torebjørk, H.E. (1974). Afferent C units responding to mechanical, thermal and chemical stimuli in human nonglabrous skin. Acta Physiol. Scand. 92:374.
32. Hensel, H., and Boman, K.K.A. (1960). Afferent impulses in cutaneous sensory nerves in human subjects. J. Neurophysiol. 23:564.
33. Konietzny, F., and Hensel, H. (1975). Warm fiber activity in human skin nerves. Pflügers Arch. 359:265.
34. Adrian, E.D., and Zotterman, Y. (1926). The impulses produced by sensory nerve endings. Part 3. Impulses set up by touch and pressure. J. Physiol. (Lond.) 61:465.
35. Zotterman, Y. (1936). Specific action potentials in the lingual nerve of cat. Scand. Arch. Physiol. 75:106.
36. Zotterman, Y. (1939). Touch, pain and tickling: an electrophysiological investigation on cutaneous sensory nerves. J. Physiol. (Lond.) 95:1.
37. Beck, P.W., and Handwerker, H.O. (1974). Bradykinin and serotonin effects on various types of cutaneous nerve fibres. Pflügers Arch. 347:209.
38. Bessou, P., and Perl, E.R. (1969). Response of cutaneous sensory units with unmyelinated fibers to noxious stimuli. J. Neurophysiol. 32:1025.
39. Burgess, P.R., and Perl, E.R. (1967). Myelinated afferent fibres responding specifically to noxious stimulation of the skin. J. Physiol. (Lond.) 190:541.
40. Iggo, A. (1962). Non-myelinated visceral, muscular and cutaneous afferent fibres and pain. *In* C.A. Keele and R. Smith (eds.), The Assessment of Pain in Man and Animals, pp. 74–87. E. & S. Livingstone Ltd., London and Edinburgh.
41. Perl, E.R. (1968). Myelinated afferent fibres innervating the primate skin and their response to noxious stimuli. J. Physiol. (Lond.) 197:593.
42. Iggo, A. (1959). Cutaneous heat and cold receptors with slowly conducting C afferent fibres. Quart. J. Exp. Physiol. 44:362.
43. Iggo, A., and Ogawa, H. (1971). Primate cutaneous thermal nociceptors. J. Physiol. (Lond.) 216:77P.
44. Iriuchijima, J., and Zotterman, Y. (1960). The specificity of afferent cutaneous C fibres in mammals. Acta Physiol. Scand. 49:267.
45. Witt, I. (1962). Aktivität einzelner C-Fasern bei schmerzhaften und nicht schmerzhaften Hautreizen. Acta Neuroveg. 24:208.
46. Hardy, I.D., Wolff, H.D., and Goodell, H. (1952). Pain Sensations and Reactions. Williams & Wilkins Company, Baltimore, Md.
47. Torebjørk, H.E., and Hallin, R.G. (1974). Identification of afferent C units in intact human skin nerves. Brain Res. 67:387.
48. Burgess, P.R., and Perl, E.R. (1973). Cutaneous mechanoreceptors and nociceptors. *In* A. Iggo (ed.), Handbook of Sensory Physiology, Vol. II, pp. 30–78. Springer, Berlin, Heidelberg, New York.

49. Iggo, A. (ed.) (1973). Handbook of Sensory Physiology, Vol. II, Somatosensory System. Springer, Berlin, Heidelberg, New York.
50. Floyd, K., Hick, V.E., and Morrison, J.F.B. (1975). The responses of visceral afferent nerves to bradykinin. J. Physiol. (Lond.) 247:53 p.
51. Dickhaus, H., Zimmermann, M., and Zotterman, Y. (1975). Neurophysiological investigations of the development of nociceptors in regenerating cutaneous nerves of the cat. Presented at the 1st Conference of Int. Assoc. Stud. Pain, Sept. 5–8, Florence, Italy.
52. Head, H. (1920). Studies in Neurology. Oxford University Press, London.
53. Trotter, W., and Davies, H.M. (1908). Experimental studies in the innervation of the skin. J. Physiol. (Lond.) 38:134.
54. Neil, E. (ed.) (1972). Handbook of Sensory Physiology, Vol. III/1, Enteroceptors. Springer, Berlin, Heidelberg, New York.
55. Iggo, A. (1966). Physiology of visceral afferent systems. Acta Neuroveget. 28:121.
56. Leek, B.F. (1972). Abdominal visceral receptors. *In* E.Neil (ed.), Handbook of Sensory Physiology, Vol. III/1, pp. 113–160, Springer, Berlin, Heidelberg, New York.
57. Iggo, A., and Leek, B.F. (1970). Sensory receptors in the ruminant stomach and their reflex effects. *In* A.T. Phillipson (ed.), Physiology of Digestion and Metabolism in the Ruminant, pp. 23–34. Oriel Press, Newcastle.
58. Head, H. (1893). On disturbances of sensation, with especial reference to the pain of visceral disease. Brain 16:1.
59. Kuntz, A., and Haselwood, L.A. (1940). Circulatory reactions in gastro-intestinal tract elicited by local cutaneous stimulation. Amer. Heart J. 20:743.
60. Wood, J.D. (1975). Neurophysiology of Auerbach's plexus and control of intestinal motility. Physiol. Rev. 55:307.
61. Paintal, A.S. (1973). Vagal sensory receptors and their reflex effects. Physiol. Rev. 53:159.
62. Fillenz, M., and Widdicombe, J.G. (1972). Receptors of the lungs and airways. *In* E. Neil (ed.), Handbook of Sensory Physiology, Vol. III, pp. 81–112. Springer, Berlin, Heidelberg, New York.
63. Paintal, A.S. (1960). Functional analysis of Group III afferent fibres of mammalian muscles. J. Physiol. (Lond.) 152:250.
64. Bessou, P., and Laporte, Y. (1960). Activation des fibres afférentes myelinisées de petit calibre d'origine musculaire (fibres du Groupe III). C.R. Soc. Biol. (Paris) 154:1093.
65. Bessou, P., and Laporte, Y. (1961). Étude des recepteurs musculaires innervés par les fibres afférentes du Groupe III (fibres myelinisées fines) chez le chat. Arch. Ital. Biol. 99:293.
66. Iggo, A. (1961). Non-myelinated afferent fibres from mammalian skeletal muscle. J. Physiol. (Lond.) 155:52.
67. Mense, S., and Schmidt, R.F. (1974). Activation of Group IV afferent units from muscle by algesic agents. Brain Res. 72:305.
68. Bessou, P., and Laporte, Y. (1958). Activation des fibres afférentes amyeliniques d'origine musculaire. C.R. Soc. Biol. (Paris) 152:1587.
69. Kalia, M., Senapati, J.M., Parida, B., and Panda, A. (1972). Reflex increase in ventilation by muscle receptors with nonmedullated fibers (C fibers). J. Appl. Physiol. 32:189.
70. Guzman, F., Braun, C., and Lim, R.K.S. (1962). Visceral pain and the pseudoaffective response to intra-arterial injection of bradykinin and other algesic agents. Arch. Int. Pharmacodyn. Ther. 136:353.
71. White, J.D. (1957). Cardiac pain. Anatomic pathways and physiologic mechanisms. Circulation 16:644.
72. Uchida, Y., and Murao, S. (1974). Excitation of afferent cardiac sympathetic nerve fibers during coronary occlusion. Amer. J. Physiol. 226:1094.
73. Brown, A.M. (1967). Excitation of afferent cardiac sympathetic nerve fibres during myocardial ischaemia. J. Physiol. (Lond.) 190:35.
74. Keele, C.A., and Armstrong, D. (1964). Substances Producing Pain and Itch. E. Arnold Ltd., London.
75. Lim, R.K.S. (1970). Pain. Annu. Rev. Physiol. 32:269.

76. Lindahl, O. (1961). Experimental skin pain. Induced by injection of water-soluble substances in humans. Acta Physiol. Scand. 51: Suppl. 179, 1.
77. Fjällbrant, N., and Iggo, A. (1961). The effect of histamine, 5-hydroxytryptamine and acetylcholine on cutaneous afferent fibres. J. Physiol. (Lond.) 156:578.
78. Uchida, Y., and Murao, S. (1974). Potassium induced excitation of afferent cardiac sympathetic nerve fibers. Amer. J. Physiol. 226:603.
79. Movat, H.Z. (1972). Chemical mediators of the vascular phenomena of the acute inflammatory reaction and of immediate hypersensitivity. Med. Clin. N. Amer. 56:541.
80. Albe-Fessard, D., and Besson, J.M. (1973). Convergent thalamic and cortical projections–the non-specific system. *In* A. Iggo (ed.), Handbook of Sensory Physiology, Vol. II, pp. 489–560, Springer, Berlin, Heidelberg, New York.
81. Bowsher, D., and Albe-Fessard, D. (1965). The anatomo-physiological basis of somatosensory discrimination. *In* C.C. Pfeiffer and J. R. Smythies (eds.), International Review of Neurobiology, pp. 35–75. Academic Press, New York, London.
82. Mountcastle, V.B. (1961). Some functional properties of the somatic afferent system. *In* W.A. Rosenblith (ed.), Sensory Communication, pp. 403–436. MIT Press, Cambridge, Mass.
83. Wall, P.D. (1970). The sensory and motor role of impulse travelling in the dorsal columns towards cerebral cortex. Brain 93:505.
84. Darian-Smith, I. (1973). The trigeminal system. *In* A. Iggo (ed.), Handbook of Sensory Physiology, Vol. II, pp. 271–314. Springer, Berlin, Heidelberg, New York.
85. Rexed, B. (1952). The cytoarchitectonic organization of the spinal cord in the cat. J. Comp. Neurol. 96:415.
86. Scheibel, M.E., and Scheibel, A.B. (1968). Terminal axonal patterns in cat spinal cord. II. The dorsal horn. Brain Res. 9:32.
87. Scheibel, M.E., and Scheibel, A.B. (1969). Terminal patterns in cat spinal cord. III. Primary afferent collaterals. Brain Res. 13:417.
88. Réthelyi, M., and Szentágothai, J. (1973). Distribution and connections of afferent fibres in the spinal cord. *In* A. Iggo (ed.), Handbook of Sensory Physiology, Vol. II, pp. 207–252. Springer, Berlin, Heidelberg, New York.
89. Christensen, B.N., and Perl, E.R. (1970). Spinal neurons specifically excited by noxious or thermal stimuli: marginal zone of the dorsal horn. J. Neurophysiol. 33:293.
90. Cajal, S.R. (1909). Histologie du Système Nerveux de l'Homme et des Vertébrés. Tome 2. Maloine, Paris.
91. Willis, W.D., Trevino, D.L., and Coulter, J.D. (1974). Responses of primate spinothalamic tract neurons to natural stimulation of hindlimb. J. Neurophysiol. 37:358.
92. Kumazawa, T., Perl, E.R., Burgess, P.R., and Whitehorn, D. (1975). Ascending projections from marginal zone (Lamina I) neurons of the spinal dorsal horn. J. Comp. Neurol. 162:1.
93. Handwerker, H.O., Iggo, A., Ogawa, H., and Ramsey, R.L. (1975). Input characteristics and rostral projection of dorsal horn neurons in the monkey. J. Physiol. (Lond.) 244:76P.
94. Mosso, J.A., and Kruger, L. (1973). Receptor categories represented in spinal trigeminal nucleus caudalis. J. Neurophysiol. 36:472.
95. Hillman, P., and Wall, P.D. (1969). Inhibitory and excitatory factors influencing the receptive fields of lamina 5 spinal cord cells. Exp. Brain Res. 9:284.
96. Kolmodin, G.M., and Skoglund, C.R. (1960). Analysis of spinal interneurons activated by tactile and nociceptive stimulation. Acta Physiol. Scand. 50:337.
97. Wall, P.D. (1960). Cord cells responding to touch, damage, and temperature of skin. J. Neurophysiol. 23:197.
98. Eisenman, J., Landgren, S., and Novin, D. (1963). Functional organization in the main sensory trigeminal nucleus and in the rostral subdivision of the nucleus of the spinal trigeminal tract in the cat. Acta Physiol. Scand. 59:214.
99. Gregor, M., and Zimmermann, M. (1972). Chacteristics of spinal neurons responding to cutaneous myelinated and unmyelinated fibres. J. Physiol. (Lond.) 221:555.
100. Wagman, I.H., and Price, D.D. (1969). Responses of dorsal horn cells of *Macaca mulatta* to cutaneous and sural nerve A and C fiber stimuli. J. Neurophysiol. 32:803.

101. Mendell, L.M. (1966). Physiological properties of unmyelinated fiber projection to the spinal cord. Exp. Neurol. 16:316.
102. Manfredi, M. (1970). Modulation of sensory projections in anterolateral column of cat spinal cord by peripheral afferents of different size. Arch. Ital. Biol. 108:72.
103. Handwerker, H.O., Iggo, A., and Zimmermann, M. (1975). Segmental and supraspinal actions on dorsal horn neurons responding to noxious and non-noxious skin stimuli. Pain 1:147.
104. Zimmermann, M. (1975). Cutaneous C-fibers: peripheral properties and central connections. *In* H.H. Kornhuber (ed.), The Somatosensory System, pp. 45–53. Thieme Stuttgart.
105. Price, D.D., and Browe, A.C. (1973). Responses of spinal cord neurons to graded noxious and non-noxious stimuli. Brain Res. 64:425.
106. Price, D.D., and Mayer, D.J. (1975). Neurophysiological characterization of the anterolateral quadrant neurons subserving pain in *Macaca mulatta*. Pain 1:59.
107. Wall, P.D. (1973). Dorsal horn electrophysiology. *In* A. Iggo (ed.), Handbook of Sensory Physiology, Vol. II, pp. 253–270, Springer, Berlin, Heidelberg, New York.
108. MacKenzie, J. (1893). Some points bearing on the association of sensory disorders and visceral disease. Brain 16:21.
109. Selzer, M., and Spencer, W.A. (1969). Convergence of visceral and cutaneous afferent pathways in the lumbar spinal cord. Brain Res. 14:331.
110. Hancock, M.B., Rigamonti, D.D., and Bryan, R.N. (1973). Convergence in the lumbar spinal cord of pathway activated by splanchnic nerve and hind limb cutaneous nerve stimulation. Exp. Neurol. 38:337.
111. Pomeranz, B., Wall, P.D., and Weber, W.V. (1968). Cord cells responding to fine myelinated afferents from viscera, muscle and skin. J. Physiol. (Lond.) 199:511.
112. Fields, H.L., Meyer, G.A., and Partridge, L.D., Jr. (1970). Convergence of visceral and somatic input onto spinal neurons. Exp. Neurol. 26:36.
113. Fields, H.L., Partridge, L.D., and Winter, D.L. (1970). Somatic and visceral receptive field properties of fibers in ventral quadrant white matter of the cat spinal cord. J. Neurophysiol. 33:827.
114. Selzer, M., and Spencer, W.A. (1969). Interactions between visceral and cutaneous afferents in the spinal cord: reciprocal primary afferent fiber depolarization. Brain Res. 14:349.
115. Hancock, M.B., Willis, W.D., and Harrison, F. (1970). Viscerosomatic interactions in lumbar spinal cord of the cat. J. Neurophysiol. 33:46.
116. Bonica, J.J. (ed.) (1974). Advances in Neurology, Vol. 4. Raven Press, New York.
117. Zieglgänsberger, W., and Herz, A. (1971). Changes of cutaneous receptive fields of spinocervical-tract neurons and other dorsal horn neurones by microelectrophoretically administered amino acids. Exp. Brain Res. 13:111.
118. Brown, A.G., and Franz, D.N. (1969). Responses of spinocervical-tract neurones to natural stimulation of identified cutaneous receptors. Exp. Brain Res. 7:231.
119. Wall, P.D. (1967). The laminar organization of dorsal horn and effects of descending impulses. J. Physiol. (Lond.) 188:403.
120. Schmidt, R.F. (1973). Control of the access of afferent acrivity to somatosensory pathways. *In* A. Iggo (ed.), Handbook of Sensory Physiology, Vol. II, pp. 151–206. Springer, Berlin, Heidelberg, New York.
121. Towe, A.L. (1973). Somatosensory cortex: descending influences on ascending systems. *In* A. Iggo (ed.), Handbook of Sensory Physiology, Vol. II, pp. 701–718. Springer, Berlin, Heidelberg, New York.
122. Brown, A.G. (1971). Effects of descending impulses on transmission through the spinocervical tract. J. Physiol. (Lond.) 219:103.
123. Fetz, E.E. (1968). Pyramidal tract effects on interneurons in the cat lumbar dorsal horn. J. Neurophysiol. 31:69.
124. Coulter, J.D., Maunz, R.A., and Willis, W.D. (1974). Effects of stimulation of sensorimotor cortex on primate spinothalamic neurons. Brain Res. 65:351.
125. Mayer, D.J., and Liebeskind, J.D. (1974). Pain reduction by focal electrical stimulation of the brain: an anatomical and behavioural analysis. Brain Res. 68:73.
126. Long, D.M., and Hagfors, N. (1975). Electrical stimulation in the nervous system: the current status of electrical stimulation of the nervous system for relief of pain. Pain 1:109.

127. Oliveras, J.L., Besson, J.M., Guilbaud, G., and Liebeskind, J.C. (1974). Behavioral and electrophysiological evidence of pain inhibition from midbrain stimulation in the cat. Exp. Brain Res. 20:32.
128. Foerster, O. (1927). Die Leitungsbahnen des Schmerzgefühls und die chirurgische Behandlung der Schmerzzustände. Urban & Schwarzenberg, Berlin, Wien.
129. Foerster, O. (1936). Symptomatologie der Erkrankungen des Rückenmarks und seiner Wurzeln. *In* O. Bumke and O. Foerster (eds.), Handbuch der Neurologie, Bd. 5, pp. 1–403. Springer, Berlin.
130. Trevino, D.L., Maunz, R.A., Bryan, R.N., and Willis, W.D. (1972). Location of cells of origin of the spinothalamic tract in the lumbar enlargement of cat. Exp. Neurol. 34:64.
131. Trevino, D.L., Coulter, J.D., and Willis, W.D. (1973). Location of cells of origin of spinothalamic tract in lumbar enlargement of the monkey. J. Neurophysiol. 36:750.
132. Fields, H.L., Wagner, G.M., and Anderson, S.D., (1975). Some properties of spinal neurons projecting to the medial brain-stem reticular formation. Exp. Neurol. 47:118.
133. Albe-Fessard, D., Levante, A., and Lamour, Y. (1974). Origin of spinothalamic and spinoreticular pathways in cats and monkeys. *In* J.J. Bonica (ed.), Advances in Neurology, Vol. 4, pp. 157–166. Raven Press, New York.
134. Kerr, F.W.L. (1975). The ventral spinothalamic tract and other ascending systems of the ventral funiculus of the spinal cord. J. Comp. Neurol. 159:335.
135. Kerr, F.W.L., and Lippman, H.H. (1974). The primate spinothalamic tract as demonstrated by anterolateral cordotomy and commissural myelotomy. *In* J.J. Bonica (ed.), Advances in Neurology, Vol. 4, pp. 147–156. Raven Press, New York.
136. Lippman, H.H., and Kerr, F.W.L. (1972). Light and electron microscopic study of crossed ascending pathways in the anterolateral funiculus in the monkey. Brain Res. 40:496.
137. Bloedel, J.R. (1974). The substrate for integration in the central pain pathways. Clin. Neurosurg. 21:194.
138. Hassler, R. (1960). Die zentralen Systeme des Schmerzes. Acta Neurochirurgica, 8:354.
139. Hassler, R. (1966). Die am Schmerz beteiligten Hirnsysteme und ihre gegenseitige Beeinflussung. Verhandlungen d.Deutschen Ges.f.innere Medizin. 72, Kongress, pp. 15–35. J.F. Bergmann, München.
140. Hassler, R. (1970). Dichotomy of facial pain conduction in the diencephalon. *In* R. Hassler and A.E. Walker (eds.), Trigeminal Neuralgia, pp. 123–138. Thieme, Stuttgart.
141. Bowsher, D. (1957). Termination of the central pain pathway in man: the conscious appreciation of pain. Brain 80:606.
142. Sweet, W.H. (1959). Pain. *In* J. Field and H.W. Magoun (eds.), Handbook of Physiology, Neurophysiology, Vol. 1, pp. 459–506. American Physiological Society, Washington, D.C.
143. Whitlock, D.G., and Perl, E.R. (1961). Thalamic projections of spinothalamic pathways in monkey. Exp. Neurol. 3:240.
144. Mayer, D.J., Price, D.D., and Becker, D.P. (1975). Neurophysiological characterization of the anterolateral spinal cord neurons contributing to pain perception in man. Pain 1:51.
145. Brown, A.G. (1973). Ascending and long spinal pathways: dorsal columns, spinocervical tract and spinothalamic tract. *In* A. Iggo (ed.), Handbook of Physiology, Vol. II, pp. 315–338. Springer, Berlin, Heidelberg, New York.
146. Bryan, R.N., Trevino, D.L., Coulter, J.D., and Willis, W.D. (1973). Location and somatotopic organization of the cells of origin of the spino-cervical tract. Exp. Brain Res. 17:177.
147. Bryan, R.N., Coulter, J.D., and Willis, W.D. (1974). Cells of origin of the spinocervical tract in the monkey. Exp. Neurol. 42:574.
148. Taub, A. (1964). Local, segmental and supraspinal interaction with a dorsolateral spinal cutaneous afferent system. Exp. Neurol. 10:357.
149. Gregor, M., and Zimmermann, M. (1976). Location of fibers of spinocervical tract in the dorsolateral funiculus of the cat. In preparation.
150. Kennard, M.A. (1954). The course of ascending fibers in the spinal cord of the cat essential to the recognition of painful stimuli. J. Comp. Neurol. 100:511.

151. Breazile, J.E., and Kitchell, R.L. (1968). A study of fiber systems within the spinal cord of the domestic pig that subserve pain. J. Comp. Neurol. 133:373.
152. Schiff, J.M. (1854). Sur la transmission des impressions sensitives dans la moelle épinière. C.R. Soc. Biol. 38:926.
153. Karplus, I.P., and Kreidl, A. (1925). Zur Kenntnis der Schmerzleitung im Rückenmark. Pflügers Arch.ges.Physiol., 207:134.
154. Basbaum, A.I. (1973). Conduction of the effects of noxious stimulation by short-fiber multisynaptic systems of the spinal cord in the rat. Exp. Neurol. 40:699.
155. Kitai, S.T., and Weinberg, J. (1968). Tactile discrimination study of the dorsal column-medial lemniscal system and spino-cervico-thalamic tract in cat. Exp. Brain Res. 6:234.
156. Norrsell, U. (1966). The spinal afferent pathway of conditioned reflexes to cutaneous stimuli in the dog. Exp. Brain Res. 2:269.
157. Foreman, R.D., Beall, J.E., Willis, W.D., Applebaum, A.E., and Trevino, D.L. (1975). Responses of primate spinothalamic tract neurons to electrical stimulation of hind-limb peripheral nerves. J. Neurophysiol. 38:132.
158. Szentágothai, J. (1964). Propriospinal pathways and their synapses. *In* J.C. Eccles and J.P. Schadé (eds.), Progress in Brain Research, Vol. 11, pp. 155–174. Elsevier, Amsterdam.
159. Denny-Brown, D., Kirk, E.J., and Yanagisawa, N. (1973). The tract of Lissauer in relation to sensory transmission in the dorsal horn of spinal cord in the macaque monkey. J. Comp. Neurol. 151:175.
160. Nakahama, H. (1975). Pain mechanisms in the central nervous system. Int. Anaesthesiol. Clin. 13:109.
161. Fussey, I.F., Kidd, C., and Whitwam, J.G. (1973). Activity evoked in the brainstem by stimulation of C fibres in the cervical vagus nerve of the dog. Brain Res. 49:436.
162. Wagman, I.H., and McMillan, J.A. (1974). Relationships between activity in spinal sensory pathways and "pain mechanisms" in spinal cord and brainstem. *In* J.J. Bonica (ed.), Advances in Neurology, Vol. 4, pp. 171–177, Raven Press, New York.
163. Burton, H. (1968). Somatic sensory properties of caudal bulbar reticular neurons in the cat (*Felis domestica*). Brain Res. 11:357.
164. Casey, K.L. (1969). Somatic stimuli, spinal pathways, and size of cutaneous fibers influencing unit activity in the medial medulary reticular formation. Exp. Neurol. 25:35.
165. Benjamin, R.M. (1970). Single neurons in the rat medulla responsive to nociceptive stimulation. Brain Res. 24:525.
166. Goldman, P.L., Collins, W.F., Taub, A., and Fitzmartin, J. (1972). Evoked bulbar reticular unit activity following delta fiber stimulation of peripheral somatosensory nerve in cat. Exp. Neurol. 37:597.
167. Nord, S.G., and Ross, G.S. (1973). Responses of trigeminal units in the monkey bulbar reticular formation to noxious and non-noxious stimulation of the face: experimental and theoretical considerations. Brain Res. 58:385.
168. Nyquist, J.K., and Greenhoot, J.H. (1974). A single neuron analysis of mesencephalic reticular formation responses to high intensity cutaneous input in cat. Brain Res. 70:157.
169. Besson, J.M., Guilbaud, G., and Lombard, M.C. (1974). Effects of bradykinin intra-arterial injection into the limbs upon bulbar and mesencephalic reticular unit activity. *In* J.J. Bonica (ed.), Advances in Neurology, Vol. 4, pp. 207–215. Raven Press, New York.
170. Becker, D.P., Gluck, H., Nulsen, F.E., and Jane, J.A. (1969). An inquiry into the neurophysiological basis for pain. J. Neurosurg. 30:1.
171. Casey, K.L. (1971). Responses of bulboreticular units to somatic stimuli eliciting escape behavior in the cat. Int. J. Neurosci. 2:15.
172. Albe-Fessard, D., and Kruger, L. (1962). Duality of unit discharges from cat centrum medianum in response to natural and electrical stimulation. J. Neurophysiol. 25:3.
173. Casey, K.L. (1966). Unit analysis of nociceptive mechanisms in the thalamus of the awake squirrel monkey. J. Neurophysiol. 29:727.
174. Perl, E.R., and Whitlock, D.G. (1961). Somatic stimuli exciting spinothalamic projections to thalamic neurons in cat and monkey. Exp. Neurol. 3:256.

175. Bowsher, D., Mallart, A., Petit, D., and Albe-Fessard, D. (1968). A bulbar relay to the centre median. J. Neurophysiol. 31:288.
176. Bowsher, D. (1974). Thalamic convergence and divergence of information generated by noxious stimulation. *In* J.J. Bonica (ed.), Advances in Neurology, Vol. 4, pp. 223–232. Raven Press, New York.
177. Collins, W.F., Nulsen, F.E., and Shealy, C.N. (1966). Electrophysiological studies of peripheral and central pathways conducting pain. *In* R.S. Knighton and P.R. Dumke (eds.), Pain, pp. 33–45. Little, Brown & Co., Boston.
178. Pompeiano, O. (1973). Reticular formation. *In* A. Iggo (ed.), Handbook of Sensory Physiology, Vol. II, pp. 381–488. Springer, Berlin, Heidelberg, New York.
179. Casey, K.L. (1971). Escape elicited by bulboreticular stimulation in the cat. Int. J. Neurosci. 2:29.
180. Wall, P.D. (1975). Editorial. Pain 1:1.
181. Mehler, W.R. (1966). Some observations on secondary ascending afferent systems in the central nervous system. *In* R.S. Knighton and P.R. Dumke (eds.), Pain, pp. 11–32. Little, Brown & Co., Boston.
182. Poggio, G.F., and Mountcastle, V.B. (1960). A study of the functional contributions of the lemniscal and spinothalamic systems to somatic sensibility: central nervous mechanisms in pain. Bull. Johns Hopkins Hosp. 106:266.
183. Poggio, G.F., and Mountcastle, V.B. (1963). The functional properties of ventrobasal thalamic neurons studied in unanaesthetized monkeys. J. Neurophysiol. 26:775.
184. Curry, M.J. (1972). The exteroceptive properties of neurons in the somatic part of the posterior group (PO). Brain Res. 44:439.
185. Curry, M.J. (1972). The effects of stimulating the somatic sensory cortex on single neurons in the posterior group (PO) of the cat. Brain Res. 44:463.
186. Curry, M.J., and Gordon, G. (1972). The spinal input to the posterior group in the cat. An electrophysiological investigation. Brain Res. 44:417.
187. Berkley, K.J. (1973). Response properties of cells in ventrobasal and posterior group nuclei of the cat. J. Neurophysiol. 36:940.
188. Boivie, J. (1970). The termination of the cervico-thalamic tract in the cat. An experimental study with silver impregnation methods. Brain Res. 19:333.
189. Boivie, J. (1971). The termination in the thalamus and the zona incerta of fibres from the dorsal column nuclei (DCN) in the cat. An experimental study with silver impregnation methods. Brain Res. 28:459.
190. Boivie, J. (1971). The termination of the spinothalamic tract in the cat. An experimental study with silver impregnation methods. Exp. Brain. Res. 12:331.
191. Rinvik, E. (1968). A re-evaluation of the cytoarchitecture of the ventral nuclear complex of the cat's thalamus on the basis of corticothalamic connections. Brain Res. 8:237.
192. Jones, E.G., and Powell, T.P.S. (1971). An analysis of the posterior group of thalamic nuclei on the basis of its afferent connections. J. Comp. Neurol. 143:185.
193. Jabbur, S.J., Baker, M.A., and Towe, A.L. (1972). Wide-field neurons in thalamic nucleus ventralis posterolateralis of the cat. Exp. Neurol. 36:213.
194. Baker, M.A. (1971). Spontaneous and evoked activity of neurons in the somatosensory thalamus of the waking cat. J. Physiol. (Lond.) 217:359.
195. Bava, A., Fadiga, E., and Manzoni, T. (1968). Extra-lemniscal reactivity and commissural linkages in the VPL nucleus of cats with chronic cortical lesions. Arch. Ital. Biol. 106:204.
196. Gaze, R.M., and Gordon, G. (1954). The representation of cutaneous sense in the thalamus of the cat and monkey. Quart. J. Exp. Physiol. 39:279.
197. Hellon, R.F., and Mitchell, D. (1975). Characteristics of neurons in the ventrobasal thalamus of the rat which respond to noxious stimulation of the tail. J. Physiol. (Lond.) 250:29 p.
198. Foerster, O. (1936). Sensible corticale Felder. *In* O. Bumke and O. Foerster (eds.), Handbuch der Neurologie, Bd. 6, pp. 358–448. Springer, Berlin.
199. Penfield, W., and Boldrey, E. (1937). Somatic motor and sensory representation in the cerebral cortex of man as studied by electrical stimulation. Brain 60:389.
200. Libet, B. (1973). Electrical stimulation of cortex in human subjects, and conscious sensory aspects. *In* A. Iggo (ed.), Handbook of Sensory Physiology, Vol. II, pp. 743–790. Springer, Berlin, Heidelberg, New York.

201. Delgado, J.M.R. (1955). Cerebral structures involved in transmission and elaboration of noxious stimulation. J. Neurophysiol. 18:261.
202. White, J.C., and Sweet, W.H. (1969). Pain and the Neurosurgeon. A Forty-Year Experience. Charles C. Thomas, Springfield, Ill.
203. Carreras, M., and Anderson, S.A. (1963). Functional properties of neurons of the anterior ectosylvian gyrus of the cat. J. Neurophysiol. 26:100.
204. Whitsel, B.L., Petrucelli, L.M., and Werner, G. (1969). Symmetry and connectivity in the map of the body surface in somatosensory area II of primates. J. Neurophysiol. 32:170.
205. Werner, G., and Whitsel, B.L. (1973). Functional organization of the somatosensory cortex. *In* A. Iggo (ed.), Handbook of Sensory Physiology, Vol. II, pp. 621–700. Springer, Berlin, Heidelberg, New York.
206. Andersson, S.A. (1962). Projection of different spinal pathways to the second somatic sensory area in the cat. Acta Physiol. Scand. 56:Suppl. 194, 1.
207. Guillery, R.W., Adrian, H.O., Woolsey, C.N., and Rose, J.E. (1966). Activation of somatosensory areas I and II of the cat's cerebral cortex by focal stimulation of the ventrobasal complex. *In* D.P. Purpura and H. Yahr (eds.). The Thalamus, pp. 197–206. Columbia University Press, New York.
208. Morse, R.W., and Vargo, R.A. (1970). Functional neuronal subsets in the forepaw focus of somatosensory area II of the cat. Exp. Neurol. 27:125.
209. Mountcastle, V.B., and Powell, T.P.S. (1959). Neural mechanisms subserving cutaneous sensibility, with special reference to the role of afferent inhibition in sensory perception and discrimination. Bull. Johns Hopkins Hosp. 105:201.
210. Andersson, S.A., and Norrsell, U. (1973). Unit activation in the post central gyrus of the monkey via ventral spinal pathways. Acta Physiol. Scand. 87:47.
211. Vyklický, L., and Keller, O. (1974). Cortical representation of Aδ tooth pulp primary afferents in the cat. *In* J.J. Bonica (ed.), Advances in Neurology, Vol. 4, pp. 233–240. Raven Press, New York.
212. Hassler, R., and K. Muhs-Clement (1964). Architektonischer Aufbau des sensomotorischen Cortex der Katze. J. Hirnforschung 6:377.
213. Wall, P.D., and Gutnick, M. (1974). Properties of afferent nerve impulses originating from a neuroma. Nature 248:740.
214. Zimmermann, M. (1975). Neurophysiological models for nociception, pain, and pain therapy. *In* H. Penzholz, M. Brock, J. Hamer, M. Klinger, and O. Spoerri (eds.), Advances in Neurology, Vol. 3, pp. 199–209. Springer, Berlin, Heidelberg, New York.
215. Guth, L. (1974). Axonal regeneration and functional plasticity in the central nervous system. Exp. Neurol. 45:606.
216. Sharpless, S.K. (1975). Disuse supersensitivity. *In* A.H. Riesen (ed.), The developmental neuropsychology of sensory deprivation, pp. 125–152. Academic Press, New York.
217. Sharpless, S.K. (1964). Reorganization of function in the nervous system–use and disuse. Annu. Rev. Physiol. 26:357.
218. Van Beusekom, G.T. (1955). Fibre analysis of the anterior and lateral funiculi of the cord in the cat. Thesis, 143 pp. Eduard Ijdo N.V., Leiden.

International Review of Physiology
Neurophysiology II, Volume 10
Edited by Robert Porter
 University Park Press Baltimore

5
Large Receptive Fields and Spatial Transformations in the Visual System

J. T. McILWAIN
Brown University, Providence, Rhode Island

INTRODUCTION

Long before the development of electrophysiology and formulation of the receptive field concept, anatomical and clinical studies firmly established the idea that the visual pathways preserve the spatial arrangement of the retina. According to Polyak (1), those who entertained this idea believed "that such an anatomical arrangement might be a factor responsible for the exquisitely spatial character of the sense of sight" (p. 335). Modern anatomical and physiological analyses have amply confirmed the high degree of spatial order in the primary visual pathways, supporting the concept that stimulus position could be coded in the physical location of the responding visual neurons. This place coding hypothesis is particularly compelling when neurons with very small receptive fields are found in retinotopically organized brain structures.

Receptive field analysis of single visual neurons in cat and primate quickly revealed that cells having the smallest receptive fields are located in parts of the retina and brain subserving the central few degrees of the visual field where spatial vision is most acute. This was consistent with the idea that neural connections in the central retina preserve the spatial arrangement of the receptor mosaic with maximum fidelity (2). The findings also reinforced the view that fine spatial information is transmitted by cells with small receptive fields.

A complementary perspective on this situation is that most of the visual field is subserved by cells having relatively large receptive fields. In the retina (3–10), lateral geniculate nucleus (LGN) (11–14), striate cortex (15–19), and superior colliculus (20–32) of cat and primate, receptive field size is observed to increase more or less monotonically with eccentricity from the fovea or area centralis. Moreover, all receptive fields, regardless of their location in the visual field, appear to increase in size as one probes progressively further along the visual pathway. This process may be observed within the striate cortex, if one holds with Hubel and Wiesel (16) that the large-field complex cells represent a stage of visual processing beyond the small-field simple cells. The phenomenon of progressive receptive field enlargement is even more striking as observations are made in areas 18 and 19 (33–36) and in yet more distant cortical stations such as the suprasylvian and splenial gyri of the cat (37–40) and the temporal cortex of the primate (41–43).

The receptive fields of cells in the retinotectal pathway exhibit the same apparently inexorable increase in size (22, 23, 25–27, 31, 32, 44–47), prompting Goldberg and Wurtz (22) to conclude that ". . . . . a major function of the superficial gray and optic layers of the monkey superior colliculus seems to be the construction of cells with large receptive fields . . . . . " (p. 556). Extrageniculate regions of the thalamus, which are related anatomically to tectal and cortical visual areas, also contain cells with characteristically large receptive fields (48–53).

The fact that large receptive fields are the rule rather than the exception in many parts of the visual system suggests that their story may be as interesting and important as that of the tiny receptive fields of the fovea and its primary projection. This review begins with a survey of current interpretations of the function of such large-field cells. There follows a discussion of certain geometric factors involved in the synthesis of large receptive fields and their implications for spatial transformations in the visual system. The last section considers a recent suggestion about how large-field cells in the superior colliculus may encode the retinal location of stimuli. The discussion focuses on studies of the cat and primate visual systems.

CURRENT INTERPRETATIONS OF LARGE-FIELD VISUAL NEURONS

The discovery of visual cells with large receptive fields within a neural structure poses an obvious question: Can the discharge of such cells convey information about the retinal location of the stimuli to which they respond? Since this question is addressed by all current conceptions of the functional roles of large-field cells, it will serve as the focal issue of the following discussion.

Spatial Information Lost

Perhaps the most common assumption is that large-field visual cells cannot transmit precise spatial information at all. Since identical discharge patterns may be evoked by stimuli in different parts of the receptive field, the output of the cell is ambiguous as to the exact location of the stimulus. Thus, Goldberg and Wurtz (22) concluded from their study that large-field cells in the macaque's superior colliculus "can only crudely indicate the location of a stimulus" (p. 556). Similar views about large collicular fields have been expressed by others (23, 31, 32, 44, 54). Considerations such as this may have prompted Kuffler's (55) concern that the "receptive field" concept could lose its usefulness if receptive fields were found to be very large.

It seems likely that some visual cells with large receptive fields are not concerned with analysis of spatial detail or with the exact spatial location of stimuli. Thus, cells in the pretectum, which probably mediate the pupillary response to light, would be expected to integrate retinal illumination over large areas (24, 56–60). Also, the spatial distribution of retinal illumination may not be important to visually responsive cells in the brainstem reticular formation which are involved in general arousal mechanisms (61). On the other hand, some visual regions, richly endowed with large-field cells, appear from other evidence to be mediating rather precise spatial transactions. In these cases, of which certain examples will be discussed below, it is difficult to avoid the conclusion that large-field cells are sometimes capable of encoding fine spatial information.

Feature Detector Hypothesis

Many visual cells in the central nervous system, though possessed of large receptive fields, are particularly sensitive to specific parameters of the stimulus. For instance, complex cells of the striate cortex often respond preferentially to oriented lines located at various positions within a large receptive field (16, 62). Cells in the inferotemporal gyrus of the macaque which have huge receptive fields may exhibit highly specific "trigger features" (42). These and similar observations have led to the notion that certain classes of central neurons detect specific features of the retinal image, as was suggested originally for neurons of the frog's retina (63). A large receptive field, according to this view, would imply that the neuron signals the presence of a particular stimulus feature, but is unconcerned with its exact position. Hubel and Wiesel (16), discussing complex cells of striate cortex, observe that "their responsiveness to the abstraction which we call orientation is thus generalized over a considerable retinal area" (p. 146). A similar theme has been pursued in discussions of large-field cells which respond preferentially to other aspects of the retinal image such as its size, shape, direction, and velocity of movement (23, 32, 37, 42–44, 53, 60, 64).

It may be observed that this conception, although positing a function for large receptive fields, does not answer the question of how the retinal location of the feature is coded. Presumably, this is accomplished by other neurons with smaller receptive fields. Furthermore, this interpretation rests on the idea that single neurons serve as detectors or extractors of specific stimulus features. In its extreme form this hypothesis holds that the discharge of some neural elements carries a fixed significance for perception, so that the perception of, say, a vertical edge simply reflects the discharge of certain neurons which respond selectively to vertical edges. Such neurons would be "labeled lines" for the feature "vertical edge" (65). The feature detector hypothesis has been advocated (66–68) and criticized (69–71) with equal vigor.

Because a feature-detecting role is so often ascribed to large-field cells, it is germane to recall several arguments which have been raised against the basic hypothesis. One frequently cited objection is that few if any visual neurons increase their discharge as a function of only one stimulus parameter (18, 69, 70, 72, 73). Since the increased discharge can have more than one interpretation, it is ambiguous. A second objection is that cortical neurons are often broadly tuned, so that widely separated values of the presumed trigger parameter produce equivalent discharge rates (73, 74). Combinatorial or ensemble processes, which could resolve the ambiguity (74), are not permitted in the pure feature detector model (66), since this would violate the axiom that a single cell's discharge has an immutable significance for perception. Related uncertainties arise from as yet unsettled questions about the inherent variability of the cell's selectivity for a given feature (18, 75–82), and the degree to which anesthetic agents contribute to this apparent specificity (18, 82–85). Another discomforting observation is that the foveal representation in the primate's striate cortex has relatively fewer narrowly tuned neurons than the parafoveal representation (18, 72, 86). Since form vision is most acute near the center of gaze, one might expect the highly tuned feature detectors to be most prominent in the foveal projection area of the cortex.

Erickson (73) reminds us that the feature detector hypothesis attempts to extend Müller's doctrine of specific nerve energies (87, 88) to the functional level of single nerve cells. Such attempts have been pursued most vigorously in the area of somesthesis, where debate continues over neuronal specificity theories and pattern theories of the submodalities (see Melzack and Wall (89) for review). It is of interest that Müller took pains to distinguish between the idea of a specific nerve energy (quality, modality) and the propensity of a nerve to be activated by a given stimulus, a property which he called "specific irritability" (87). Although he was unsure of the substrate of the unique sensation associated with a given nerve, suggesting that it was either in the "nerve itself or in the parts of the brain and spinal cord with which it is connected" (88) (p. 1072), he clearly believed that this "specific energy" did not arise from the "specific irritability" of the peripheral sensory organ. In current terminology, Müller concluded that the optic nerve is a labeled line for vision, and that the fixed perceptual significance of its activity does not arise from the photoreceptors' selectivity for light quanta. This may be contrasted with the current tendency to attribute feature detector properties to single neurons *because* of their stimulus preferences or "specific irritabilities."

Ensemble Coding Hypothesis

Unlike the experimenter, whose attention is restricted to the activity of the one or two cells discharging near his microelectrode, the brain has immediate access to all the cells responding to a sensory input. In fact, these cells *are* the brain, or a considerable fraction of it. Recognition of this led long ago to the idea that sensory information could be represented in the pattern of discharge in populations of neurons. For instance, Adrian, Cattel, and Hoagland (90) invoked the process, often called ensemble or distributed coding, to explain how cutaneous afferents of the frog could encode the position of a point stimulus:

> There is no reason to suppose that the widespread distribution of the sensory endings of a single fiber will necessarily interfere with the exact localization of a stimulus. Owing to the overlapping of the area of distribution of different fibers the stimulation of any point on the skin will cause impulse discharges in several fibers and the particular combination of fibers in action, together with the relative intensity of the discharge in each, would supply all the data needed for localization (p. 384).

This idea has been used in the literatures of somesthesis (90–94), olfaction (95, 96), gustation (97, 98), and audition (99) to account for the apparent ability of neurons with large or broadly tuned receptive fields to mediate fine discriminations along some sensory dimension. The concept is also central to models of distributed or holographic memory which have appeared in recent years (100–107). Erickson (73) points out that Thomas Young exploited the idea in his classic explanation of how small regions of retina encode many different wavelengths using a limited number of sensitive elements. Although several recent discussions emphasize the possibility of ensemble coding in visual function (69, 86, 108–110), it is perhaps fair to say that an early concern for the importance of patterns of neural activity in the retina (111) and visual cortex (112) has been supplanted by current interest in the stimulus specificities of single visual neurons.

The concepts of ensemble coding and neuronal feature detection are joined in the so-called Fourier or spatial frequency theory of visual analysis, which has attracted wide attention in recent years.[1] According to one form of this theory, visual cells are differentially responsive to sinusoidal luminance contours of specified periodicity. Hence, the cells act as spatial filters and their pattern of discharge represents the spatial spectral composition of part of the retinal image. Large receptive fields arise quite naturally in such systems because a cell's frequency selectivity is in part related to the size of its receptive field (35, 114, 115). Although the discharge of a given cell has a fixed significance in terms of a preferred spatial frequency, its output is relatively meaningless without reference to the activity of other cells. This may be contrasted with the pure feature detector models in which the familiar elements of perception, such as oriented edges, correspond isomorphically to the discharge of individual neurons. An ensemble code of the spatial frequency variety would represent the presence and position of the edge in the discharge of many cells. Other combinations of feature detection and ensemble coding have been proposed (74).

It is one thing to speculate that large-field visual cells may participate in some process of ensemble coding and quite another to understand how such codes are interpreted, or as Perkel and Bullock (65) put it, "Who reads ensemble codes?" These authors and others (116) point out that there is, in fact, no need to postulate a process, analogous to decoding, which "reads" the ensemble code and returns the message to its original form. An external observer might simply see the pattern of neural activity undergo repeated transformations, some contingent upon past experience, until it emerged as behavior.

The study of ensemble coding in populations of visual cells requires knowledge of the spatial distribution of neural responses to visual stimuli. Although several investigators have considered the problem in its mammalian context (117–120), few experimental studies have dealt with it directly. It has been suggested (121) that a major obstacle to research on the question is the great success enjoyed by the technique of receptive field analysis, which generally asks, "Here is a cell, what and where are the stimuli which it sees?" To approach the problem of ensemble coding, the question must be inverted: "Here is a stimulus, where are the cells which see it?" The distribution of cells which see a particular stimulus depends on geometric factors which are intimately related to the construction of large receptive fields, and the following section examines some of these.

INGREDIENTS OF LARGE RECEPTIVE FIELDS: SOME GEOMETRIC FACTORS AND THEIR IMPLICATIONS

Dendritic Fields and Afferent Maps: The Convergence Perspective

Convergence processes in the retina occur in the familiar spherical coordinate system of visual space, where distance across the retina may be directly and

[1] Consideration of the extensive physiological and psychophysical literature which has grown up around this theory is beyond the scope of this review. The reader will find detailed discussion and references in a recent review by Sekuler (113).

simply related to visual angle. The magnification in this system is space invariant, in the sense that a degree of visual angle is equivalent to the same distance across the retina at all points. In the primary central visual pathways, the linear distance allotted to a given visual angle (the Magnification Factor of Daniel and Whitteridge (122)) is space variant, decreasing with distance from the projection of the primate's fovea or the cat's area centralis. This fact has important consequences for the synthesis of large receptive fields in the central nervous system and for the spatial organization of central neural structures containing large-field cells. Of particular interest are the potential interactions of dendritic field size with this space variant magnification and certain of these effects are illustrated in Figure 1.

The cells of Figure 1*A* lie directly beneath one another on a line normal to the surface of a hypothetical central visual structure. The dendritic fields increase in diameter with depth in this structure, which is assumed to receive a point-to-point afferent projection from the retina. The afferent terminals rise perpendicular to the surface through the dendritic fields of the resident neurons. In Figure 1, *B* and *C*, this set of neurons is shown embedded in two different visual nuclei, one considerably larger than the other. In Figure 1*B*, the retina projects to an extensive area and the scale of the afferent map is relatively large compared to the dimensions of the cells' dendritic fields. The visual nucleus of Figure 1*C* is small and the dendritic fields of the cells occupy sizeable portions of the afferent map.[2]

If we assume that the input is purely excitatory, the receptive fields mapped with a point stimulus might appear as in Figure 1, *D* and *E*. Several effects may be seen. The receptive fields of identical cells differ in size if their dendrites sweep out regions of the afferent map representing visual space at different magnification. This situation may be contrasted with the retina, where cells with identical dendritic fields tend to have receptive field centers of comparable size (3, 9, 123). It may also be seen that the receptive fields of cells located immediately above or below one another are centered near the same point only when the dendrites are small relative to the scale of the map. When this condition is met, cells lying along a line normal to the surface appear to form topographically organized cell columns. When the dendritic field dimensions are large with respect to the scale of the afferent map, as in Figure 1*C*, neighboring cells no longer have receptive fields nested concentrically in visual space. Depending on the coordinate system of the afferent map, cells with concentrically nested receptive fields may not even lie along straight lines drawn through the neural tissue. Note also that, as receptive field diameter increases, the location of the receptive field center becomes a poorer estimator of the retinal origin of the afferent fibers projecting to the vicinity of the cell body. In the case of Figure 1*C*, a map of the afferent projection based on the average field center position

---

[2] The relative spacings of the iso-eccentricity lines in the polar coordinate system of Figure 1, *B* and *C*, were measured from the 270° meridian of Figure 5 in Daniel and Whitteridge (122).

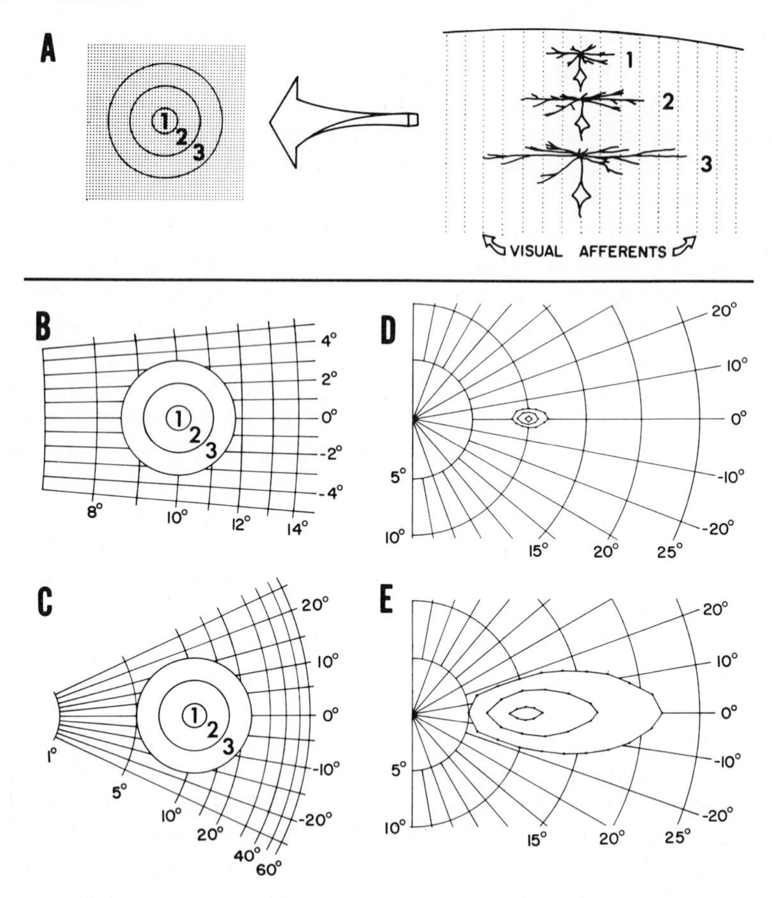

Figure 1. Relationships among dendritic trees, afferent maps, and receptive fields. *A, right,* neurons 1, 2, and 3 viewed from the side. *Dotted lines,* visual afferents ascending to surface through dendritic trees. *Left,* schematic representation of dendritic trees of same cell trio viewed from surface of visual structure. *Stipple,* termination of point-to-point afferent projection from retina. *B* and *C,* dendritic trees of neurons 1, 2, and 3 in two regions where the afferent visual maps differ in scale. *D* and *E,* receptive fields resulting from excitatory action of afferents on cell trios in B and C. Fields obtained by projecting the concentric profiles of *B* and *C* into the coordinate systems of *D* and *E,* respectively. Graphic projection executed on grids of higher resolution than those illustrated.

would be biased significantly by the inclusion of large-field cells. The visual coordinates assigned to the recording site would reflect a more eccentric origin of the retinal afferents than was actually the case.

The situation in Figure 1*B* may be compared roughly to that of the cat's or monkey's striate cortex, where the total area of the map is large relative to the dimensions of the neural elements (122, 124). The residual scatter in receptive field position reported for cell columns oriented normal to the cortical surface appears to be more or less random about some average center position (16, 17, 125). Consistent drifts across the visual field are revealed only when the electrode traverses a series of neighboring columns. The eccentrically nested receptive fields of Figure 1*C* are similar to those found in the cat's superior colliculus (25) where many cells have dendritic arborizations which are relatively large with respect to the scale of the afferent map (126).

Colonnier (127) has suggested that the oriented receptive fields of striate cortex neurons may reflect regular asymmetries in the dendritic fields of cortical cells, such as are seen in Golgi preparations cut tangential to the cortical surface. This possibility appears to have received little further attention, but it is clear that dendritic geometry must be considered in the context of the afferent map and any anisotropies which it exhibits.

The interaction of dendritic field dimensions and scale of the afferent map may have an interesting consequence in structures receiving two or more topographically organized visual projections, such as the superior colliculus. Degeneration studies following punctate lesions suggest that a small region of the colliculus is innervated from restricted and topographically corresponding regions of retina and striate cortex (128). On the other hand, electrical recordings from single cells in the cat's colliculus reveal that neighboring neurons have quite disparate receptive fields and may be driven electrically from cortical areas differing greatly in size (54). In fact, the cortical area in functional connection with the collicular cell appears to be that which "sees" the same region of the visual field. This somewhat curious finding is easy to understand in terms of the factors operating in Figure 1. Because the colliculus is relatively small, the retinal and cortical connections of a given collicular cell are determined as much by the size of the cell's dendritic tree as by the location of its perikaryon. Thus, neighboring cells of different size may make quite incongruent connections within the space of corticotectal and retinotectal afferent terminals. Since these afferent projections to the superficial gray layer are retinotopically organized and in register, the cortical connections to a given cell will be highly correlated with its own visual receptive field, but not necessarily with the cortical connections and receptive fields of its neighbors.

Yet another consequence of dendritic dimension is suggested by the hypothetical situation of Figure 1*C*, in which the dendrites of the largest neuron ramify in regions of the afferent map differing considerably in magnification factor. Consider the effects which a small visual stimulus might have on this neuron. Because of the space variant magnification of the afferent projection, the stimulus would activate a larger region of the afferent map when presented near the visual axis than when introduced into the peripheral visual field. From this, one might expect the stimulus to produce a brisker response at the central edges than at the peripheral edges of the receptive field. This is observed in studies of large-field cells of the cat's superior colliculus (25, 44).

Point Images and Receptive Field Images: The Divergence Perspective

Fischer's Relationship The illustrations used in the preceding discussion assumed that the retina projects in point-to-point fashion to some hypothetical central visual structure. It is well known that real retinal ganglion cells have overlapping receptive fields of finite dimensions, so no point in visual space is "represented" by a single ganglion cell or optic tract axon. This is true even of the primate's midget ganglion cells which are said to receive their input from single cones (1, 2). Since the point spread function[3] of the eye is large with respect to the diameters and separation of foveal cones (1, 2, 129), several cones are illuminated by a point stimulus and a given cone is illuminated from neighboring points in a finite region of visual space. Electrical recordings made from ganglion cells presumably in the midget cell pathway reveal receptive fields with finite dimensions and a center-surround organization (131).

Since a point in the visual field is "seen" by more than one retinal ganglion cell, a point stimulus will produce a distribution of ganglion cell activity, which can be called the neural *point image* (25).[4] Fischer (132) has proposed that the point image in the space of retinal ganglion cells in the cat exhibits an interesting invariance property. He argues that any point on the retina falls within the field centers of a constant *number* of ganglion cells, because the rate at which ganglion cell density decreases with distance from the area centralis is just matched by the concomitant increase of receptive field center diameter. In consequence, the collection of optic nerve axons viewing a point through their receptive field centers will occupy a cross-sectional *area* of the nerve which is roughly invariant with translation of the point. Fischer's relationship acquires added significance from recent findings that the afferent axons which are decisive in determining responses in the LGN are those which have been stimulated through their receptive field centers (133–135). This implies that the component of the retinal point image which is relayed to cortex is largely that which occupies a constant area of the optic nerve.

Is it possible that such translation invariance of the neural point image is preserved in the geniculocortical projection and that the volume of cortex receiving information from a point in the visual field is independent of the location of the point? This would indeed be an interesting consequence of receptive field enlargement and might confer advantageous properties on the visual system. For instance, if the afferent neural activity concerning a visual point were distributed to a large enough region of cortex, the point could be viewed by cells representing a full spectrum of feature selectivities and preferences (orientation, color, etc.). Furthermore, a neighboring visual point, separated from the first by a vanishingly small distance, could also be served by a full analytic array of the same dimensions, provided that the various tuned sensory elements were appropriately distributed. Thus, a complete range of feature-selective cortical neurons would survey each point in the visual field. Neighboring visual points would, of course, share cortical elements to a degree determined by

---

[3] The point spread function of an optical system describes the distribution of light in the image of a luminous point object (138).

[4] Fischer (117) uses the German work Punktbild to describe the distribution of excitation in a population of ganglion cells due to activity of a single photoreceptor.

the magnification factor, since this would dictate how far apart the cortical point images would be located. An idea closely related to this has been proposed by Hubel and Wiesel (17) and is discussed below.

The data on which Fischer based his conclusions did not distinguish between the retinal neurons which project to the tectum and those which participate in the retino-cortical pathway. The numerically dominant input to striate cortex is via the sustained or X-cell population (3, 11, 36, 136–139), so it is of interest to know whether Fischer's relationship holds for this class of cells. One form of the relationship states that

$$R = (k/\mathrm{d}\cdot\pi)^{1/2}$$

where R is receptive field center radius, d is ganglion cell density, and k is the number of cells whose field centers contain a given retinal point. R and d are functions of eccentricity. Since $d^{-1/2}$ estimates the distance between neighboring ganglion cells (i.e., the reciprocal of linear density (140)), this expression requires that receptive field diameter be a linear function of cell spacing, if k is to be constant at all values of eccentricity.

Figure 2*A* shows how inter-cell distance varies with eccentricity for X-cells of the cat's retina. The points represent the reciprocal square roots of the X-cell densities given in Figure 7 of Fukuda and Stone (137). The distribution is well described by the straight line ($r = 0.99$). Figure 2*B*, from Stone and Fukuda (9), shows that, except for the immediate vicinity of $0°$, the center diameters of X-cells increase approximately linearly with eccentricity, a finding also reported by others (3, 6). Therefore, cell spacing and field center diameter are approximately linear functions of eccentricity and thus of each other, supporting Fischer's proposal for the case of X-cells.[5]

Translation Invariance and Receptive Field Images Is there any evidence, then, that the point image in the primary visual pathway retains some kind of translation invariance as it ends in the striate cortex? The number of geniculo-cortical afferents terminating beneath a unit surface area of striate cortex appears to be approximately constant, at least in the primate (142), so if the point image were to form in an invariant number of LGN relay cells it might occupy a constant area of the terminal projection into cortex. Unfortunately, the connectional geometry of the LGN is not known in sufficient detail to permit a direct estimation of the size of the point image in the postsynaptic elements. However, another kind of evidence bearing on this question has been obtained in a recent study of the macaque's striate cortex. Hubel and Wiesel (17) report that a microelectrode must move a relatively constant distance of 2–3 mm parallel to the cortical surface before an entirely new region of visual

---

[5] The observation that ganglion cell spacing varies linearly with eccentricity was first made by Weymouth (141) using data from the human retina. The same relationship for extrafoveal retina of the monkey may be observed in Figure 5 of Rolls and Cowey (140).

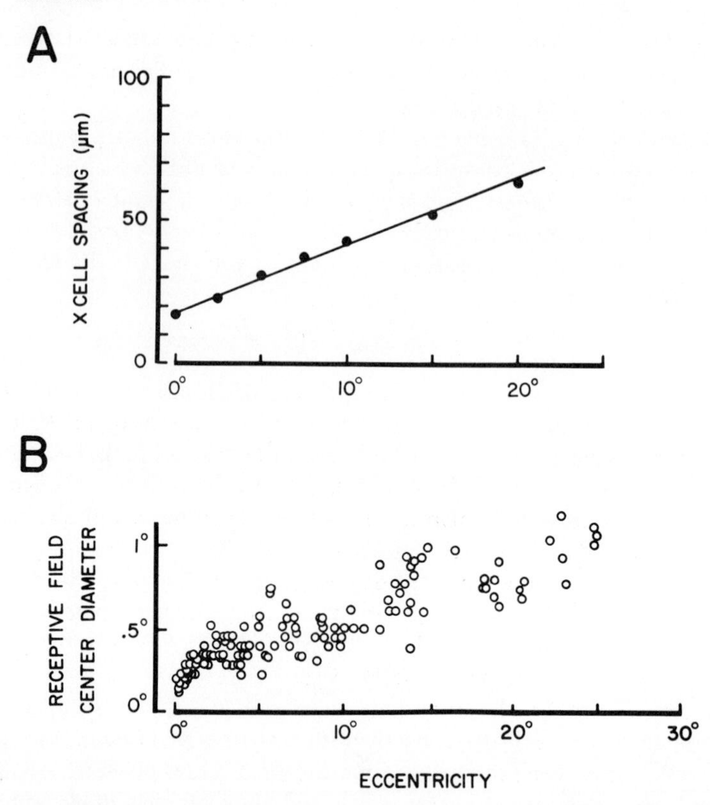

Figure 2. Relationship of cell spacing and receptive field center diameter to eccentricity in retinal X-cells of the cat. *A, ordinate:* inter-cell spacing estimated as reciprocal square roots of area densities given by Fukuda and Stone (137). *Abscissa,* distance from area centralis. *B, ordinate:* X-cell receptive field center diameter. *Abscissa,* distance from area centralis. From Stone and Fukuda (9), courtesy of American Physiological Society.

field is represented in the receptive fields of the recorded cells. Within this 2–3 mm distance, the electrode crosses several organizational modules which the authors call *hypercolumns.* A hypercolumn comprises a set of adjacent cell columns or slabs in which the cells' preferred stimulus orientations rotate 180° as the set is traversed. The term also refers to a pair of adjacent right and left eye dominance columns. The orientation hypercolumns and the eye dominance hypercolumns are co-extensive and hence are the same size, 0.5–1.0 mm thick. From these findings, Hubel and Wiesel (17) propose that "a 2–3 mm region of cortex can be said to contain by a comfortable margin the machinery it needs to analyze the region of visual field which it subserves" (p. 303).

This idea and that of the space invariant point image, suggested by Fischer's relationship, are clearly related, so we may ask if the evidence obtained by Hubel and Wiesel supports the latter concept. The question may be approached through a relationship illustrated in Figure 3 and discussed in more detail elsewhere (25). In Figure 3*A,* a point stimulus has produced a small region of

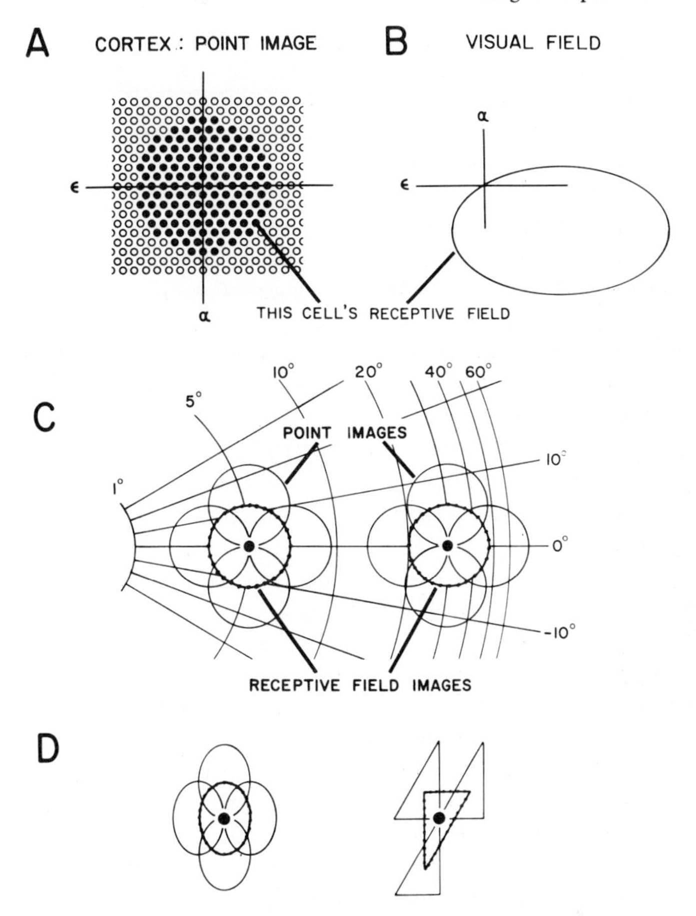

Figure 3. Reciprocity of receptive field images and point images. *A, schematic point image* in a class of cortical cells. *Filled circles,* excited cells. *Open circles,* quiescent cells. *B,* receptive field of cell at margin or cortical point image. Coordinates α, ϵ refer to conjugate loci in visual field and cortical sensory map. *C,* two cells (*filled circles*) located in different regions of cortical map. Each cell lies at margin of four identical circular point images (*light profiles*). *Dark profile,* locus of centers of all such point images having these cells at their margins. Coordinates of this profile correspond to those of receptive field. *D,* reciprocity of non-circular point images and receptive field images. Receptive field image (*dark profile*) obtained by sliding point image (*light profile*) around cell (*filled circle*), keeping cell at edge of point image and point image in same orientation. During this maneuver any fixed point in the point image describes a locus which has the shape of the point image.

activity (*filled circles*) in a hypothetical class of cortical neurons. This point image is assumed to be centered at the coordinates α,ϵ corresponding to the location of the point stimulus in visual space. For simplicity of illustration, it is also assumed that all cells in the point image are excited, though not necessarily to the same degree. The cell designated by the heavy line lies at the extreme edge of the cortical point image, which is equivalent to saying that the point stimulus at α,ϵ lies at the edge of the cell's receptive field. Because of the conjugate relationship between point image center and point stimulus location, the centers of all point images having this cell at their very edges will lie at map coordinates corresponding to the boundaries of the cell's receptive field. A line connecting these coordinates is the geometric projection of the receptive field's boundaries onto the cortical sensory map or the *receptive field image* in that coordinate system (25). Figure 3*C* shows that, if the point images in this cell class are circular and of constant diameter, the receptive field images for the class have exactly the same size and shape as the point images. This reciprocity also obtains when the point images are not circular, as long as their orientation with respect to the index cell is constant and the map projection of the point stimulus occurs at some fixed location in the point image's profile (Figure 3*D*). The geometry of this relationship imposes a 180° rotation in going from point images to receptive field images and vice versa, which becomes evident when these profiles are not symmetric (Figure 3*D*). Were the receptive field images for a cell class heterogeneous in size over a small region of tissue, the boundaries of the point image would be related to the largest receptive field images and the point image would contain quiescent members of the class. An observation of particular interest here is that, if the point image is translation invariant in a class of cells such as those of Figure 3, the receptive field images for the class will also be translation invariant.

In their study of the macaque's striate cortex, Hubel and Wiesel (17) observed that average receptive field size and cortical magnification factor vary inversely across the cortex, which causes the "cortical projection of the boundaries of an average cortical receptive field" (p. 304) to enclose a relatively constant area of about 0.66 mm^2. In other words, the area of the receptive field image of the "average receptive field" is invariant with translation, a finding which is clearly consistent with the possibility that the area of the point image in the cortex is also translation invariant (Figure 3).

Unfortunately, several reservations about such an inference come immediately to mind. First, the point image, as defined above, refers to a distribution of neural activity in a population of cells which actually respond to a point stimulus. Although a point image will form in classes of cortical cells which respond to point or spot stimuli, such as simple cells and corticotectal complex cells of the cat (16, 143) and cells in layer IVc of the primate (34), the notion would appear to have little concrete meaning for populations of neurons which respond poorly to point-like stimuli. In this case, it may prove useful to broaden the definition of point image to designate the locus of all cells whose receptive fields contain a given point in the visual field. Knowledge of this would permit one to locate the set of cells potentially capable of responding to some stimulus at a given visual point.

A second difficulty arises with respect to the meaning of the receptive field image or the "cortical projection of the boundaries" of a receptive field. A key assumption in the scheme of Figure 3 is that the cortical sensory coordinates corresponding to the location of the point stimulus occur at a fixed position in the point image. It is not clear to what degree this assumption is valid in striate cortex, particularly in view of the scatter of receptive fields belonging to neighboring cortical cells (34). The continued separation in cortex of visual inputs from the two eyes also complicates the picture. In the macaque, the separation is most pronounced in layer IVc, where the smallest receptive fields are observed (17, 144). Thus, a point stimulus viewed binocularly may normally yield two point images in this layer, rather than one. Hubel and Wiesel (17) excluded cells from this layer when they demonstrated the reciprocal relationship between average receptive field size and magnification factor. The situation in deeper and more superficial layers may be comparable to the scheme of Figure 3, since the separation of binocular inputs is less marked in these regions and the representation of the visual field does not exhibit the discontinuities which are evident in layer IVc (125, 144).

Visual Acuity and Translation Invariant Point Images These difficulties indicate that caution is required when making inferences from receptive field images. For this reason it is encouraging to find an old observation, which does not involve receptive field measurements at all, yet suggests that the boundaries of cortical point images may exhibit translation invariance. Daniel and Whitteridge (122) called attention to the fact that when the minimal angle of resolution (minimum separable), as measured at various eccentricities for the human eye, is plotted on the visual map of the monkey's striate cortex, the distance between the plotted points is the same at all eccentricities. Subsequent reports (140, 145) indicate that this relationship holds if the monkey's or the human's minimum separable are plotted on their own cortical visual maps. Daniel and Whitteridge (122) suggested that "presumably two peaks of excitation would have to be separated by this distance and by the corresponding number of cortical cells for them to give rise to separate sensations" (p. 218). It is also possible that when the neural images of two point stimuli overlap sufficiently, the points are no longer discriminable as separate. In other words, visual acuity tests of the minimum separable type may measure the capacity of the nervous system to resolve two overlapping neural point images in striate cortex. The observations of Daniel and Whitteridge (122) and others (140, 145) suggest that the critical overlap occurs when the point images are centered about the same distance apart, wherever they occur in the cortex. It follows that the diameter of the point images in the critical cell class must vary little with translation, since otherwise the degree of overlap and consequently the "resolvable" separation would depend on the location of the point images in the cortical map.

Point Images and Hypercolumn Clusters We have seen how Fischer's relationship leads to the notion of a translation invariant region of cortex which receives input from a point in the visual field and that this idea, though still speculative, is nonetheless consistent with certain experimental observations. Hubel and Wiesel (17) have proposed that a cluster of hypercolumns also has the property of translation invariance and comprises a full array of neuronal elements for analyzing visual stimuli. One would like next to ask how a point image and a cluster of hypercolumns differ, since both concepts refer to translation invariant cortical regions which may function as analytic modules. They clearly differ with respect to their reference elements in the visual field. The hypercolumn cluster surveys a finite region of visual space which must vary in size as a function of its location, because of the space variant magnification factor of cortex. The point image is referred to a point in visual space, which has by definition no dimensions, only location.

A less obvious distinction between the two ideas lies in their conception of the operational fabric of the visual cortex. From Figure 3 one sees that the size of the point image in a cell class is reflected in the dimensions of the receptive field images of the member neurons. If a given cortical region contains several classes of neurons with receptive fields of different dimensions, then there may exist multiple point images of different size within that region. Since complex cells generally have larger receptive fields than simple cells (16, 62), receptive field images and point images could be larger in the former than in the latter class. Also, the size of the point image in the space of afferent terminals is a function of the dimensions of the receptive fields of the afferent neurons (25). Thus, if a given visual area receives input from two classes of afferents having, respectively, small and large receptive fields, the point images in the two afferent systems can occupy regions of different size. For instance, a point stimulus on the cat's retina may influence a larger region of striate cortex via the Y-cell than via the X-cell projections, since at any eccentricity retinal and geniculate Y-cells have larger receptive field centers than do X-cells (3, 9, 11). These possibilities are linked by recent evidence that the input to complex cells is through the Y-cell system, whereas simple cells are activated principally by X-cell inputs (36). Furthermore, there is evidence that LGN axons terminate in cortical layers V and VI (146, 147) as well as in layer IV, thus providing a structural basis for distinct point images in the afferent input to various cortical strata.

The receptive field image of the "average cortical receptive field" was estimated by Hubel and Wiesel to have an area of 0.66 mm (2, 17). This would imply an "average" point image of the same area, perhaps a square about 0.8 mm on a side. Thus, the "average" point image inferred in this way would be approximately the thickness of a single hypercolumn (17). On the other hand, Hubel and Wiesel estimated that the translation invariant hypercolumn cluster was 2–3 mm in diameter, because "a certain constant distance along the cortex, amounting to 2–3 mm, must be traversed in order to obtain a shift in field position comparable to the size of the fields plus their scatter" (p. 303). Their observation implies that even the largest receptive fields mapped at two cortical loci will not overlap if the recording sites are about 2–3 mm apart. In a system like that of Figure 3, this would mean that the receptive field images, and consequently the point images, for the class of cortical cells with the largest receptive fields are about 2–3 mm in diameter, are translation invariant and span several hypercolumns. In contrast, the "average" receptive field image, and by inference the "average" point image, span only one hypercolumn. This again raises the possibility that point images of different size form in different classes of cortical cells.

The picture of cortical organization which emerges from this development of Fischer's relationship differs from that of a mosaic of cell columns extending through the cortical thickness (16, 34). While the columnar perspective emphasizes the anatomical arrangement of cells which respond to similar stimulus features, the point image notion suggests that neighboring loci in the visual field may be surveyed by discrete sets of neurons in some cell classes and by strongly overlapping sets of other cell classes. The elements of cortical organization derived from this perspective need not be constrained by walls running normal to the surface from pia to white matter, nor are cells located directly above and below one another necessarily linked in an invariant functional relationship.

Note Added in Proof In two papers which have appeared since submission of this manuscript, Albus (162, 163) develops arguments virtually identical with those offered here. He concludes that "Each retinal point . . . is functionally represented by the same number of cortical cells, irrespective of its position within the retina." (Ref. 162, p. 176).

LARGE RECEPTIVE FIELDS AND SENSORY-MOTOR TRANSFORMATIONS IN THE SUPERIOR COLLICULUS

Although point stimuli may elicit neural responses which display invariant properties under translation, such stimuli are generally of less interest than the complex patterns associated with form vision. There are instances, however, in which the response of the visual system to punctate stimulation may reveal significant details of functional organization. One such case is the highly integrated visual orienting response to a novel stimulus, which, it will be argued, is mediated by large-field cells.

The role of the superior colliculus in visual orienting behavior has been studied actively in the last two decades and reviews of this extensive literature are available (148–150). Of interest here is the evidence that saccadic eye movements, which are part of the visual orienting response, may also be elicited by focal electrical stimulation of the colliculus in cat and primate (29, 151–155). These electrically evoked saccades deflect the visual axes toward the region of visual space represented in the sensory projection to the stimulus site (29, 151, 152, 154). This congruence of the sensory and motor maps of the colliculus supports the view that the colliculi are involved in the production of saccadic eye movements to novel stimuli, a process called "foveation" by Schiller and Koerner (28).

In primates, the direction and amplitude of the electrically evoked saccades are relatively independent of the initial position of the eyes (29, 151, 152), but there has been some controversy as to whether this is true in the cat (152, 153). A recent report by Straschill and Rieger (154) suggests that saccades elicited by stimulation in the superficial gray layer of the cat's superior colliculus resemble those of the primate in being independent of initial eye position.

The output pathway from superficial gray to brain stem oculomotor systems very probably involves cells of the deeper collicular strata (148, 156). Hence, the conclusion seems virtually inescapable that the discharge of these cells encodes the equivalent retinal location of the electrical stimulus applied to the colliculus. Yet the experimental evidence indicates that these cells have large receptive fields (22, 23, 29, 44, 46, 157). Furthermore, to the extent that the colliculus mediates saccadic eye movements to novel visual stimuli, these same cells must be capable of encoding retinal location, despite their large receptive fields. The hypothesis sketched below (25) suggests that the key to the puzzle lies in the requirements of the oculomotor system, in this instance the ultimate consumer of the collicular output.

The anatomy of the oculomotor system requires that the neural message evoked by a novel visual stimulus eventually take the form of a distribution of activity in twelve sets of motoneurons serving as many extraocular muscles. Single unit recording in the motor nuclei indicates that the amplitude of the resulting saccade is correlated with the duration of the motoneuronal discharge (158, 159). The direction of the saccade must be determined by the relative amount of activity in the different sets of motoneurons. Thus, the saccadic system requires information from the colliculus about the amplitude or duration of the required saccade and about the distribution of activity to the several extraocular muscles.

The results of focal electrical stimulation indicate that there exist distinct spatial gradients in the functional connections made by the colliculus to subdivisions of the oculomotor system. Systems moving the eyes upward, presumably via the superior rectus and the inferior oblique muscles, are increasingly represented toward the medial border of the colliculus, since a stimulating electrode moved in this direction evokes saccades of increasing vertical amplitude (29, 151–154). One form which this functional gradient might take is illustrated in Figure 4*B* on an outline of the cat's superior colliculus. A population of cells with pre-motor connections to the brainstem "up-system" forms the majority cell type in the medial part of the colliculus (*upward directed arrows*). Analogous pre-motor representations of the systems moving the eyes downward and laterally are located in Figure 4*B* as suggested by the stimulation studies. These distributions may be compared to the sensory map in Figure 4*A*. The pre-motor cells of Figure 4*B* are assumed here to be in the superficial gray layer, since this stratum provides spatially organized visual input to the deeper output layers (148, 160).

It has been shown (25) that a point stimulus can potentially excite cells within an oval zone of the cat's superficial gray layer about 1.5 mm (anteroposterior) by 2.5 mm (mediolateral) in size. Cells with the largest receptive fields are excited throughout the zone, which is centered on the coordinate projection of the point stimulus in the sensory map. Within the central part of the colliculus, the largest receptive field images, and consequently the largest point images, are relatively invariant with translation (25). If we assume that the pre-motor cells have the largest receptive fields in the superficial gray layer, then a point stimulus in the right lower visual quadrant might activate cells in the lower of the two overlapping shaded zones in Figure 4*B*. Note that the resultant

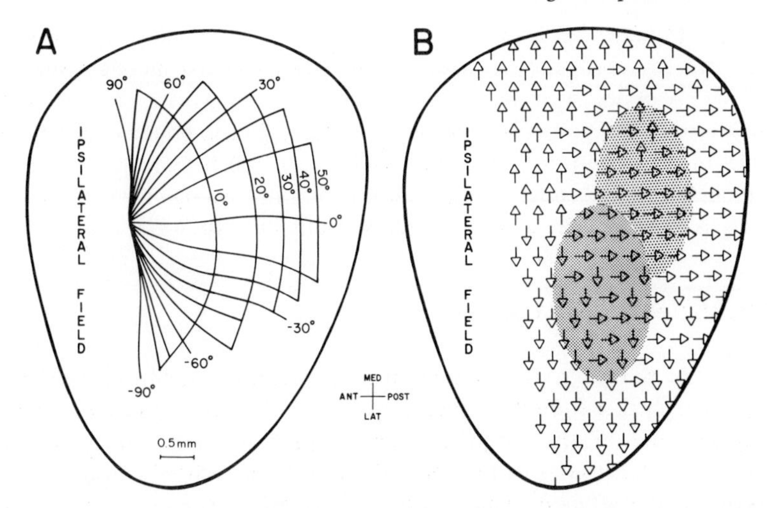

Figure 4. Sensory-motor relationships in the cat's superior colliculus. *A*, polar representation of visual projection to cat's left superior colliculus (25). O° is horizontal meridian. *B*, hypothetical distribution of pre-motor cells in superficial gray layer. Cells symbolized by downward and upward directed arrows connected to systems moving the eyes downward and upward, respectively. Arrows pointing to right signify cells connected to systems which deflect eyes laterally and to the right. *Stippled ellipses*, two point images in the superficial gray layer involving the class of cells having the largest receptive fields.

pre-motor signal from this sample of cells is predominantly "down" and "to-the-right," the appropriate direction for a saccade. Displacement of the point stimulus upward and toward the right results in excited populations of pre-motor cells which include increasingly larger contingents of "up" and "to-the-right" elements and fewer "down" elements (*upper shaded oval zone*). If the conflicting signals of the "up" and "down" elements are summed algebraically by brainstem mechanisms, such an arrangement of large-field pre-motor cells would transform a point stimulus into an output signal containing finely graded amplitude and directional information. The distribution of the excited population formats the output signal to suit the anatomy of the oculomotor system and the signal is independent of the initial position of the eyes. Precise information about stimulus position, not evident in the sensory receptive fields, is nonetheless present in the topography of the motor connections made by the large-field neurons.

This sketch of how large-field cells might encode spatial information does not incorporate many known facts about collicular physiology. For instance, the variation in excitability across receptive fields must surely result in a non-uniform distribution of excitation in the collicular point image. There are also other ways to build directional and amplitude gradients into the collicular networks. The scheme of Figure 4 has arranged the pre-motor elements so that no net signal occurs when a point stimulus appears in the area centralis, but this is only one of many possible explanations for the absence of saccades under these conditions. Further brainstem processing must be assumed to convert the amplitude signal (number of active cells) to a duration of discharge in the motoneurons and to coordinate the action of yoked extraocular muscles. Nevertheless, a neural device such as this would be consistent with earlier views that ensemble codes are particularly suited to sensory-motor transformations (65) and with the suggestion of Wurtz and Goldberg (161) that " . . . . . ensembles of these (collicular) cells could transmit information fine enough for guidance that our analysis of single cell discharge patterns would miss" (p. 447).

SUMMARY AND CONCLUSIONS

In the cat's or primate's visual system, neuronal receptive fields increase in size as they occur farther from the center of gaze. Although this is usually attributed to increasing size of the receptive fields of afferent fibers, other important factors are the dimensions of the dendritic trees of the central neurons and the local scale of the afferent maps in which they reside. The decrease in central magnification factor which parallels the growth of receptive field size may mean that a relatively invariant amount of neural tissue receives input from any point in visual space. Assessment of this potential invariance and its functional significance depends importantly on establishing concrete structural referants for the sensory coordinate maps of central visual pathways, since there appear to exist reciprocity relationships between neural point images and receptive field images. It may be hoped that developments in this area will eventually permit reconstruction of the neural activity patterns resulting from complex visual stimuli.

Large receptive fields occur with increasing frequency in more central stations of the visual pathway and it is difficult to avoid the conclusion that this sometimes reflects spatial transformations supporting complex visual functions. Several hypotheses about such transformations are reviewed here. For the most part, these theories deal with visual neuronal activity in a purely sensory context, often from a strongly psychophysical point of view. It is also possible to consider large-field cells solely as participants in a sequence of neural transformations which map a sensory event into a behavioral event. A simple example from the superior colliculus illustrates how the significance of large receptive fields may be more apparent from the putative pre-motor functions of the cells than from their sensory behavior. Although it may be unwarranted to extrapolate from this situation to higher visual functions, the example nonetheless supports the view that one is at a serious disadvantage, when interpreting the output signals of central visual neurons, if the neural systems receiving that output are not understood (33, 65).

ACKNOWLEDGMENTS

I thank Kent M. Chapman, James A. Anderson, and Menasche Nass for helpful criticism. Support of the United States Public Health Service (Grant NS 09997) and an A. P. Sloan Research Fellowship in Neurophysiology are gratefully acknowledged.

REFERENCES

1. Polyak, S. (1975). The Vertebrate Visual System. University of Chicago Press, Chicago.
2. Polyak, S. (1941). The Retina, 389 pp. University of Chicago Press, Chicago.
3. Cleland, B.G., and Levick, W.R. (1974). Brisk and sluggish concentrically organized ganglion cells in the cat's retina. J. Physiol. (Lond.) 240:421.
4. Enroth-Cugell, C., and Robson, J.G. (1966). The contrast sensitivity of retinal ganglion cells in the cat. J. Physiol. (Lond.) 187:517.
5. Fischer, B., and May, H.V. (1970). Invarianzen in der Katzenretina. Gesetzmässige Beziehungen zwischen Emfindlichkeit, Grösse und Lage receptiver Felder von Ganglionzellen. Exp. Brain Res. 11:448.
6. Hammond, P. (1974). Cat retinal ganglion cells: size and shape of receptive fields. J. Physiol. (Lond.) 242:99.
7. Hubel, D.H., and Wiesel, T.N. (1960). Receptive fields of optic nerve fibers in the spider monkey. J. Physiol. (Lond.) 154:572.
8. Ikeda, H., and Wright, M.J. (1972). Differential effects of refractive errors and receptive field organization of central and peripheral ganglion cells. Vision Res. 12:1465.
9. Stone, J., and Fukuda, Y. (1974). Properties of cat retinal ganglion cells: a comparison of W-cells with X- and Y-cells. J. Neurophysiol. 37:722.
10. Wiesel, T.N. (1960). Receptive fields of ganglion cells in the cat's retina. J. Physiol. (Lond.) 153:583.
11. Hoffmann, K.-P., Stone, J., and Sherman, S.M. (1972). Relay of receptive-field properties in dorsal lateral geniculate nucleus of the cat. J. Neurophysiol. 35:518.
12. Hubel, D.H., and Wiesel, T.N. (1961). Integrative action in the cat's lateral geniculate body. J. Physiol. (Lond.) 155:385.
13. Sanderson, K.J. (1971). Visual field projection columns and magnification factors in the lateral geniculate nucleus of the cat. Exp. Brain Res. 13:159.
14. Wiesel, T.N., and Hubel, D.H. (1966). Spatial and chromatic interactions in the lateral geniculate body of the rhesus monkey. J. Neurophysiol. 24:1115.
15. Joshua, D.E., and Bishop, P.O. (1970). Binocular single vision and depth discrimination. Receptive field disparities for central and peripheral vision and binocular interactions of peripheral single units in cat striate cortex. Exp. Brain Res. 10:389.
16. Hubel, D.H., and Wiesel, T.N. (1962). Receptive fields, binocular interaction and functional architecture in the cat's visual cortex. J. Physiol. (Lond.) 160:106.
17. Hubel, D.H., and Wiesel, T.N. (1974). Uniformity of monkey striate cortex: a parallel relationship between field size, scatter and magnification factor. J. Comp. Neurol. 158:295.
18. Poggio, G.F. (1972). Spatial properties of neurons in striate cortex of unanesthetized macaque monkey. Invest. Opthalmol. 11:368.
19. Wilson, J.R., and Sherman, S.M. (1974). Receptive field characteristics in cat striate cortex: changes with visual eccentricity. Soc. Neurosci. 4:480.
20. Cynader, M., and Berman, N. (1972). Receptive field organization of monkey superior colliculus. J. Neurophysiol. 35:187.
21. Dreher, B., and Hoffmann, K.-P. (1973). Properties of excitatory and inhibitory regions in the receptive fields of single units in the cat's superior colliculus. Exp. Brain Res. 16:333.
22. Goldberg, M.E., and Wurtz, R.H. (1972). Activity of superior colliculus in behaving monkey. I. Visual receptive fields of single neurons. J. Neurophysiol. 35:542.
23. Hoffmann, K.-P. (1970). Retinotopische Beziehungen und Strukture rezeptiver Felder im Tectum opticum und Praetectum der Katze. Z. Vergl. Physiol. 67:26.
24. Kadoya, S., Wolin, L.R., and Massopust, L.C., Jr. (1971). Photically evoked unit activity in the tectum opticum of the squirrel monkey. J. Comp. Neurol. 142:495.
25. McIlwain, J.T. (1975). Visual receptive fields and their images in superior colliculus of the cat. J. Neurophysiol. 38:219.
26. McIlwain, J.T., and Buser, P. (1968). Receptive fields of single cells in the cat's superior colliculus. Exp. Brain Res. 5:314.
27. Rosenquist, A.C., and Palmer, L.A. (1971). Visual receptive field properties of cells of the superior colliculus after cortical lesions in the cat. Exp. Neurol. 33:629.

28. Schiller, P.H., and Koerner, F. (1971). Discharge characteristics of single units in superior colliculus of the alert rhesus monkey. J. Neurophysiol. 34:920.
29. Schiller, P.H., and Stryker, M. (1972). Single unit recording and stimulation in superior colliculus of the alert rhesus monkey. J. Neurophysiol. 35:915.
30. Sterling, P. and Wickelgren, B.G. (1969). Visual receptive fields in the superior colliculus of the cat. J. Neurophysiol. 32:1.
31. Updyke, B.V. (1974). Characteristics of unit responses in superior colliculus of the *Cebus* monkey. J. Neurophysiol. 37:896.
32. Humphrey, N.K. (1968). Responses to visual stimuli of units in the superior colliculus of rats and monkeys. Exp. Neurol. 20:312.
33. Hubel, D.H., and Wiesel, T.N. (1965). Receptive fields and functional architecture in two non-striate visual areas (18 and 19) of the cat. J. Neurophysiol. 28:229.
34. Hubel, D.H., and Wiesel, T.N. (1974). Sequence regularity and geometry of orientation columns in the monkey striate cortex. J. Comp. Neurol. 158:267.
35. Pollen, D.A., and Ronner, S.F. (1975). Periodic excitability changes across the receptive fields of complex cells in the striate and parastriate cortex of the cat. J. Physiol. (Lond.) 245:667.
36. Stone, J., and Dreher, B. (1973). Projection of X- and Y-cells of the cat's lateral geniculate nucleus to areas 17 and 18 of visual cortex. J. Neurophysiol. 36:551.
37. Dow, B.M., and Dubner, R. (1969). Visual receptive fields and responses to movement in an association area of cat cerebral cortex. J. Neurophysiol. 32:773.
38. Dow, B.M., and Dubner, R. (1971). Single unit responses to moving visual stimuli in the middle suprasylvian gyrus of the cat. J. Neurophysiol. 34:47.
39. Kalia, M., and Whitteridge, D. (1973). The visual areas in the splenial sulcus of the cat. J. Physiol. (Lond.) 232:275.
40. Hubel, D.H., and Wiesel, T.N. (1969). Visual area of the lateral suprasylvian gyrus (Clare-Bishop area) of the cat. J. Physiol. (Lond.) 202:251.
41. Dubner, R., and Zeki, S.M. (1971). Response properties and receptive fields of cells in an anatomically defined region of the superior temporal sulcus in the monkey. Brain Res. 35:528.
42. Gross, C.G., Rocha-Miranda, C.E., and Bender, D.B. (1972). Visual properties of neurons in inferotemporal cortex of the macaque. J. Neurophysiol. 35:96.
43. Zeki, S.M. (1974). Functional organization of a visual area in the posterior bank of the superior temporal sulcus of the rhesus monkey. J. Physiol. (Lond.) 236:549.
44. Gordon, B. (1973). Receptive fields in deep layers of cat superior colliculus. J. Neurophysiol. 36:157.
45. Sprague, J.M., Marchiafava, P.L., and Rizzolatti, G. (1968). Unit responses to visual stimuli in the superior colliculus of the unanesthetized, mid-pontine cat. Arch. Ital. Biol. 106:169.
46. Stein, B.E., and Arigbede, M.O. (1972). Unimodal and multimodal response properties of neurons in the cat's superior colliculus. Exp. Neurol. 36:179.
47. Straschill, M., and Hoffmann, K.-P. (1969). Functional aspects of localization in the cat's tectum opticum. Brain Res. 13:274.
48. Godfraind, J.-M., Meulders, M., and Veraart, C. (1969). Visual receptive fields of neurons in pulvinar, nucleus lateralis posterior and nucleus suprageniculatus thalami of the cat. Brain Res. 15:552.
49. Godfraind, J.-M., Meulders, M., and Veraart, C. (1972). Visual properties of neurons in pulvinar, nucleus lateralis posterior and nucleus suprageniculatus thalami in the cat. I. Qualitative investigation. Brain Res. 44:503.
50. Mathers, L.H., and Rapisardi, S.C. (1973). Visual and somatosensory receptive fields in the squirrel monkey pulvinar. Brain Res. 64:65.
51. Suzuki, H., and Kato, H. (1969). Neurons with visual properties in the posterior group of the thalamic nuclei. Exp. Neurol. 23:353.
52. Veraart, C., Meulers, M., and Godfraind, J.-M. (1972). Visual properties of neurons in pulvinar, nucleus lateralis posterior and nucleus suprageniculatus thalami in the cat. II. Quantitative investigations. Brain Res. 44:527.
53. Wright, M.J. (1971). Responsiveness to visual stimuli of single neurons in the pulvinar and lateral posterior nuclei of the cat's thalamus. J. Physiol. (Lond.) 219:32P.
54. McIlwain, J.T. (1973). Retinotopic fidelity of striate cortex-superior colliculus interactions in the cat. J. Neurophysiol. 36:702.

55. Kuffler, S.W. (1952). Neurons in the retina: organization, inhibition and excitation problems. Cold Spring Harbor Symp. Quant. Biol. 17:281.
56. Cavaggioni, A., Madarasz, I., and Zampollo, A. (1968). Photic reflex and pretectal region. Arch. Ital. Biol. 106:227.
57. Harutiunian-Kozak, B., Kozak, W., and Dec, K. (1968). Single unit activity in the pretectal region of the cat. Acta Biol. Exp. Warsaw 28:333.
58. Harutiunian-Kozak, B., Kozak, W., and Dec, K. (1970). Analysis of visually evoked activity in the pretectal region of the cat. Acta Neurobiol. Exp. 30:233.
59. Smith, J.D., Ichinose, L.Y., Masek, G.A., Watanabe, T., and Stark, L. (1968). Midbrain single units correlating with pupil response to light. Science 162:1302.
60. Straschill, M., and Hoffmann, K.-P. (1969). Response characteristics of movement-detecting neurons in pretectal region of the cat. Exp. Neurol. 25:165.
61. Bell, C., Sierra, G., Buendia, N., and Segundo, J.P. (1964). Sensory properties of units in mesencephalic reticular formation. J. Neurophysiol. 27:961.
62. Pettigrew, J.D., Nikara, T., and Bishop, P.O. (1968). Responses to moving slits by single units in cat striate cortex. Exp. Brain Res. 6:373.
63. Maturana, H.R., Lettvin, J.Y., McCulloch, W.S., and Pitts, W.H. (1960). Anatomy and physiology of vision in the frog (*Rana pipiens*). J. Gen. Physiol. 43: Suppl. 2, 129.
64. Zeki, S.M. (1974). Cells responding to changing image size and disparity in the cortex of the rhesus monkey. J. Physiol. (Lond.) 242:827.
65. Perkel, D.H., and Bullock, T.H. (1968). Neural Coding. Neurosci. Res. Prog. Bull. 6, No. 3.
66. Barlow, H.B. (1972). Single units and sensation: a neuron doctrine for perceptual psychology. Perception 1:371.
67. Blakemore, C. (1974). Developmental factors in the formation of feature extracting neurons. *In* F.O. Schmitt and F.G. Worden, (eds.), The Neurosciences, Third Study Program, p. 105. MIT Press, Cambridge, Mass.
68. Ganz, L., and Fitch, M. (1968). The effect of visual deprivation of perceptual behavior. Exp. Neurol. 22:638.
69. Hoeppner, T.J. (1974). Stimulus analyzing mechanisms in the cat visual cortex. Exp. Neurol. 45:257.
70. Spinelli, D.N., and Barrett, T.W. (1969). Visual receptive field organization of single units in the cat's visual cortex. Exp. Neurol. 24:76.
71. Uttal, W.R. (1971). The psychobiological silly season—or what happens when neurophysiological data become psychological theories. J. Gen. Psychol. 84:151.
72. Bartlett, J.R., and Doty, R.W., Sr. (1974). Response of units in striate cortex of squirrel monkey to visual and electrical stimuli. J. Neurophysiol. 37:621.
73. Erickson, R.P. (1968). Stimulus coding in topographic and non-topographic afferent modalities. Psychol. Rev. 75:447.
74. Erickson, R.P. (1974). Parallel "population" neural coding in feature extraction. *In* F.O. Schmitt and F.G. Worden (eds.), The Neurosciences, Third Study Program, p. 155. MIT Press, Cambridge, Mass.
75. Bear, D.M., Sasaki, H., and Ervin, F.R. (1971). Sequential changes in receptive fields of striate neurons in dark adapted cats. Exp. Brain Res. 13:256.
76. Donaldson, I.M.L., and Nash, J.R.G. (1975). Variability of the relative preference for stimulus orientation and direction of movement in some units of the cat visual cortex (areas 17 and 18). J. Physiol. (Lond.) 245:305.
77. Henry, G.H., Bishop, P.O., Tupper, R.M., and Dreher, B. (1973). Orientation specificity and response variability of cells in the striate cortex. Vision Res. 13:1771.
78. Horn, G., and Hill, R.M. (1969). Modifications of receptive fields of cells in the visual cortex occurring spontaneously and associated with bodily tilt. Nature 221:186.
79. Rose, D., and Blakemore, C. (1974). An analysis of orientation selectivity in the cat's visual cortex. Exp. Brain Res. 20:1.
80. Sasaki, H., Saito, Y., Baer, D.M., and Ervin, F.R. (1971). Quantitative variation in striate receptive fields of cats as a function of light and dark adaptation. Exp. Brain Res. 13:273.
81. Schwartzkroin, P.A. (1972). The effect of body tilt on the directionality of units in cat visual cortex. Exp. Neurol. 36:498.
82. Ikeda, H., and Wright, M.J. (1974). Sensitivity of neurons in visual cortex (area 17) under different levels of anesthesia. Exp. Brain Res. 20:471.

83. Lee, B.B. (1970). Effect of anesthetics upon visual responses of neurons in the cat's striate cortex. J. Physiol. (Lond.) 207:74P.
84. Pettigrew, J.D. (1974). The effect of visual experience on the development of stimulus specificity by kitten cortical neurons. J. Physiol. (Lond.) 237:49.
85. Robertson, A.D.J. (1965). Anesthesia and receptive fields. Nature 205:80.
86. Spinelli, D.N., Pribram, K.H., and Bridgeman, B. (1970). Visual receptive field organization of single units in the visual cortex of monkey. Int. J. Neurosci. 1:67.
87. Müller, J. (1826). Über die phantastischen Gesichtserscheinungen, p. 6, Hölscher, Coblenz.
88. Müller, J. (1842). Elements of Physiology, Vol. 2, p. 1072. W. Baly (translator). Taylor and Walton, London.
89. Melzack, R., and Wall, P.D. (1962). On the nature of cutaneous sensory mechanisms. Brain 85:331.
90. Adrian, E.D., Cattel, M., and Hoagland, H. (1931). Sensory discharges in single cutaneous nerve fibers. J. Physiol. (Lond.) 72:25.
91. Hahn, J.F. (1971). Stimulus response relationships in first order sensory fibers from cat vibrissae. J. Physiol. (Lond.) 213:215.
92. Mountcastle, V.B. (1966). The neural replication of sensory events in the somatic afferent system. *In* J.C. Eccles (ed.), Brain and Conscious Experience, p. 85. Springer, New York.
93. Sinclair, D.C. (1955). Cutaneous sensation and the doctrine of specific energy. Brain 78:584.
94. Tower, S.S. (1940). Unit of sensory reception in cornea. J. Neurophysiol. 3:486.
95. Gesteland, R.C., Lettvin, J.Y., and Pitts, W.H. (1965). Chemical transmission in the nose of the frog. J. Physiol. (Lond.) 181:525.
96. O'Connell, R.J., and Mozell, M.M. (1969). Quantitative stimulation of the frog olfactory receptors. J. Neurophysiol. 32:51.
97. Ganchrow, J.R., and Erickson, R.P. (1970). Neural correlates of gustatory intensity and quality. J. Neurophysiol. 33:768.
98. Pfaffman, C. (1944). Gustatory afferent impulses. J. Cell. Comp. Physiol. 17:243.
99. Whitfield, I.C. (1967). The Auditory Pathway, p. 147. Edward Arnold, London.
100. Anderson, J.A. (1972). A simple neural network generating an interactive memory. Math. Biosci. 14:197.
101. Cooper, L.N. (1973). A possible organization of animal memory and learning. *In* B. Lundquist and S. Lundquist (eds.), Nobel Sym. Med. Nat. Sci., Collective Properties of Physical Systems, p. 252. Academic Press, New York.
102. Van Heerden, P.J. (1968). The Foundations of Empirical Knowledge. N.V. Vitgeverij Wistik, Wassenaar.
103. Julesz, B., and Pennington, K.S. (1965). Equidistributed information mapping: an analogy to holograms and memory. J. Opt. Soc. Amer. 55:604.
104. Kabrisky, M. (1966). A Proposed Model for Visual Information Processing in the Human Brain. University of Illinois, Urbana.
105. Longuet-Higgens, H.C. (1968). The non-local storage of temporal information. Proc. Roy. Soc. Lond. Ser. B **171**:327.
106. Pribram, K., Nuwer, M., and Baron, R. (1974). The holographic hypothesis of memory structure in brain function and perception. Contemp. Dev. Math. Psychol. 2:416.
107. Westlake, P.R. (1970). The possibilities of neural holographic processes within the brain. Kybernetik 7:129.
108. Burns, B.D. (1968). The Uncertain Nervous System, p. 28. Edward Arnold, London.
109. Ewert, J.-P., and Borchers, H.-W. (1971). Reaktionscharakteristik von Neuronen aus dem Tectum opticum und Subtectum der Erdkröte *Bufo bufo* (L). Z. Vergl. Physiol. 71:165.
110. Mandl, G. (1970). Localization of visual patterns by neurons in cerebral cortex of the cat. J. Neurophysiol. 33:812.
111. Hartline, H.K. (1940). The receptive fields of optic nerve fibers. Amer. J. Physiol. 130:690.
112. Marshall, W.H., and Talbot, S.A. (1942). Recent evidence for neural mechanisms in vision leading to a general theory of sensory acuity. Biol. Symp. 7:117.
113. Sekuler, R. (1974). Spatial Vision. Annu. Rev. Psychol. 25:195.

114. Legéndy, C.R. (1975). Can the data of Campbell and Robson be explained without assuming Fourier Analysis? Biol. Cybernet. 17:157.
115. MacLeod, I.D.G., and Rosenfeld, A. (1974). The visibility of gratings: spatial frequency channels of bar-detecting units. Vision Res. 14:909.
116. Uttal, W.R. (1973). The Psychobiology of Sensory Coding, p. 208. Harper and Row, New York.
117. Fischer, B. (1972). Optische und neuronale Grundlagen der visuellen bildübertragung: Einheitliche mathematische Behandlung des retinalen Bildes und der Erregbarkeit von retinalen Ganglienzellen mit Hilfe der Linearen Systemtheorie. Vision Res. 12:1125.
118. Kaji, S., Yamane, S., Yoshimura, M., and Sugie, N. (1974). Contour enhancement of two-dimensional figures observed in the lateral geniculate cells of cats. Vision Res. 14:113.
119. Marko, H. (1969). Die Systemtheorie der homogenen schichten. I. Mathematische Grundlagen. Kybernetik 6:221.
120. von Seelen, W. (1970). Zur Informationsverarbeitung im visuellen System der Wirbeltiere. I. Kybernetik 7:43.
121. Uttal, W.R. (1969). Emerging principles of sensory coding. Perspect. Biol. Med. 12:344.
122. Daniel, P.M., and Whitteridge, D. (1961). The representation of the visual field on the cerebral cortex in monkeys. J. Physiol. (Lond.) 159:203.
123. Boycott, B.B., and Wässle, H. (1974). The morphological types of ganglion cells of the domestic cat's retina. J. Physiol. (Lond.) 240:397.
124. Bilge, M., Bingle, A., Seneviratne, K.N., and Whitteridge, D. (1967). A map of the visual cortex in the cat. J. Physiol. (Lond.) 191:116P.
125. Hubel, D.H., and Wiesel, T.N. (1968). Receptive fields and functional architecture of monkey striate cortex. J. Physiol. (Lond.) 195:215.
126. Sterling, P. (1971). Receptive fields and synaptic organization of the superficial gray layer of the cat superior colliculus. Vision Res. Suppl. 3:309.
127. Colonnier, M. (1964). The tangential organization of the visual cortex. J. Anat. 98:327.
128. Garey, L.J., Jones, E.G., and Powell, T.P.S. (1968). Interrelationships of striate and extrastriate cortex with the primary relay sites of the visual pathway. J. Neurol. Neurosurg. Psychiat. 31:135.
129. Gubisch, R.W. (1967). Optical performance of the human eye. J. Opt. Soc. Amer. 57:407.
130. Westeimer, G. (1971). Optical properties of vertebrate eyes. *In* M.G.F. Fuortes (ed.), Handbook of Sensory Physiology, Vol. 7, p. 449. Springer, Berlin, New York.
131. Gouras, P. (1968). Identification of cone mechanisms in monkey ganglion cells. J. Physiol. (Lond.) 199:533.
132. Fischer, B. (1973). Overlap of receptive field centers and representation of the visual field in the cat's optic tract. Vision Res. 13:2113.
133. Hammond, P. (1972). Chromatic sensitivity and spatial organization of LGN neuron receptive fields in cat: cone-rod interactions. J. Physiol. (Lond.) 225:391.
134. Hammond, P. (1973). Contrasts in spatial organization of receptive fields at geniculate and retinal levels: centre, surround and outer surround. J. Physiol. (Lond.) 228:115.
135. Maffei, L., and Fiorentini, A. (1972). Retinogeniculate convergence and analysis of contrast. J. Neurophysiol. 35:65.
136. Cleland, B.G., Dubin, M., and Levick, W.R. (1971). Sustained and transient neurons in the cat's retina and lateral geniculate nucleus. J. Physiol. (Lond.) 217:473.
137. Fukuda, Y., and Stone, J. (1974). Retinal distribution and central projections of Y-, X- and W-cells of the cat's retina. J. Neurophysiol. 37:749.
138. Fukuda, Y., and Saito, H. (1972). Phasic and tonic cells in the cat's lateral geniculate nucleus. Tohuku J. Exp. Med. 106:209.
139. Singer, W., and Bedworth, N. (1973). Inhibitory interaction between X- and Y-units in the cat LGN. Brain Res. 49:291.
140. Rolls, E.T., and Cowey, A. (1970). Topography of the retina and striate cortex and its relationship to visual acuity in rhesus monkeys and squirrel monkeys. Exp. Brain Res. 10:298.

141. Weymouth, F.W. (1958). Visual sensory units and the minimal angle of resolution. Amer. J. Ophthalmol. 46:102.
142. Clark, W.E. LeGros (1941). The laminar organization and cell content of the lateral geniculate body in the monkey. J. Anat. 75:419.
143. Palmer, L.A., and Rosenquist, A.C. (1974). Visual receptive fields of single striate cortical units projecting to the superior colliculus in the cat. Brain Res. 67:27.
144. Hubel, D.H. Wiesel, T.N., and LeVay, S. (1974). Visual field representation in layer IVc of monkey striate cortex. Soc. Neurosci. 4:264.
145. Cowey, A., and Rolls, E.T. (1974). Human cortical magnification factor and its relation to visual acuity. Exp. Brain Res. 21:447.
146. Rosenquist, A.C., Edwards, S.B., and Palmer, L.A. (1974). An autoradiographic study of the projections of the dorsal lateral geniculate nucleus and the posterior nucleus in the cat. Brain Res. 80:71.
147. Rossignol, S., and Colonnier, M. (1971). A light microscopic study of degeneration patterns in cat cortex after lesions of the lateral geniculate nucleus. Vision Res. Suppl. 3:329.
148. Ingle, D., and Sprague, J.M. (1975). Sensorimotor function of the midbrain tectum. Neurosci. Res. Prog. Bull. 3:169.
149. McIlwain, J.T. (1972). Central vision: visual cortex and superior colliculus. Annu. Rev. Physiol. 34:291.
150. Sprague, J.M., Berlucchi, G., and Rizzolatti, G. (1973). The role of the superior colliculus and pretectum in vision and visually guided behavior. *In* R. Jung (ed.), Handbook of Sensory Physiology, Vol. 7/III/B, 27. Springer, Berlin.
151. Robinson, D.A. (1972). Eye movements evoked by collicular stimulation in the alert monkey. Vision Res. 12:1795.
152. Schiller, P.H. (1972). The role of the monkey superior colliculus in eye movement and vision. Invest. Ophthalmol. 11:451.
153. Straschill, M., and Rieger, P. (1972). Optomotor integration in the superior colliculus of the cat. *In* J. Dichgans and E. Bizzi (eds.), Cerebral Control of Eye Movements and Motion Perception, p. 130. Karger, Basel.
154. Straschill, M., and Rieger, P. (1973). Eye movements evoked by focal stimulation of the cat's superior colliculus. Brain Res. 59:211.
155. Syka, J., and Radil-Weiss, T. (1971). Electrical stimulation of the tectum in freely moving cats. Brain Res. 28:567.
156. Kanaseki, T., and Sprague, J.M. (1974). Anatomical organization of pretectal nuclei and tectal laminae in the cat. J. Comp. Neurol. 158:319.
157. Wurtz, R.H., and Goldberg, M.E. (1972). Activity of superior colliculus in behaving monkey. III. Cells discharging before eye movements. J. Neurophysiol. 35:575.
158. Fuchs, A.F., and Luschei, E.S. (1970). Firing patterns of abducens neurons of alert monkeys in relationship to horizontal eye movement. J. Neurophysiol. 33:382.
159. Robinson, D.A. (1970). Oculomotor unit behavior in the monkey. J. Neurophysiol. 33:393.
160. Kawamura, S., Sprague, J.M., and Niimi, K. (1974). Corticofugal projections from the visual cortices of the thalamus, pretectum and superior colliculus in the cat. J. Comp. Neurol. 158:339.
161. Wurtz, R.H., and Goldberg, M.E. (1972). The primate superior colliculus and the shift of visual attention. Invest. Ophthalmol. 11:441.
162. Albus, K. (1975). A quantitative study of the projection area of the central and the paracentral visual field in area 17 of the cat. I. The precision of the topography. Exp. Brain Res. 24:159.
163. Albus, K. (1975). A quantitative study of the projection area of the central and the paracentral visual field in area 17 of the cat. II. The spatial organization of the orientation domain. Exp. Brain Res. 24:181.

International Review of Physiology
Neurophysiology II, Volume 10
Edited by Robert Porter
 University Park Press Baltimore

6
Tonotopic Organization at Higher Levels of the Auditory Pathway

L. M. AITKIN
Monash University, Clayton, Victoria, Australia

INTRODUCTION

The orderly relationships between the frequency of a tonal stimulus and the region of a particular brain nucleus activated—"tonotopic organization"—may be regarded as the analog in the brain auditory system of the representations of the body surface in somatosensory and motor areas (somatotopic organization) and of the visual world upon the brain visual areas (visuotopic organization). This analogy requires the assumption that the frequency of a tonal stimulus is correlated with a place on the cochlear partition.

The existence of tonotopic organization has been an accepted fact of the organization of parts of the auditory system for several decades, but it has been considered largely as a by-product of anatomical connections rather than as being a reflection of a precise anatomical framework of potentially great functional significance. However, publications giving structural and physiological details of brain auditory areas have demonstrated that a need exists for a critical evaluation of tonotopic organization, particularly in relationship to the higher nuclei of the brain auditory pathway.

The word "tonotopic" first appeared in the neurophysiological literature when it was used by McCulloch and his colleagues in their study of the functional organization of the temporal lobe in primates in 1942 (1). They used "tonotopic localization" to mean that different sound pitches evoked potentials which varied in magnitude across the supra-temporal plane of the temporal cortex. In the same year the classic study of Woolsey and Walzl (2) laid the basis for later studies of the topographic relationship between the cochlear partition and brain auditory areas.

Shortly after this time the refinement of acoustic stimulation techniques was associated with the demonstration of precise audiofrequency localization in the auditory cortex of the dog by Tunturi (3). His method of strychnine-induced spikes for enhancing electrical activity of the cortex was later successfully used by Hind (4) in his electrophysiological determination of tonotopic localization in the auditory cortex of the cat.

In the 30-odd years since these pioneering experiments were initiated, numerous studies have been carried out upon single neurons which have given more and more refined accounts of the precise relationships between the region of basilar membrane and the volumes of brain tissue activated by a pure tone. The modern concept of *tonotopic organization* has largely arisen from and may be defined by the now-classic description of the orderly relationship between the *best* (also called *characteristic*) *frequencies* of neurons and their locations in the cat cochlear nuclear complex, given by Rose and his colleagues in 1959 (5). The existence of precise tonotopic organization throughout the auditory pathway is likely to be basic to any theory of pitch perception which depends on the orderly correlation between tone frequency, region of brain auditory nucleus activated, and perceived pitch.

In this review I examine tonotopic organization in several ways in an attempt to define its importance as an organizational principle in the higher regions of the mammalian auditory system. Most of the material that is discussed has been drawn from studies of the cat inferior colliculus, medial geniculate body, and auditory cortex since these studies have mostly been carried out under very similar conditions of anesthesia, electrical recording, and acoustic stimulation.

After a review of the methodology used in experiments on tonotopicity, I briefly review studies of tonotopic organization in submammalian species and describe its development in a mammalian brainstem nucleus. A detailed account of the anatomical and physiological bases of tonotopicity is then given, succeeded by a brief evaluation of the significance of tonotopic organization in frequency discrimination.

It is worthwhile to define some of the terms which will be encountered. A "unit" is the neural element, usually a neuron soma, whose action potentials are monitored extracellularly by the recording electrode. "Best frequency" is the tone frequency which activates a unit at some intensity below which no other frequencies are effective (27); this intensity is usually called "best frequency threshold." The curve described by joining threshold points for different frequencies which activate a unit is called the "tuning curve" for that unit. "Tonotopic organization" is the orderly spatial relationship between the best frequencies of units and their locations within a nucleus or across a surface. "Cochleotopic organization" is the orderly spatial relationship between the cochlear partition and a nucleus and may be deduced from a knowledge of the area of basilar membrane and the volume of a particular brain region activated by a series of tone stimuli.

EXPERIMENTAL CONSIDERATIONS

Three major experimental difficulties face the investigator of topographic relationships in the brain auditory system. First, many experiments concerned with tonotopic organization have employed anesthetized animals yet it is clear that anesthetic agents may profoundly alter or depress neural activity. Second, it is necessary to be able to precisely relate the physiological "unit" under study to an accurate location in its parent auditory nucleus. Third, it may be important to be able to distinguish between pre- and postsynaptic axon spikes and the action potentials of neuron somas. These three problems may be as important as more technical issues such as methods of tonal stimulation and sound intensity measurement.

The differences in the responses of neurons in auditory areas in anesthetized and unanesthetized animals have been the subjects of a number of publications (6–17). Generally, the depressing effects of anesthesia become more profound at successively higher levels of the auditory pathway and are most noticeable upon discharge patterns of single units rather than upon their thresholds. Thus, anesthetic agents appear to have little effect upon discharge patterns at the ventral cochlear nucleus (17), while, at the auditory cortex, a "one-quarter surgical dose" of pentobarbitone sodium may block response activity (8).

Although it seems likely that discharge patterns at higher auditory levels are severely modified by anesthetic agents, it is not yet clear whether frequency response areas are similarly altered. The studies by Bock and his colleagues (9) and Whitfield and Purser (10) suggest that anesthesia may influence the tuning ranges of some neurons, but their conclusions are weakened by the uncertain histological controls in both studies. Furthermore, the presence of much spontaneous activity of neurons in unanesthetized animals, although undoubtedly important in auditory coding processes, severely hampers the measurement of thresholds which are easily obtained in anesthetized animals.

The receptive field of an auditory neuron, defined as the frequency and intensity pairings that influence the discharges of that neuron (18), is determined by the anatomical connections between the cochlear partition and the unit under study, and the best frequency of that neuron would be a reflection of the central core in the topography of these connections. Viewed in this way, it seems unlikely that anesthetic agents would alter the best frequency of a neuron at threshold, even if the temporal sequences of excitation and inhibition are modified. Consequently it is appropriate to consider that tonotopic organization is a "static" (19) property of the auditory system.

The correlation between unit best frequency and depth within a nucleus or location across a cortical surface is dependent for its accuracy on a number of variables. First, discharges from a given neural element may be detected extracellularly over many microns of electrode movement—in one published case, for 200 μm (20). The size and constancy of this location "error" are difficult to establish but its magnitude is likely to be considerable where unit "clusters" are studied. Second, histological procedures invariably lead to some tissue shrinkage which will vary from brain to brain, nucleus to nucleus and may be differential within a given nucleus. This factor can be partly overcome by the use of marking electrolytic lesions at various depths along a particular track. Third, combinations of tonotopic maps for an area may produce errors because of possible individual anatomical differences between nuclei in different animals. This factor is of major importance where composite auditory cortex maps have been used to describe the details of the organization of the primary auditory area (AI) (7, 21, 22). The failure in these experiments to find strict tonotopic organization has been attributed to individual differences between boundaries of the primary field in relation to prominent sulci on the cortical surface which are generally used as guidelines for composite maps (13, 14).

A third experimental consideration which may require evaluation is the nature of the neural element from which spikes are recorded extracellularly. Three main candidates exist—pre- and postsynaptic axons and neurons somas. This consideration is of only small import if the experimenter is able to show an orderly tonotopic relationship since, whatever the nature of the neural elements, the demonstration of such an organization is beyond dispute. However, the presence of inconsistencies in a tonotopic map may be indicative of the sampling of discharges of fibers of passage, destined for parts of the nucleus other than that in which the electrode tip is presumed to lie. Most experiments in this field have used metal electrodes—tungsten, indium, platinum-iridium or stainless steel—and it is generally accepted that these relatively large-tipped electrodes are poorly suited for recording the activity of fibers and tend to detect the discharges of larger cell bodies (23, 24). In contrast to this, fine saline or KCl-filled micropipettes give satisfactory extracellular recordings from myelinated fibers (23), and it is probable that the use of micropipettes by workers such as Katsuki and his colleagues may have been the reason for their inability to discern orderly tonotopic arrangements in the cat auditory pathway (25, 26).

Finally, differing results may also be attributable to the location of the stimulating sound source in the free-field about the head of the animal (6, 7, 9–11) compared with monaural stimulation used in other studies. The binaural interactions produced by free-field stimulation may interfere with the interpretation of the influence of tone frequency per se, so greatest emphasis will be placed in this review on the results of experiments employing monaural stimulation. Similarly, the intensive criteria used to define best frequency vary from study to study. In view of the fact that, for some neurons, the best frequency shifts from its threshold value to higher or lower values with increases in sound intensity (18, 113), most weight will be assigned to experiments in which tonotopic organization has been assessed at threshold best frequency.

TONOTOPIC ORGANIZATION IN SUBMAMMALIAN SPECIES

Very few studies have been carried out to determine the topography of the relationship between the auditory receptor and the brain auditory nuclei in submammalian species. The avian cochlear nucleus consists of two major parts, *nucleus magnocellularis* and *nucleus angularis* (28, 29), and similar structures exist in the reptile, the caiman (*Caiman crocodilus*) (30). Dorsoventral microelectrode penetrations in these nuclei in various songbirds reveal a tonotopic organization in which increasing best frequency occurs in medial-to-lateral, rostral-to-caudal and ventral-to-dorsal directions in *n. angularis* and a less clear organization occurs in *n. magnocellularis* (29). The cochlear nuclei in the caiman show the same tonotopic arrangement except that the medial-to-lateral frequency increase in *n. angularis* is reversed in the caiman (30).

The midbrain auditory region in the avian, the *nucleus mesencephalicus lateralis pars dorsalis* (NMLD), lies beneath the optic tectum (Figure 1) and is believed to be the homolog of the mammalian inferior colliculus and the reptilian *torus semicircularis* (31, 114). Recent microelectrode studies by R. Coles in this laboratory have shown that NMLD in the chicken is organized in a strict tonotopic fashion (32) and contains neurons responding to the full audible frequency range for the chicken (33).

One such penetration is illustrated in Figure 1. A micropipette, filled with the marking dye Pontamine sky blue, dissolved in 2 M NaCl, was inserted through the optic tectum (OT) at an angle of approximately 20° to the vertical plane. Light-evoked potentials were detected in the optic tectum, and, following this, 22 units were studied in NMLD which responded from frequencies as low as 100–200 Hz in the upper part of the nucleus to 4,500 Hz at the deepest point in the penetration, confirmed by dye marking. The sequence of best frequencies consisted of an initial drop from 500 to 210 Hz, and then, with one main exception (160 Hz), a steady increase in best frequency to reach 4,500 Hz at the

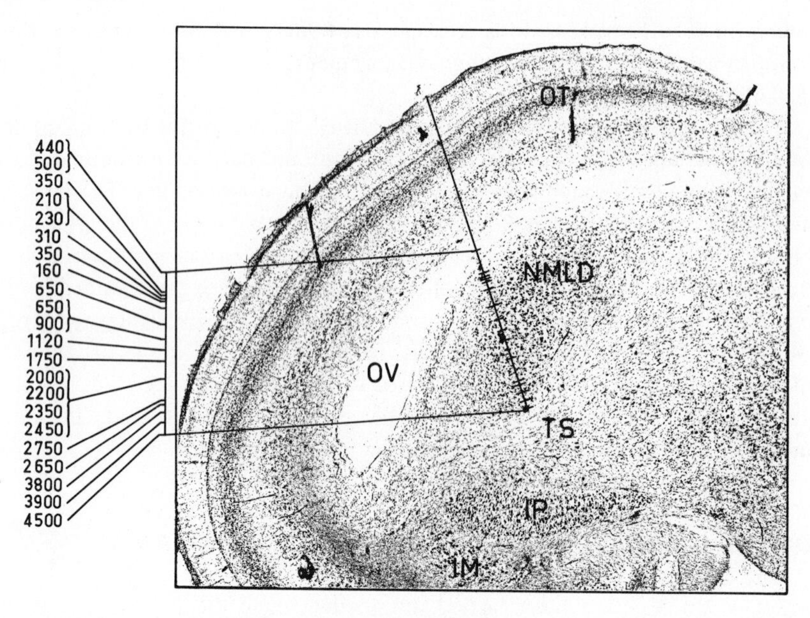

Figure 1. Photomicrograph of frontal section through the midbrain of the domestic chicken (*Gallus domesticus*) revealing tonotopic organization in the *nucleus mesencephalicus lateralis pars dorsalis* (*NMLD*). Numbers to left of figure are the best frequencies of single units; *OT*, optic tectum; *OV*, optic ventricle; *TS*, tectospinal tract; *IP*, isthmus nucleus, parvocellular division; *IM*, isthmus nucleus, magnocellular division. Thionine stain, × 20. (From R. Coles, unpublished observation.)

ventral-most point of the penetration. The initial auditory units were located in a narrow band of small cells and fibers just dorsal to NMLD and ventral to the optic ventricle (OV). These units, whose best frequencies did not conform to the main sequence, were located in a region homologous to those regions of the inferior colliculus immediately surrounding the central nucleus in mammals (48).

It is difficult to make generalizations about the evolutionary significance of tonotopic organization in the light of so few measurements in mammalian and submammalian species. It may be that the degree and precision of sound-frequency organization in the nervous system of species more primitive than cats or monkeys may depend very greatly on the importance of hearing to these species. However, it is also clear that tonotopic organization is not a specialized feature of the brain organization of higher animals alone.

DEVELOPMENT OF TONOTOPIC ORGANIZATION

The changes in the sensitivity and organization of auditory neurons to tonal stimuli as a function of age have recently been studied in the central nucleus of inferior colliculus of the kitten (34). During the first week of life discharges are evoked only by very high intensity stimuli and show little frequency specificity. As a result it is very difficult to specify a best frequency for a given unit in this period although a tendency appears to exist for higher frequencies to cause maximal excitation ventrally in the central nucleus relative to more dorsally located low frequency responses.

During the second and third postnatal weeks tuning becomes sharper and a drop occurs in threshold at best frequency. The best frequencies of units in these animals show an orderly relationship to the depth in the penetration at which they are encountered. As an example, in Figure 2 best frequencies recorded in a 14-day cat become higher with increasing depth of penetration. At this age frequencies between 1 and 4 kHz are commonly encountered and frequencies in excess of 10 kHz occupy a relatively small proportion of the penetration compared with that in adults (p. 259). The 20 kHZ best frequency unit encountered at the beginning of the penetration was located in the pericentral nucleus of this structure. The small sizes of the cells in both divisions of the inferior

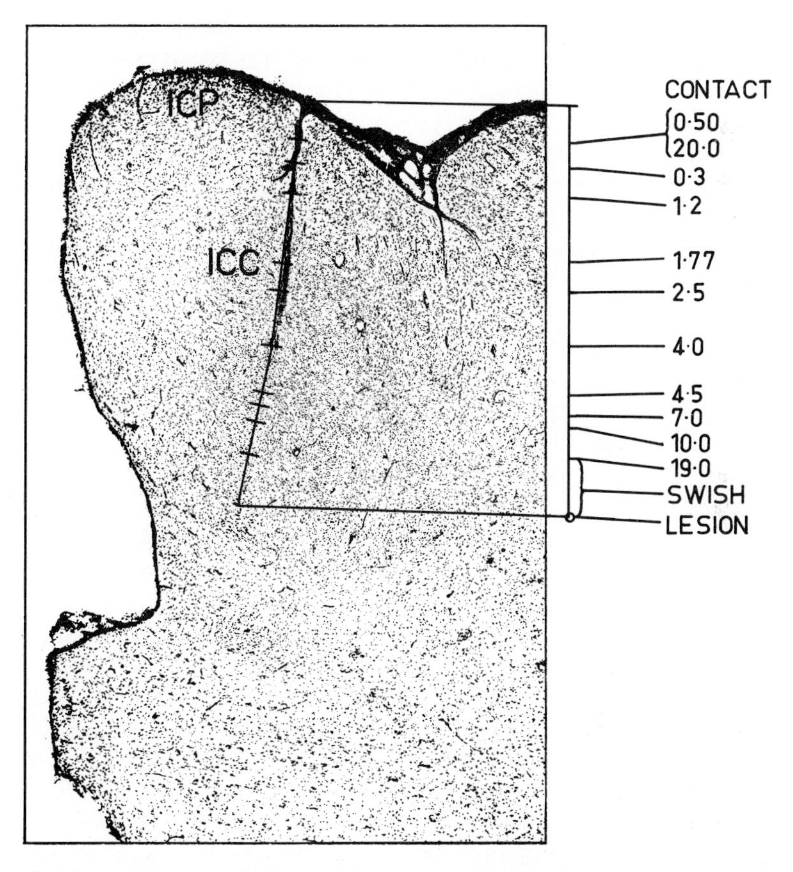

Figure 2. Photomicrograph of sagittal section through the brainstem of a 14-day-old kitten. Numbers to right are best frequencies of single units, as for Figure 1. *Swish,* background response to tones; *ICP,* pericentral nucleus of inferior colliculus; *ICC,* central nucleus of inferior colliculus. Thionine stain, × 20. (From D.R. Moore and L.M. Aitkin, unpublished observations.)

colliculus of the immature kitten may be contrasted with the much larger cells in NMLD of the adult chicken (Figure 1).

The results of this study suggest that the topography of the projection to the central nucleus of the inferior colliculus may be present at birth since, as soon as neurons show frequency specificity, they appear to be arranged in a tonotopic array. This conclusion is supported by the observation that tonotopic organization is unmodified when kittens are reared from birth to 50–79 days in an environment in which a dominantly loud, continuous tone is presented (35). If tonotopic organization was experience-dependent, some alterations might be expected following such a rearing procedure.

ANATOMICAL AND PHYSIOLOGICAL BASES OF TONOTOPIC ORGANIZATION

The prime target for this review is the evaluation of tonotopic organization at higher levels in the auditory pathway, but it is first necessary to describe the topography of the connections of the lower auditory centers to the cochlear partition.

The frequency-selective properties of the basilar membrane have been well known since the pioneering experiments of von Békésy (36). Recent publications using the Mössbauer technique (37, 38) and the capacitance probe (39) have supported the basic concept of the mechanical tuning of the basilar membrane in which high frequency sounds produce maximal vibration at the basal regions and low frequency sounds at the apical regions of the cochlea. Although the precise details of the transduction processes at the basilar membrane-hair cell-cochlear nerve fiber junctions are still unresolved, it would appear that the tuning curves of cochlear nerve fibers are sharper than the mechanical tuning curves of the cochlear partition to which they attach (40, 41).

The spatial segregation of sound frequencies in the cochlea is maintained by the orderly projection of the cochlear nerve upon the cochlear nuclei (42). Owing to spiraling of the cochlear nerve in its path to the medulla, apical fibers make approximately 1¾ turns about the axis of the nerve trunk to occupy a medial location while extreme basal turn fibers enter the cochlear nerve trunk laterally and inferiorly to terminate more distally and dorsally in the cochlear nuclei than do the apical fibers. Such a representation of the cochlear spirals in the cochlear nerve has been confirmed in the tonotopicity demonstrated in the nerve trunk using microelectrodes (40).

The conclusions of Sando (42) support the earlier observations of Lewy and Kobrak (43) that fibers from the basal coil bifurcate and distribute most dorsomedially, while those from the apical turn bifurcate and distribute most ventrolaterally. This distribution of cochlear nerve endings is in general agreement with the detailed study of tonotopic organization in the cochlear nucleus by Rose, Galambos, and Hughes (5), and in particular agreement with their description of the ventral cochlear nucleus.

The relationships between basal and apical regions of the cochlea and the various auditory nuclei are schematically summarized in Figure 3. If fibers reaching the cochlear nucleus are considered as first order, second order fibers depart from the dorsal, posteroventral, and anteroventral cochlear nucleus to the inferior colliculi via either a second medullary locus in the superior olivary complex or a direct pathway (44, 45). No information at present exists as to the proportions of afferent fibers reaching the inferior colliculus by the two routes. However, it is possible that the proportion from the bilaterally innervated olivary regions is substantial since a high proportion of units in the central nucleus of the inferior colliculus have binaural properties similar to those of superior olivary units (46–48), yet a direct projection from the ipsilateral cochlear nucleus to the inferior colliculus is either minor or absent (49–51).

Descriptions of tonotopic organization have been published for the lateral superior olive (52, 53), medial superior olive (53, 54), nucleus of the trapezoid body (53) and nuclei of the lateral lemniscus (18). The nuclei of the superior olivary complex have been particularly well studied in the report of Guinan and his colleagues (53), who have shown that isofrequency lines for the tonotopic organization in the medial nucleus of the trapezoid body are arranged with low

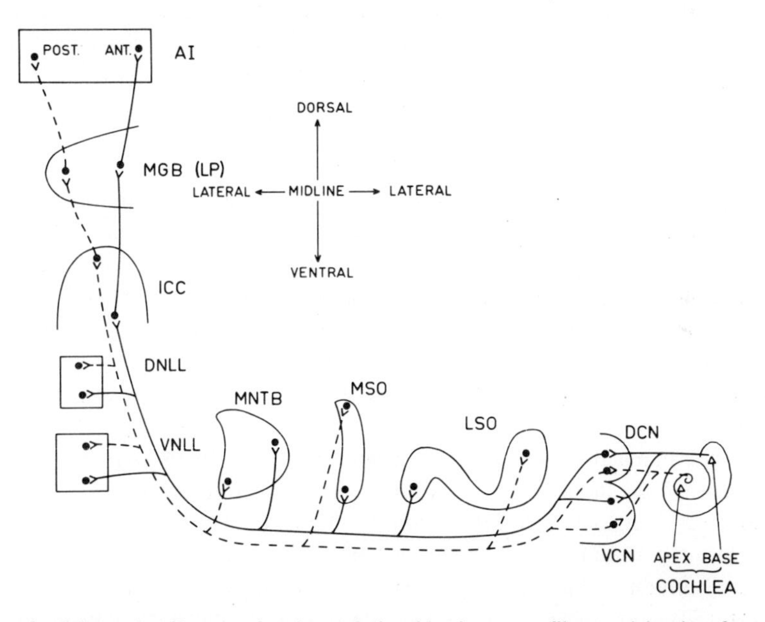

Figure 3. Schematic diagram showing relationship between fibers originating from the apical (*dashed lines*) and basal (*solid lines*) regions of the cochlea and the various auditory nuclei. *DCN, VCN,* dorsal and ventral cochlear nuclei; *LSO, MSO,* lateral and medial superior olivary nuclei; *NMTB,* medial nucleus of trapezoid body; *DNLL, VNLL,* dorsal and ventral nuclei. *DCN, VCN,* dorsal and ventral cochlear nuclei; *LSO, MSO,* lateral and medial superior lateralis of medial geniculate body; *AI,* primary auditory cortex; *POST., ANT.,* posterior and anterior.

frequencies placed dorsolaterally and high frequencies ventromedially. Their observations on the lateral superior olive concur with those of Tsuchitani and Boudreau (52) in that the lateral arm of this S-shaped nucleus contains low frequency units while the medial arm contains high frequencies. For the medial superior olive and the nuclei of the lateral lemniscus, a low-to-high, dorsal-to-ventral sequence is the simplest interpretation of the available data (18, 54).

These arrangements, and those of the higher nuclei, are schematically depicted in the sketch of Figure 3. It should be noted, in viewing this figure, that no attempt has been made to dissect the different pathways leading from the cochlear nuclei to the inferior colliculus since the sole purpose of the illustration is to relate the orientations of the tonotopic arrangements in the different auditory nuclei. A further qualification that should be noted is that evidence has accumulated for a disproportionate representation of low frequencies in the medial superior olive (53, 54) and of high frequencies in the lateral superior olive (52, 53). These distributions are qualitatively different from the greater representation of higher octaves in the upper nuclei, which will shortly be discussed, and are possibly related to the role of the nuclei of the superior olivary complex in sound localization (55).

The organizations of neurons in medullary nuclei of the auditory pathway thus reflect the connections of the auditory nerve to the cochlear partition but, with the exception of the cochlear nuclear complex, three-dimensional maps of the representation of the cochlea have not been published for these nuclei. Such material is, however, available for the central nucleus of the inferior colliculus.

Central Nucleus of the Inferior Colliculus

A recent study of Osen (51) utilizing anterograde fiber degeneration and retrograde cell degeneration techniques has delineated the region-to-region relationship between the dorsal and ventral cochlear nuclei and the central nucleus of the inferior colliculus. Lesions involving the dorsal (high frequency) portions of both divisions of the cochlear nucleus result in restricted degeneration rostroventrally in the central nucleus, while more ventrally placed lesions (low frequencies) produce corresponding changes in the dorsolateral regions of the central nucleus (Figure 4). Osen further noted that fibers from the ventral and dorsal cochlear nuclei intermingled in the central nucleus and that there were no apparent differences in the modes of termination of the two categories of fiber.

Since the original brief description by Morest (56) of the Golgi-impregnated cat inferior colliculus in which he described obliquely vertical layers of nerve cells and afferent fibers, two major studies have appeared. That of Geniec and Morest (57) described the neuronal architecture of the human posterior colliculus observed with Golgi-Cox stains, and many of its findings bear directly on the cat and other mammals. Even under relatively low magnification the impregnated neuronal processes may be seen to be arranged in parallel webs (Figure 5) which sharply differentiate the central nucleus (CN) from the adjacent cortex regions (CC,DC). The study of Rockel and Jones (58) of the inferior colliculus of the cat utilized both the Golgi-Kopsch procedure, which stains many axons as

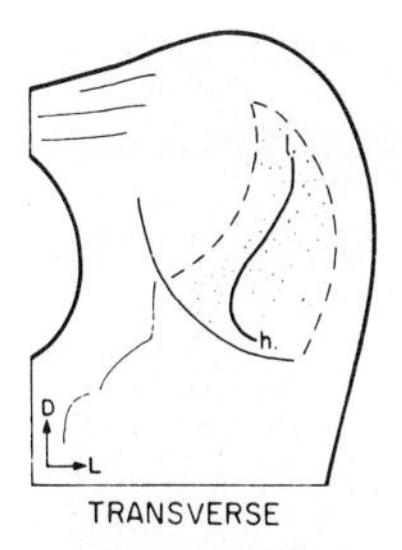

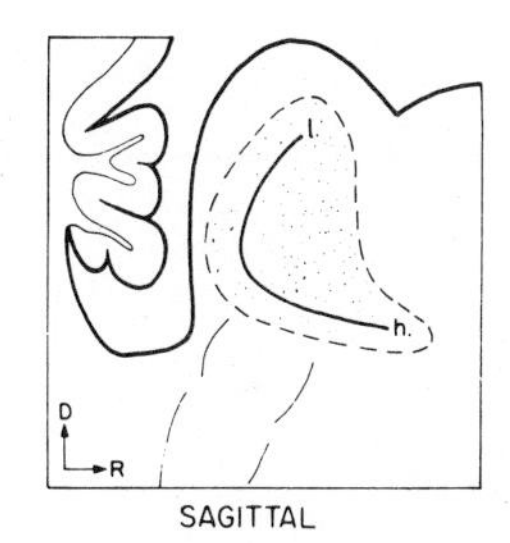

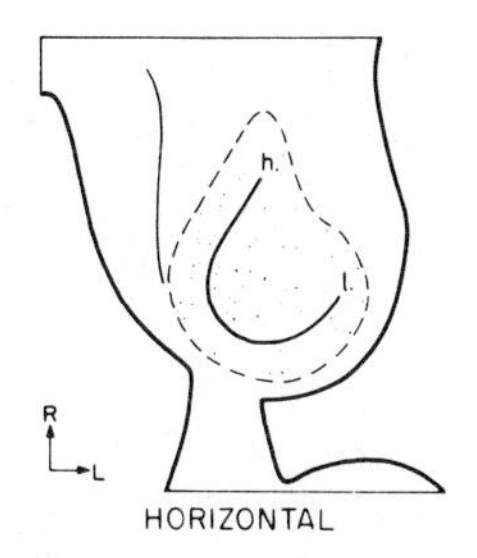

Figure 4. Diagrams of the right inferior colliculus in the transverse, sagittal, and horizontal planes to show the termination to second order auditory fibers (*dotted area*). The curved line interconnecting low frequency region (*l*) and high frequency region (*h*) indicates the suggested direction of the rise or fall in tonal frequencies within the central nucleus. *D,* dorsal; *R,* rostral; *L,* lateral. (From Osen, Ref. 51.)

well as cells, and experimental degeneration techniques which enable correlations between afferents to the central nucleus and Golgi-stained elements to be made.

Within the central nucleus Rockel and Jones have distinguished two divisions. The *ventrolateral division* has a pronounced laminar arrangement of cells, dendrites, and axons which form an "onion-like series of concentric curved shells" and which receive fibers from the lateral lemniscus only (Figure 6). The smaller *dorsomedial division* is not laminated and appears to receive afferents from other sources, including the auditory cortex (58, 110). The lamination of the ventrolateral division is not always clear in Golgi sections since there is often substantial overlap between the rows of cells and fibers. This factor makes it difficult to assign a thickness to a given lamina and Rockel and Jones consider that the best index of lamina thickness is the height of the dendritic field of the principal cells which make up the lamina, viz, approximately 100 μm.

Comparisons between the direction of tonotopic organization suggested by the degeneration study of Osen (51) and the Golgi studies of Morest (56) and Rockel and Jones (58) led to the conclusion that the fibrocellular laminae may each subserve the transmission of a restricted range of frequencies, with laminae devoted to low frequencies situated dorsolaterally and those concerned with high frequencies ventrally placed in the central nucleus of the inferior colliculus.

The presence of a tonotopic organization in the central nucleus has been accepted since the publication of the experiments of Rose and his colleagues (59), and more recent studies have confirmed these findings in the cat (18, 63) and other mammals (60–62). It is only in the last few years that a detailed physiological correlation between the layers of cells and fibers and "frequency-band laminae" has been presented (63). Merzenich and Reid (63) have shown that parallel penetrations entering over nearly the entire dorsal surface of the central nucleus reveal a similar change in best frequency as a function of depth (Figure 7). They were able to reconstruct from multiple penetrations "iso-frequency contours" which were lines joining points of equal best frequency plotted on photographs of sagittal sections of the central nucleus (Figure 8).

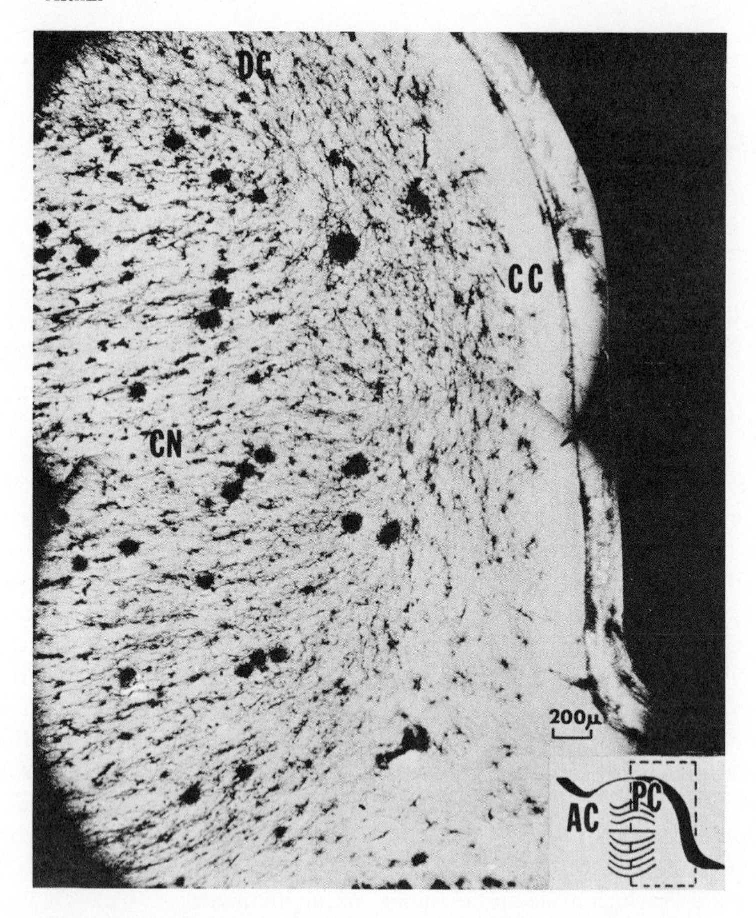

Figure 5. Photomicrograph of a parasagittal section at the junction of the middle and lateral thirds of the posterior colliculus showing impregnation of the dendritic laminae of the central nucleus and the cellular layers of the cortex. *Inset,* orientation of the dendritic laminae in the central nucleus. The *dashed lines* demarcate the sector of the posterior colliculus shown in the photomicrograph. *CN,* central nucleus; *CC, DC,* caudal and dorsal cortex, respectively; *AC, DC,* anterior (superior) and posterior (inferior) colliculi. Golgi-Cox, 53-year-old man. (From Geniec and Morest, Ref. 57.)

The isofrequency contours so obtained were approximately parallel to each other but their orientations varied somewhat, depending on the part of the central nucleus sampled. In the medial part of the central nucleus, as for the example illustrated (Figure 8), the contours were nearly horizontal; in the central and lateral regions the contours were tilted upwards rostrally and downwards laterally. These "physiological laminae" differed from those suggested by the anatomists (56, 58) only in their orientation at the margins of the central nucleus—for example, an upward tilt of the laminae might be expected laterally from anatomical findings (Figure 6), but a downward tilt is observed physiologically (Figure 8).

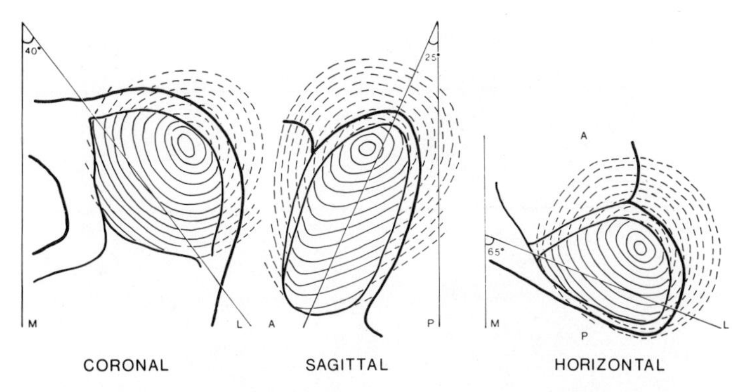

Figure 6. Semi-schematic representations of the laminar planes of the central nucleus as they appear in coronal (*left*), sagittal (*middle*), and horizontal (*right*) sections. The approximate orientations of the axes of the central nucleus in relation to the three standard planes are also indicated. *A*, anterior; *P*, posterior; *M*, medial; *L*, lateral. (From Rockel and Jones, Ref. 58.)

The results of Merzenich and Reid may thus be viewed as indicating that each fibrocellular lamina relays a discrete range of frequencies, but this view must be tempered by one qualification. The localization "errors" described under "Experimental Considerations" may be significant in the above study since the authors included in their tonotopic maps the responses of numerous

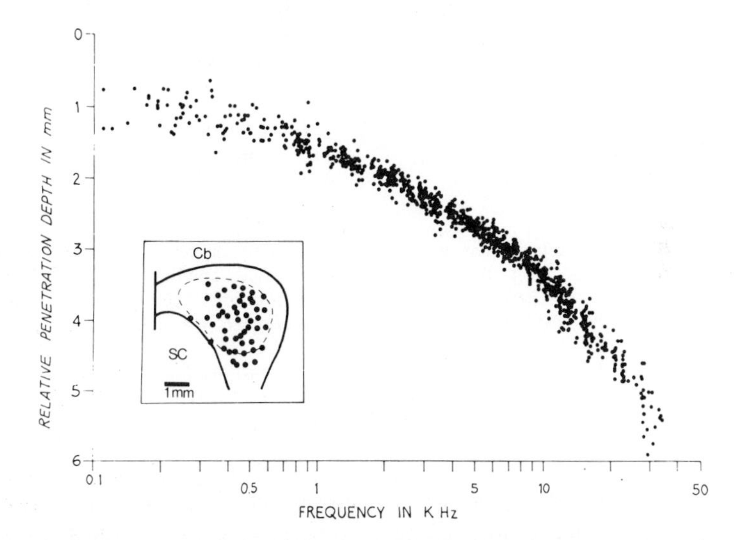

Figure 7. Best frequencies for neurons at 976 sites along 49 penetrations across the central nucleus plotted as a function of penetration depth. Individual axes were scaled along the ordinate so that they overlay one another. All penetrations were in the sagittal plane, directed caudalwards at angles ranging from 20° to 35° out of the front stereotaxic plane. The approximate locations of the penetrations are indicated on the schematic outline of a dorsal view of the inferior colliculus shown in the *inset*. The *broken line* in the *inset* defines the approximate limits of the deep lying central nucleus. *Cb*, cerebellum; *SC*, superior colliculus. (From Merzenih and Reid, Ref. 63.)

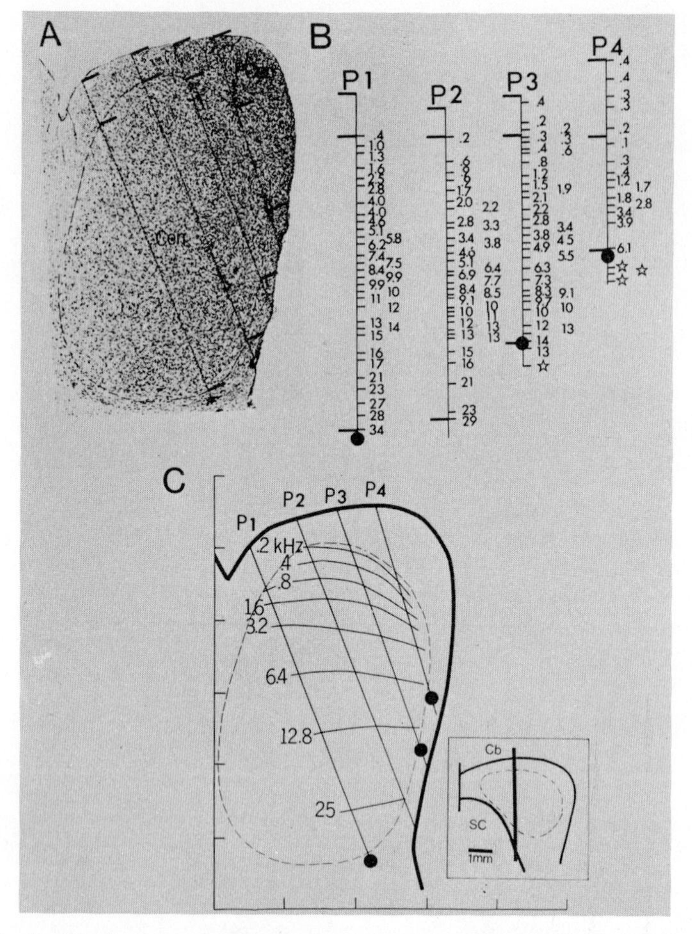

Figure 8. *A,* sagittal section through inferior colliculus at the approximate level indicated in the *inset* in *C.* The caudal aspect of the nucleus is to the right. The lines of four penetrations across the central nucleus (*Cen*) are indicated. *B,* reconstruction of the penetration shown in *A.* Each number is the best frequency in kHz for units or groups of units, and lesions are shown as filled circles near the end of each penetration. *C,* approximate iso-frequency contours, drawn for frequencies representing octave separations, from consideration of the penetrations shown in *A* and *B. P. Cen,* pericentral nucleus. (From Merzenich and Reid, Ref. 63.)

unit clusters, which are likely to be composed of discharge of many neurons and fibers in the vicinity of the recording electrode. This fact opens the possibility that, in addition to the laminar grouping of best frequency described, a progression of best frequency may also exist *along* a given fibrocellular lamina. This possibility can only be resolved by electrode penetrations made at right angles to those used by Merzenich and Reid.

Figures 7 and 8 also suggest that lower frequency octaves receive proportionally less space than do the higher frequency octaves. This may be partly accounted for by the fact that higher frequencies are represented along greater lengths of basilar membrane than are lower frequencies (64, 65), but the cochleotopic organization of the central nucleus calculated by Merzenich and Reid (63) still shows nonlinearity in that the basal-most 5 mm of basilar membrane occupy more than twice the length of a penetration than does the apical-most 5 mm of membrane.

The greater representation of basal regions of the basilar membrane could imply that more neural tissue is needed to adequately signal higher frequencies (5–30 kHz) than frequencies below 5 kHz. This suggestion is in harmony with the observation that the discharges of peripheral neurons are usually phase-locked to the cycles of low-frequency sinusoids (66), but this mechanism of frequency coding may not be available for the discrimination of higher pitches. However, phase-locked discharges are not commonly observed for units in either the anesthetized (117) or unanesthetized (9) cat inferior colliculus. A second possible implication of the disproportionate representation of high frequencies is that cats may make more use of frequencies in the higher octaves than they do of low frequency information (67).

The layered structure of the inferior colliculus is likely to be responsible for the observation that changes in best frequency are in discrete steps in a given penetration and usually no more than 30–40 separate best frequencies may be encountered during such a penetration (63). The maximum length of a microelectrode penetration through the central nucleus rarely exceeds approximately 5 mm, giving an average thickness to a physiological layer of from 130 to 170 μm. Since the relatively large-tipped platinum-iridium microelectrodes (2–6 μm tip diameter) used by Merzenich and Reid may not have detected the action potentials of smaller cell bodies (24), this result is in good agreement with the estimate of 100 μm by Rockel and Jones for the thickness of an anatomical lamina (58).

Tonotopic organization also appears to be present in the pericentral (48, 63) and external nuclei of the inferior colliculus (48, 59) but little precise information exists as to the mapping of the cochlear partition upon these structures. Furthermore, the broad tuning and rapid habituation exhibited by neurons in both nuclei argue against a function in frequency discrimination for these regions (48).

In conclusion, it may be seen that much precise information exists relating to the tonotopic organization of the central nucleus of the inferior colliculus. Fibers entering the central nucleus arising from the cochlear nucleus terminate in discrete somatodendritic laminae. The dorsoventral region of their termination is related in an orderly fashion to their region of origin in the cochlear nucleus. Microelectrode penetrations suggest that each lamina is related to a restricted range of best frequencies and that the best frequencies increase in a systematic way as a function of dorsoventral depth in the central nucleus.

Given this highly ordered framework, several further problems require solution before the role of the central nucleus in pitch discrimination may be determined. First, it will be important to discover the common and disparate properties of cells along each lamina. Second, several different morphological types of neuron comprise the central nucleus and the possibilities which exist for local interactions to occur between neurons within and across laminae need to be investigated. Finally, the organization of the central nucleus in relationship to binaural processing must be assimilated into the tonotopic organization described using monaural stimulation.

Ventral Division of the Medial Geniculate Body

At the time of writing no study of the topography of the projection of the central nucleus of the inferior colliculus upon the ventral division of the medial geniculate body has been published, although it is well known that the latter nucleus receives a dense projection from the former (68–70). Such a study is, however, under way in this laboratory and results to date suggest that lesions of the central nucleus produce localized terminal degeneration in the ventral division of the medial geniculate body which is arranged in bands roughly parallel to the lateral convex surface of this nucleus.

This banded pattern of degeneration correlates with the laminar pattern of cells and fibers observed by Morest (72, 73) to typify the ventral division. The fibrodendritic patterns are complex (Figure 9) and are composed of two coils of dendrites in the ventromedial part of the ventral division, named by Morest *pars ovoidea,* and lines of dendrites arranged parallel to the lateral surface of the medial geniculate body in a region referred to as *pars lateralis.* Between these two areas short horizontal lines of dendrites are orientated in an oblique longitudinal direction. These details relate particularly to the posterior and middle thirds of the medial geniculate and, although the anterior part of this structure was not discussed in detail, Morest has indicated that the parallel lines of dendrites become dominant at more and more anterior levels of the medial geniculate body (72, 73).

The dimensions of the laminae and their number in *pars lateralis* are difficult to assess because, as in the inferior colliculus, much overlapping occurs between adjacent dendritic trees. The radius of the dendritic field of a principal cell is, on the average, 120 μm but if overlapping is taken into account the distance between cell bodies in adjacent laminae is closer to 50 μm (73). Since the thickness of *pars lateralis,* estimated to be 2.4 mm in some sections, is only about half that of the central nucleus of the inferior colliculus, the number of lateromedial cell layers in *pars lateralis* is likely to be of the same order as the number of dorsoventral laminae in the central nucleus, viz. 40–50. Thus it is possible that each "frequency-band lamina" in the inferior colliculus has its counterpart in the medial geniculate body.

The complex structural features of the laminae in the ventral division are probably responsible for the inability of early studies to demonstrate tonotopic organization in this nucleus (26, 74). Microelectrode penetrations from dorsal-to-ventral may traverse the dorsal division, parts of *pars lateralis* and parts of *pars ovoidea,* depending on their rostrocaudal point of entry, giving a rather chaotic appearance to the sequence of best frequencies so obtained. Even

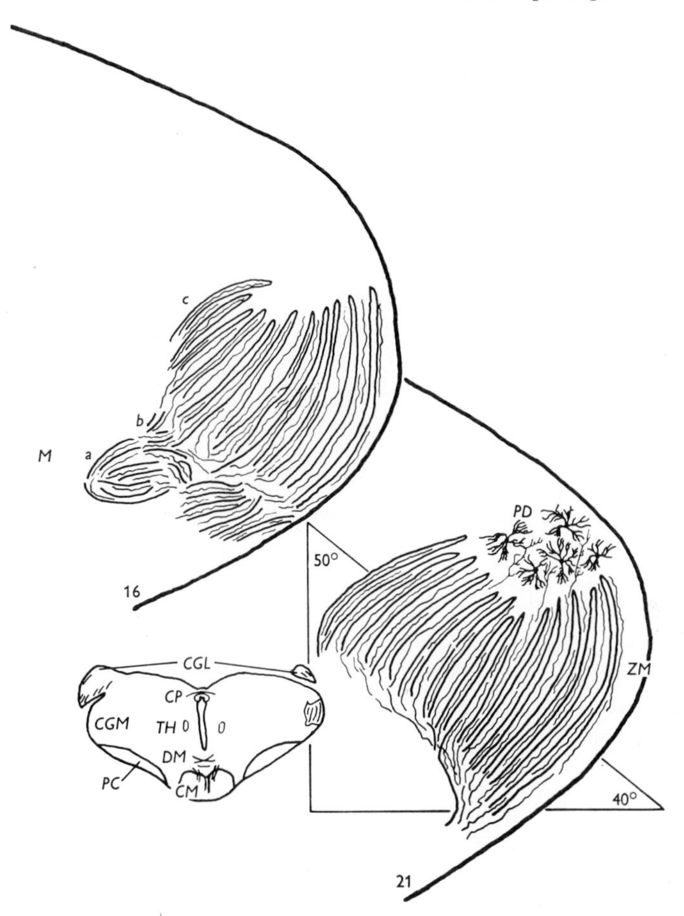

Figure 9. Transverse reconstructions (× 15.5) of the fibrodendritic layers of the ventral nucleus near the junction of the anterior and middle one-third of the medial geniculate body in a 15-day-old cat. *a,b,* spiral and transitional zones of pars ovoidea; *c,* columnar zone of pars lateralis; *m,* medial division; *z,* marginal zone; *PD,* dorsal division of medial geniculate body (*CGM*). *Inset,* outline drawing of section 16, showing plane of sectioning through the posterior pole of the lateral geniculate nucleus (*CGL*) and mammillary bodies (*CM*). *PC,* cerebral peduncle; *DM,* supramammillary decussation; *TH,* habenulo peduncular tract; *CP,* posterior commissure. (From Morest, Ref. 73.)

carefully reconstructed tracks passing in this manner would prove difficult to relate to the fibrodendritic laminae only apparent in Golgi material.

The first positive evidence for a lateral-to-medial, low-to-high frequency arrangement in the ventral division of the medial geniculate body was obtained when potentials evoked by electrical stimulation of the cochlea were related to the lateromedial location of the recording electrode (75). In these experiments basal turn stimulation in the cochlea was only found to be effective in the medial parts of the ventral division, while the effects of apical turn stimulation were localized to the lateral boundary of the medial geniculate body. These findings have been substantiated at the unit level by Adrian, Gross, Ley, and Lifschitz, as cited by Woolsey (76), who used microelectrodes to demonstrate that neurons with progressively higher best frequencies were located from lateral to medial in the ventral division.

Three recent studies have defined the pattern of tonotopic organization in the ventral division although they have not been carried far enough to allow the details of cochleotopic organization to be determined (77–79). In the cat, units responding to low frequencies are first encountered at the lateral margins of *pars lateralis* and best frequencies become higher as the electrode is located more and more medially (77, 78). The best frequencies of 44 units from 4 penetrations show a more or less linear relationship to distance from the lateral margin (Figure 10) but whether this pattern is an expression of a real difference between *pars lateralis* and the central nucleus of the inferior colliculus (Figure 7) awaits a larger sample of measurements in the medial geniculate.

Penetrations through *pars ovoidea* have not revealed its organization, but this is probably due to the intricate windings of the laminae in this structure (73, 78). As for the central nucleus of the inferior colliculus, high best frequency units were more commonly encountered in the ventral division than were low frequency neurons. Approximately twice as many units were recorded with best frequencies in the octave 12.8–25.6 kHz as in any other octave (78).

The auditory thalamus of the squirrel monkey has also been shown to be organized tonotopically (79) and it is possible to recognize a small-cell division, presumably analogous to the ventral division of the cat, and two divisions with large, darkly stained cells, perhaps equivalent to the medial division of the cat. A microelectrode traversing from lateral to medial detects units of rising best frequency in the small-cell division, with a reversal and a new progression from low to high in the darkly stained regions.

If the latter areas are, as the authors suggest, homologs of the medial division of the cat medial geniculate body they would appear to differ from the cat in that tonotopic organization has not been demonstrated in this nucleus in the cat (80). This area is very similar to the pericentral and external nuclei of the cat inferior colliculus in that it contains units of very broad tuning and labile discharge characteristics (80).

The complete organization of the medial geniculate body remains to be elucidated but it is clear that *pars lateralis* of the ventral division consists of neurons with best frequencies encompassing the total audible spectrum. These are compressed in a band which has half the thickness of the dorsal-to-ventral dimensions of the inferior colliculus but which extends for nearly 4 mm from posterior to anterior in the form of "isofrequency sheets." It seems likely that each isofrequency lamina of the central nucleus of the inferior colliculus is related to one isofrequency sheet in the ventral medial geniculate body.

Primary Auditory Cortex

The mammalian auditory cortex was the original site for the investigation of tonal localization (2–4) and it has recently received attention in two excellent

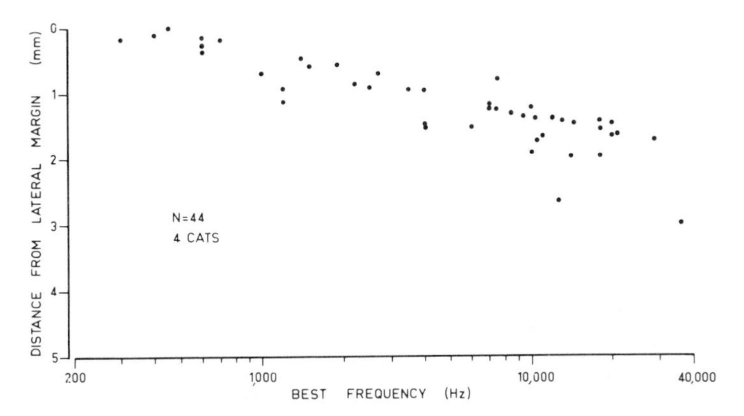

Figure 10. Best frequencies in Hz of 44 units from four cats plotted as a function of penetration distance in mm from the lateral margin of the medial geniculate body, at an anterior-posterior level corresponding to the middle one-third of this nucleus. (From Aitkin and Webster, unpublished observations.)

detailed reports on the cat and monkey auditory cortices (14, 81). Additional material is also available which reveals precise tonotopic organization in the primary auditory areas of the cortices of the owl monkey (115) and gray squirrel (116). The substance of the studies on cat and macaque monkey will be considered following a description of the topographic relationships between the ventral division of the medial geniculate body and the primary auditory cortex.

The existence of a topographic relationship between these two areas has been known since the retrograde degeneration experiments of Rose and Woolsey (82) in which the removal of small areas of auditory cortex in the cat led to circumscribed zones of cell death in the medial geniculate body. This study, combined with the cytoarchitectural study of Rose (83), defined the limits of the primary auditory cortex in the cat (AI) in relationship to its neighboring secondary areas. These areas have not received the detailed attention which has been devoted to AI in recent years and will not be considered in this review. However, the pattern of projection suggested by Rose and Woolsey (75, 82) was later substantiated by electrical stimulation of different parts of the medial geniculate body and recording of evoked potentials over the surface of the auditory cortex (76).

More recent studies involving anterograde fiber degeneration procedures have tended to obscure rather than illuminate the nature of the projection patterns, partly because the insertion of the lesion-making electrodes into the thalamus often interrupted cortical association pathways and also partly because of the different staining procedures used by different experimenters (84–86). The detailed study of Sousa-Pinto (85), using variants of the Nauta technique, indicated that *pars lateralis* of the medial geniculate body may be the only division of that nucleus to send fibers to AI in the cat. However, experiments in which horseradish peroxidase has been injected into discrete sites in AI and traced back to the cells of origin in the medial geniculate body indicate that there is a spatially ordered convergent projection from all three divisions of the medial geniculate body to a point in AI (118). Sousa-Pinto has also suggested, on the basis of lesion reconstruction and examination of the resulting degeneration in AI, that caudal regions of *pars lateralis* project to ventrocaudal regions of AI, while rostral regions of each area are connected (85). Histological studies in this laboratory suggest that a representation of latero-medial *pars lateralis* along a caudorostral axis in AI is nearly compatible with Sousa-Pinto's observations and very compatible with the electrophysiology of *pars lateralis* (78) and area AI (13, 14).

The different locations of the ventral medial geniculate lesions in the five brains illustrated in Figure 11 were associated with degeneration which encompassed most of the surface of AI when the results were combined. The boundary lines for AI and AII, derived from Woolsey (87), are shown for brain 74–13. Lesions in brains 73–39 and 74–6 encompassed much of the lateromedial extent of the middle third of the ventral division and led to a swathe of terminal degeneration across AI in a caudorostral direction, most extensive in the case of brain 73–39. The more anterior and medial lesion of brain 73–35 produced degeneration in a more anterior sector of AI, while the more posterior and ventral lesions of brains 73–52 and 74–13 were associated with smaller patches of terminal degeneration about and within the posterior ectosylvian sulcus. The overall pattern of lesion and degeneration agrees with a scheme in which lateral, caudal, and ventral regions of the ventral division of MGB project posteriorly in AI, while medial and rostral sectors project to anterior AI. Such a scheme is nearly identical to that proposed by Merzenich and Colwell (118) on the basis of experiments utilizing horseradish peroxidase.

The pattern of intracortical termination of thalamic fibers in AI revealed following lesions in the ventral division is depicted in Figure 12. The locations of the layers indicated by Roman numerals to the right of the figure were determined following consideration of adjacent thionin-stained sections and closely follow the descriptions given by Rose (83) and Sousa-Pinto (88). A dense aggregation of thalamic axon terminals is present in layer IV and a lighter scattering of degeneration is also present in layer III. Degeneration in deeper layers is mostly that of preterminal axons. Terminal degeneration may also be observed in some sections in layer I, as originally demonstrated by Wilson and Cragg (84).

The cell types observed in area AI in Golgi sections have been described by Sousa-Pinto (88) and similar observations have been made in this laboratory (Figure 13). The majority of neurons in layer II are small pyramidal cells with apical dendrites which branch after a short apical shaft (Figure 13*A*). Layers III (Figure 13*B*) and IV (Figure 13*C*) are populated by pyramidal cells with apical dendrites extending into layer I and these are associated with numerous stellate neurons. Layer V contains few cells and appears as a light band in Nissl sections. In Golgi sections large pyramidal neurons may be observed (Figure 13*E*) while cells in layer VI are diverse in dendritic conformation and some neuron types, including large fusiform cells and inverted pyramidal cells (Figure 13*D*) are observed to occur only in this layer.

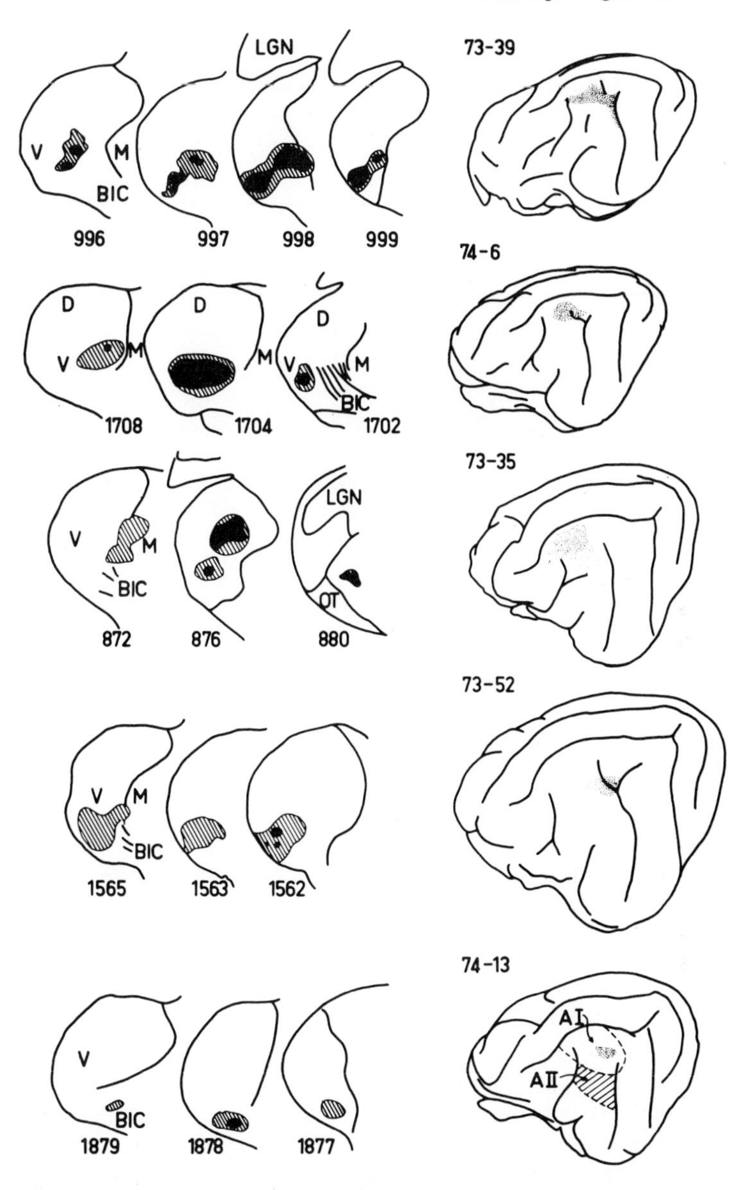

Figure 11. The distribution of degenerated axon terminals in the auditory cortex (*dotted*) in five hemispheres with lesions in the medial geniculate body (*MGB*). The latter is depicted for each animal in three or four sections which correspond to the anterior and posterior limits of the lesion and at least one section near the region of maximum damage. *Black regions* are those where brain tissue has disappeared, *hatched regions* (excluding hemisphere 74–13) are areas of tissue which stained abnormally and are probably damaged. Numbers below each 25 μm section are for reference purposes with each step 250 μm. 73–39, etc. refer to the year and the cat number. *V,* ventral division of MGB; *M,* medial or magnocellular division of MGB; *BIC,* brachium of inferior colliculus; *LGN,* lateral geniculate nucleus; *OT,* optic tract. (From Aitkin and Farrington, unpublished observations.)

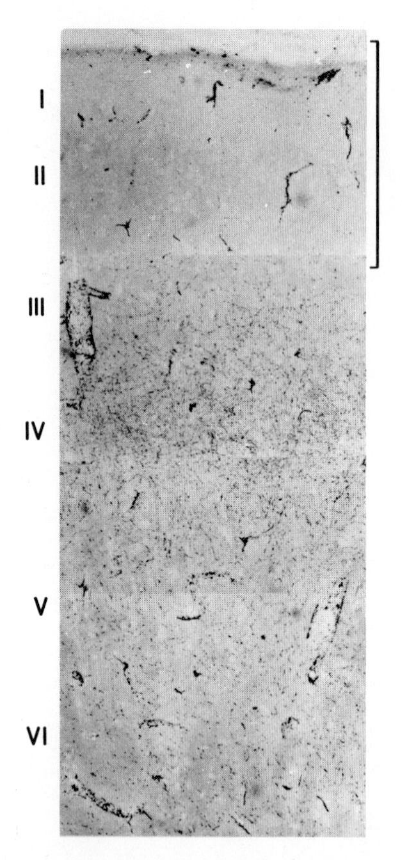

Figure 12. 25 μm Fink-Heimer section from cat 73–39, showing well defined lamination in the location of degenerating thalamic terminals within AI. Bar is 500 μm. (From Aitkin and Farrington, unpublished observations.)

Microelectrode mapping experiments in the cat have, until recently, shown only a weak general trend for tonotopic organization to be present across the primary auditory cortex (6, 7, 20–22) and theories have been proposed which presume that "the organization of the primary cortex is not tonotopic in any sense that could have the functional implication of frequency analysis" (Ref. 89, p. 285). These theories have been partly supported by ablation studies in cats in which removal of the auditory cortex does not completely remove the ability of the animal to make behavioral responses to changes in tone frequency (90–92). However, not all responses to tones remain after such ablations (93, 94) and studies in humans suggest that pitch perception is intimately related to the temporal lobe (95–97).

Merzenich, Knight, and Roth (13, 14) have recently described a highly ordered tonotopic representation in AI of the cat. They have ascribed their positive findings to the use of detailed maps of AI drawn from each of a series of animals and they have not used composite maps obtained by combination of

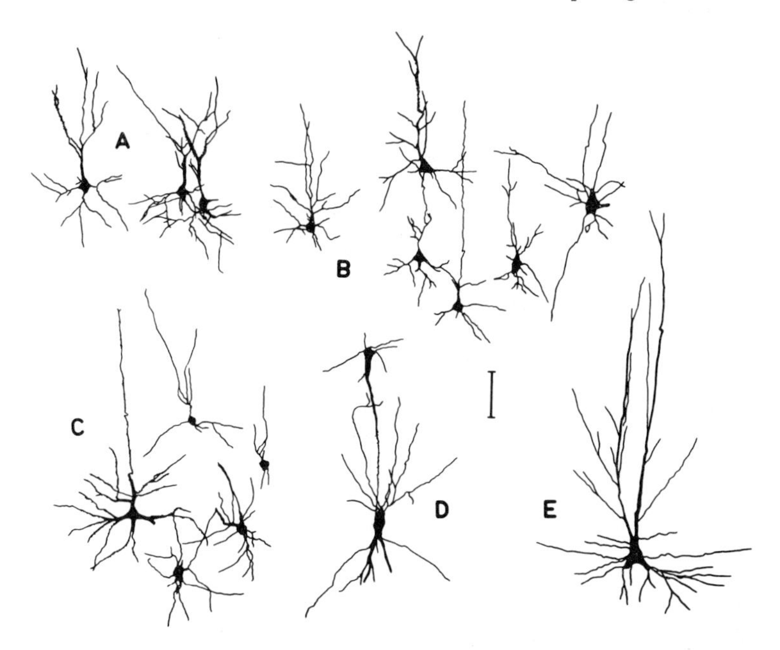

Figure 13. Golgi-Cox impregnated cell profiles from layer II (*A*), layer III (*B*), layer IV (*C*), layer V (*E*), and layer VI (*D*). Bar is 100 μm. Drawings have been arranged with the apical dendrites of pyramidal cells pointing upwards (pial-wards). (From Aitkin and Farrington, unpublished observations.)

smaller amounts of data from many experiments. The latter methods have been used in most of the previous studies which have also frequently used unanesthetized animals. The drawbacks of these procedures have been outlined previously (p. 251).

The details from two experiments of Merzenich and his colleagues (14) are shown in Figure 14. These surface maps show a simple sequence of ascending best frequencies for any line drawn through penetrations across the caudal-to-rostral (right-to-left) dimension of the primary field. They also show that any given frequency band is represented across a belt of cortex, aligned in a nearly straight mediolateral axis and extending across the width of the primary field. The dashed lines in Figure 14 thus represent "isofrequency contours."

The representation of the higher frequency octaves was found to occupy a higher proportion of AI than that of subcortical sites (63, 78), but this was difficult to quantify because much of the very low frequency representation is buried within the banks of the posterior ectosylvian sulcus.

Penetrations normal to the surface of AI, particularly in the middle and deep layers of cortex, detected units which had best frequencies usually within 0.1 octave of each other. Taken in conjunction with the observations that stepwise changes in best frequency were observed in long penetrations down the banks of the posterior ectosylvian sulcus, the results of Merzenich and his colleagues (14) suggest a vertical unit of organization in primary auditory cortex similar to those

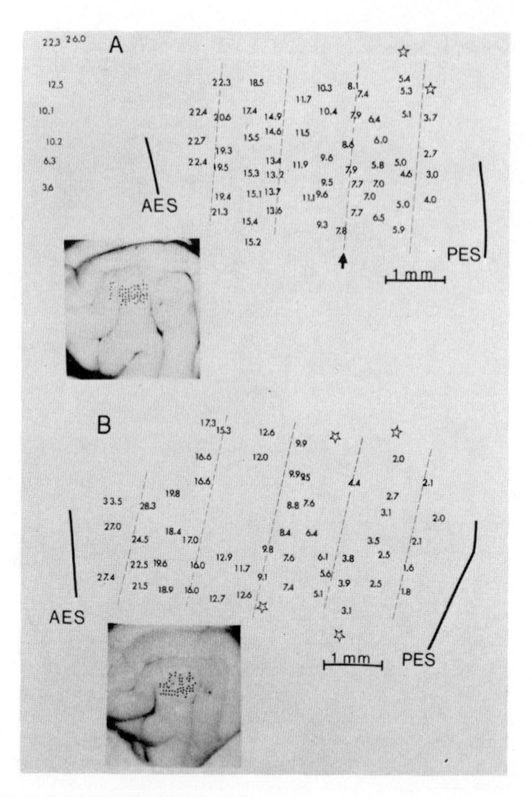

Figure 14. *A,* surface map showing the distribution of best frequencies determined within 66 parallel electrode penetrations into auditory cortex. The sites of these electrode penetrations are represented by *dots* in the brain photograph in the *inset.* Each number in the drawing is the best frequency (in kHz) of neurons encountered within a penetration at that site. *Dashed lines* are approximately parallel to "isofrequency contours." *Stars* represent electrode penetrations in which auditory responses were encountered, but in which no best frequency determination could be made. *AES,* anterior ectosylvian sulcus; *PES,* posterior ectosylvian sulcus. *B,* surface map showing the distribution of best frequencies determined in 57 electrode penetrations into auditory cortex. *Dashed lines* approximately parallel "isofrequency contours." (From Merzenich, Knight, and Roth, Ref. 14.)

demonstrated in the somatosensory (98) and visual (99) cortices. The findings for the auditory cortex are compatible with the earlier study of Abeles and Goldstein (100).

The cortical auditory area in many primates is located on the superior temporal plane of the superior temporal gyrus, and an orderly representation of the cochlear partition has been accepted in this area for at least 30 years (101, 102). The recent study of Merzenich and Brugge (81) has given exact definition to tonotopic organization in the cortex of the macaque monkey, particularly to the field which is the homolog of AI of the cat. In this area best frequencies of from 100 Hz to 32 kHz, nearly the full range of frequencies which these monkeys are known to hear (103), are represented with low frequencies rostral and lateral and high frequencies caudal and medial. It is probably significant, in view of the higher upper limit of frequency to which the cat is able to respond (104) compared with the monkey (103), that the representation of the apical regions of the cochlea is at least as pronounced as that for basal regions of the cochlea in the auditory cortex of the macaque monkey.

It is much more difficult to make the detailed correlations between neuronal architecture and fiber connections and tonotopic organization for AI that are possible for subcortical nuclei. This difficulty is probably due to the great complexity of the internal structure and connections of AI and to the probability that ascending auditory connections provide only a small proportion of the fibers terminating in a given region of the primary cortex. Although no details are available for the auditory cortex, electron microscope studies of the visual cortex suggest that only 2–10% of the terminals in layer IV of area 17 are the endings of axons arising in the lateral geniculate nucleus (105–107). Further complication also occurs because fibers from the association cortices and from contralateral AI in the cat terminate at different levels of AI to those from the thalamus (108–109).

In contrast to AI, ascending auditory fibers to the medial geniculate body and inferior colliculus are likely to be the dominant input to those areas (68, 69). Descending fibers from the auditory cortex terminate in a laminar fashion which parallels the ascending termination (110). From the point of view of tonotopic organization, subcortical sites may thus be considered as simple relay nuclei but this cannot be assumed for AI.

In conclusion, neurons of the primary auditory cortex of the cat and the monkey are tonotopically organized in a systematic fashion. This feature is particularly pronounced for neurons encountered at depths corresponding to layers III and IV (14) where most thalamocortical axons terminate. More complex responses of cortical neurons, such as those to frequency-modulated tones and to natural calls (11, 89), may be representative of further integration by auditory cortex processes for which a vast anatomical substrate exists.

CONCLUSION

This review has emphasized the organization of single neurons according to their *threshold* best frequencies in the central nucleus of the inferior colliculus, ventral division of the medial geniculate body, and area AI of the auditory cortex, particularly in the cat for which most physiological and anatomical details are available. Perhaps the most significant observations common to these studies are that tonotopic organization is very precise in all three regions, and that optimally placed microelectrode penetrations reveal step-like shifts in best frequency which correlate with the electrode traversing cell and fiber laminae or columns. If very low intensities, such as those used in the production of tonotopic maps, were the only sound intensities to which the behaving animal was exposed it would not be difficult to attempt an analysis of frequency discrimination based upon the frequency-resolving power available in a matrix of discretely sensitive neural elements. However, it has long been known that the sharpness of tuning of particular neurons to sound frequencies depends considerably upon the sound intensity employed and that most neurons respond to a substantial range of frequencies at higher intensities. Recent measurements of tuning curves of units in the inferior colliculus (48) and medial geniculate body (78) have additionally shown that sharpness of tuning is a function of best frequency as well as sound intensity. These physiological findings would appear to contrast with the fact that, for man, differential frequency limens are poorer at threshold intensities than they are at higher levels, and poorer at high compared with low sound frequency (119).

The increases in width of tuning curves for individual units at high intensities suggest that suprathreshold sound stimulation at a given frequency could produce neural activity in many "frequency-band laminae" adjoining and including that defined at threshold for that frequency. It is difficult to postulate, therefore, that activity at a particular locus in an auditory nucleus is a *sufficient* condition for defining the frequency of a sound stimulus producing that activity, although it may be sufficient for defining a *range* of frequencies.

How then is accurate frequency discrimination possible? Let it be assumed that tonotopic organization, although not a sufficient condition, is a necessary condition for the occurrence of frequency discrimination in the behaving animal. This assumption need not preclude the possibility that tonotopic organization may have additional functions which may differ for each auditory area. For example, it has been suggested that tonotopicity may subserve the apparent elevation of a sound in space when its frequency is changed from low to high (120).

A number of additional neural cues are available to distinguish between sounds of different frequency. First, for low frequency tones, the phenomenon of phase-locking may be utilized in frequency discrimination (66). Second, the number of discharges is usually greater at best frequency and the firing pattern for a particular neuron often differs at best frequency compared with frequencies on either side of this value (10, 12, 15, 18, 78, 111, 113). Third, the latent period to the first discharge to a tonal stimulus is usually shorter at best frequency than at adjacent frequencies (18, 113).

A feature of the neuronal organizations of the inferior colliculus, medial geniculate body, and area AI is that, in addition to neurons arranged with their dendrites forming the laminae previously discussed, neurons are also observed with dendrites which lie across fibrocellular laminae and these may provide connections between adjacent laminae (58, 72, 88). For example, Rockel and Jones (58) describe, for the central nucleus of the inferior colliculus, large multipolar cells whose dendrites cross several laminae. Morest (72) has similarly described Golgi type II cells in the ventral division of the medial geniculate body whose function is likely to be inhibitory (112).

If the translaminar processes present in the inferior colliculus also have an inhibitory function, they may inhibit neurons in laminae adjacent to that in which their cell bodies lie. The presence of shorter latencies and more vigorous responses at best frequency for a given neuron compared with adjacent frequencies may be translated to earlier activation in the lamina in which that neuron lies compared with adjacent laminae. Inhibitory cross-connections may then suppress activity in those adjacent laminae, providing a contrast between discharges in flanking and core laminae. These local inhibitory mechanisms, interacting with descending influences of cortical origin, are thus in a position to produce an excitation of neurons in a more limited portion of a tonotopic framework than might be suggested by the width of individual tuning curves. These influences may provide the necessary neural machinery based upon a tonotopically organized auditory pathway to allow a behaving animal to make an accurate discrimination of sound frequency.

REFERENCES

1. McCulloch, W.S., Garol, H.W., Bailey, P., and von Bonin, G. (1942). The functional organization of the temporal lobe. Anat. Rec. 82:430.
2. Woolsey, C.N., and Walzl, E.M. (1942). Topical projections of nerve fibers from local regions of the cochlea to the cerebral cortex of the cat. Bull. Johns Hopkins Hosp. 71:315.
3. Tunturi, A.R. (1944). Audio frequency localization in the acoustic cortex of the dog. Amer. J. Physiol. 141:397.
4. Hind, J.E. (1953). An electrophysiological determination of tonotopic organization in auditory cortex of cat. J.Neurophysiol. 16:475.
5. Rose, J.E., Galambos, R., and Hughes, J.R. (1959). Microelectrode studies of the cochlear nuclei of the cat. Bull. Johns Hopkins Hosp.104:211.
6. Evans, E.F., and Whitfield, I.C. (1964). Classification of unit responses in the auditory cortex of the unanesthetized and unrestrained cat. J.Physiol.(Lond.) 171: 476.
7. Evans, E.F., Ross, H.F., and Whitfield, I.C. (1965). The spatial distribution of unit characteristic frequency in the primary auditory cortex of the cat. J.Physiol.(Lond.) 179:238.
8. Goldstein, M.H. (1968). Single unit studies of cortical coding of simple acoustic stimuli. *In* Francis D. Carlson (ed.), Physiological and Biochemical Aspects of Nervous Integration, p. 131. Prentice-Hall, Inc., Englewood Cliffs, N.J.
9. Bock, G.R., Webster, W.R., and Aitkin, L.M. (1972). Discharge patterns of single units in inferior colliculus of the alert cat. J.Neurophysiol. 35:265.
10. Whitfield, I.C., and Purser, D. (1972). Microelectrode study of the medial geniculate body in unanesthetized free-moving cats. Brain Behav.Evol. 6:311.
11. Funkenstein, H.H., and Winter, P. (1973). Responses to acoustic stimuli of units in the auditory cortex of awake squirrel monkeys. Exp.Brain Res. 18:464.
12. Brugge, J.F., and Merzenich, M.M. (1973). Responses of neurons in auditory cortex of the macaque monkey to monaural and binaural stimulation. J.Neurophysiol. 36:1138.
13. Merzenich, M.M., Knight, P.L., and Roth, G.L. (1973). Cochleotopic organization of primary auditory cortex in the cat. Brain Res. 63:343.
14. Merzenich, M.M., Knight, P.L., and Roth, G.L. Representation of the cochlea within primary auditory cortex in the cat. J.Neurophysiol. In press.
15. Aitkin, L.M., and Prain, S. (1974). Medial geniculate body: unit responses in the awake cat. J.Neurophysiol. 37:512.
16. Bock, G.R., and Webster, W.R. (1974). Spontaneous activity of single units in the inferior colliculus of anesthetized and unanesthetized cats. Brain Res. 76:150.
17. Webster, W.R. (1973). Single unit studies of the cochlear nucleus in the awake cat. Proc. Australian Physiol. Pharmacol. Soc. 4:89.
18. Aitkin, L.M., Anderson, D.J., and Brugge, J.F. (1970). Tonotopic organization and discharge characteristics of single neurons in nuclei of the lateral lemniscus of the cat. J.Neurophysiol. 33:421.

19. Poggio, G.F., and Mountcastle, V.B. (1963). The functional properties of ventrobasal thalamic neurons studied in unanesthetized monkeys. J.Neurophysiol. 26:775.
20. Erulkar, S.D., Rose, J.E., and Davies, P.W. (1956). Single unit activity in the auditory cortex of the cat. Bull. Johns Hopkins Hosp. 99:55.
21. Goldstein, M.H., Abeles, M., Daly, R.L., and McIntosh, J. (1970). Functional architecture in cat primary auditory cortex: tonotopic organization. J.Neurophysiol. 33:188.
22. Hind, J.E. (1960). Unit activity in the auditory cortex. *In* G.L. Rasmussen and W.F. Windle (eds.), Neuronal Mechanisms of the Auditory and Vestibular Systems, p. 201. Charles C. Thomas, Springfield, Ill.
23. Kiang, N.Y-S. (1965). Stimulus coding in the auditory nerve and cochlear nucleus. Acta Otolaryng. 59:186.
24. Stone, J. (1973). Sampling properties of microelectrodes assessed in the cat's retina. J.Neurophysiol. 36:1071.
25. Katsuki, Y., Sumi, T., Uchiyama, H., and Watanabe, T. (1958). Electric responses of auditory neurons in the cat to sound stimulation. J.Neurophysiol. 21:569.
26. Katsuki, Y., Watanabe, T., and Maruyama, M. (1959). Activity of auditory neurons in upper levels of brain of cat. J.Neurophysiol. 22:343.
27. Galambos, R., and Davis, H. (1943). The response of single auditory nerve fibers to acoustic stimulation. J. Neurophysiol. 6:39.
28. Boord, R.L., and Rasmussen, G.L. (1963). Projection of the cochlear and lagena nerves on the cochlear nuclei of the pigeon. J.Comp.Neurol. 120:463.
29. Konishi, M. (1970). Comparative neurophysiological studies of hearing and vocalizations in songbirds. Z.vergl. Physiologie. 66:257.
30. Manley, G. (1970). Frequency sensitivity of auditory neurons in the caiman cochlear nucleus. Z. vergl.Physiologie. 66:251.
31. van Tienhoven, A., and Juhasz, L.P. (1962). The chicken telencephalon, diencephalon and mesencephalon in stereotaxic co-ordinates. J. Comp.Neurol. 118:185.
32. Coles, R. (1974). Responses to acoustic stimuli of single units in the midbrain of the adult chicken (*Gallus domesticus*). Proc. Australian Physiol. Pharmacol. Soc. 5:186.
33. Saunders, J.C., Coles, R.B., and Gates, G.R. (1973). The development of auditory evoked responses in the cochlea and cochlear nuclei of the chick. Brain Res. 63:59.
34. Aitkin, L.M., and Moore, D.R. (1975). Inferior colliculus. II. Development of tuning characteristics and tonotopic organization of the neonatal cat. J. Neurophysiol. 38:1208.
35. Moore, D.R., and Aitkin, L.M. Rearing in an acoustically unusual environment–effects on neural auditory responses. Neuroscience Letters. In press.
36. von Békésy, G. (1960). Experiments in Hearing, p. 403. McGraw-Hill, New York.
37. Johnstone, B.M., Taylor, K.J., and Boyle, A.J. (1970). Mechanics of the guinea pig cochlea. J.Acoust.Soc.Amer. 47:504.
38. Rhode, W.S. (1971). Observations of the vibration of the basilar membrane in squirrel monkeys using the Mössbauer technique. J.Acoust.Soc.Amer. 49:1218.
39. Wilson, J.P., and Johnstone, J.R. (1972). Capacitive probe measures of basilar membrane vibration. *In* Symposium on Hearing Theory, p. 172. IPO, Eindhoven.
40. Kiang, N.Y.-S. (1965). Discharge patterns of single fibers in the cat's auditory nerve. M.I.T. Press, Cambridge, Mass.
41. Evans, E.F. (1972). "Does frequency-sharpening occur in the cochlea?" *In* Symposium on Hearing Theory, p. 27. IPO, Eindhoven.
42. Sando, I (1965). The anatomical interrelationships of the cochlear nerve fibers. Acta Otolaryng. 59:417.
43. Lewy, F.H., and Kobrak, H. (1936). The neural projection of the cochlear spirals on the primary acoustic centers. Arch.Neurol.Psychiat. (Chic.) 35:839.
44. Stotler, W.A. (1953). An experimental study of the cells and connections of the superior olivary complex of the cat. J.Comp.Neurol. 98:401.
45. Woollard, H.H., and Harpman, J.A. (1940). The connexions of the inferior colliculus and of the dorsal nucleus of the lateral lemniscus. J.Anat. 74:441.
46. Webster, W.R., and Veale, J.L. (1970). Binaural response patterns in the inferior colliculus. Proc.Australian Physiol.Pharmacol.Soc. 1:60.

47. Webster, W.R., and Veale, J.L. (1971). Patterns of binaural discharge of cat inferior colliculus units. Proc. Australian Physiol.Pharmacol.Soc. 2:84.
48. Aitkin, L.M., Webster, W.R., Veale, J.L., and Crosby, D.C. (1975). Inferior colliculus. I. Comparison of response properties of neurons in central, pericentral and external nuclei of adult cat. J. Neurophysiol. 38:1196.
49. Warr, W.B. (1966). Fiber degeneration following lesions in the anterior ventral cochlear nucleus. Exp. Neurol. 14:453.
50. Warr, W.B. (1969). Fiber degeneration following lesions of the posteroventral cochlear nucleus of the cat. Exp. Neurol. 23:140.
51. Osen, K.K. (1972). Projection of the cochlear nuclei on the inferior colliculus in the cat. J.Comp.Neurol. 144:355.
52. Tsuchitani, C., and Boudreau, J.C. (1966). Single unit analysis of cat superior olive S-segment with tonal stimuli. J.Neurophysiol. 29:684.
53. Guinan, J.J., Norris, B.E., and Guinan, S.S. (1972). Single units in the superior olivary complex. II. Locations of unit categories and tonotopic organization. Intern.J.Neuroscience 4:147.
54. Goldberg, J.M., and Brown, P.B. (1968). Functional organization of the dog superior olivary complex: an anatomical and electrophysiological study. J.Neurophysiol. 31:639.
55. Erulkar, S.D. (1972). Comparative aspects of spatial localization of sound. Physiol. Rev. 52:237.
56. Morest, D.K. (1964). The laminar structure of the inferior colliculus of the cat. Anat.Rec. 148:314.
57. Geniec, P., and Morest, D.K. (1971). The neuronal architecture of the human posterior colliculus. Acta Otolaryng. Suppl. 295:1.
58. Rockel, A.J., and Jones, E.G. (1973). The neuronal organization of the inferior colliculus of the adult cat. I. The central nucleus. J.Comp.Neurol. 147:11.
59. Rose, J.E., Greenwood, D.D., Goldberg, J.M., and Hind, J.E. (1963). Some discharge characteristics of single neurons in the inferior colliculus of the cat. I. Tonotopical organization, relation of spike counts to tone intensity and firing patterns of single elements. J.Neurophysiol. 26:294.
60. Aitkin, L.M., Fryman, S., Blake, D.W., and Webster, W.R. (1972). Responses of neurons in the rabbit inferior colliculus. I. Frequency-specificity and tonotopic arrangement. Brain Res. 47:77.
61. Clopton, B.M., and Winfield, J.A. (1973). Tonotopic organization in the inferior colliculus of the rat. Brain Res. 56:355.
62. Clopton, B.M., Winfield, J.A., and Flammino, F.J. (1974). Tonotopic organization: review and analysis. Brain Res. 76:1.
63. Merzenich, M.M., and Reid, M.D. (1974). Representation of the cochlea within the inferior colliculus of the cat. Brain Res. 77:397.
64. Schuknecht, H.F. (1960). Neuroanatomical correlates of auditory sensitivity and pitch discrimination in the cat. *In* G.L. Rasmussen and W.F. Windle (eds.), Neural Mechanisms of the Auditory and Vestibular Systems, p. 76. Charles C. Thomas, Springfield, Ill.
65. Greenwood, D.D. (1961). Critical band width and the frequency co-ordinates of the basilar membrane. J.Acoust.Soc. Amer. 33:1344.
66. Rose, J.E., Brugge, J.F., Anderson, D.J., and Hind, J.E. (1968). Patterns of activity in single auditory nerve fibers of the squirrel monkey. *In* A.V.S. de Reuck and J. Knight (eds.), Hearing mechanisms in vertebrates. Churchill, London. 144 pp.
67. Rose, J.E. (1960). Organization of frequency sensitive neurons in the cochlear nuclear complex of the cat. *In* G.L. Rasmussen and W.F. Windle (eds.), Neural Mechanisms of the Auditory and Vestibular Systems, p. 116. Charles C. Thomas, Springfield, Ill.
68. Moore, R.Y., and Goldberg, J.M. (1963). Ascending projections of the inferior colliculus in the cat. J.Comp. Neurol. 121:109.
69. Goldberg, J.M., and Moore, R.Y. (1967). Ascending projections of the lateral lemniscus in the cat and monkey. J.Comp.Neurol. 129:143.
70. Powell, E.W., and Hatton, J.B. (1969). Projections of the inferior colliculus in the cat. J.Comp.Neurol. 136:183.

71. Fink, R.P., and Heimer, L. (1967). Two methods for selective silver impregnation of degenerating axons and their synaptic endings in the central nervous system. Brain Res. 4:369.
72. Morest, D.K. (1964). The neuronal architecture of the medial geniculate body of the cat. J.Anat. 98:611.
73. Morest, D.K. (1965). The laminar structure of the medial geniculate body in the cat. J.Anat. 99:143.
74. Galambos, R. (1952). Microelectrode studies on medial geniculate body of cat. III. Responses to pure tones. J.Neurophysiol. 15:381.
75. Rose, J.E., and Woolsey, C.N. (1958). Cortical connections and functional organization of the thalamic auditory system in the cat. *In* H.F. Harlow and C.N. Woolsey (eds.), Biological and biochemical bases of behavior, p. 127. University of Wisconsin Press, Madison.
76. Woolsey, C.N. (1964). Electrophysiological studies on thalamocortical relations in the auditory system. *In* A. Abrams, H.H. Garner, and J.E.P. Toman (eds.), Tasks in the Behavioral Sciences, p. 45. Williams and Wilkins, Baltimore, Md.
77. Aitkin, L.M., and Webster, W.R. (1971). Tonotopic organization in the medial geniculate body of the cat. Brain Res. 26:402.
78. Aitkin, L.M., and Webster, W.R. (1972). Medial geniculate body of the cat: organization and responses to tonal stimuli of neurons in the ventral division. J.Neurophysiol. 35:365.
79. Gross, N.B., Lifschitz, W.S., and Anderson, D.J. (1974). The tonotopic organization of the auditory thalamus of the squirrel monkey (*Saimiri sciureus*). Brain Res. 65:323.
80. Aitkin, L.M. (1973). Medial geniculate body of cat: responses to tonal stimuli of neurons in medial division. J.Neurophysiol. 36:275.
81. Merzenich, M.M., and Brugge, J.F. (1973). Representation of the cochlear partition on the superior temporal plane of the macaque monkey. Brain Res. 50:275.
82. Rose, J.E., and Woolsey, C.N. (1949). The relations of thalamic connections, cellular structure and evocable electrical activity in the auditory region of the cat. J.Comp. Neurol. 91:441.
83. Rose, J.E. (1949). The cellular structure of the auditory region of the cat. J.Comp. Neurol. 91:409.
84. Wilson, M.E., and Cragg, B.G. (1969). Projections of the medial geniculate body to the cerebral cortex in the cat. Brain Res. 13:462.
85. Sousa-Pinto, A. (1973). Cortical projection of the medial geniculate body in the cat. Adv.Anat.Embryol.Cell Biol. 48:1.
86. Niimi, K., and Naito, F. (1974). Cortical projections of the medial geniculate body of the cat. Exp. Brain Res. 19:326.
87. Woolsey, C.N. (1960). Organization of cortical auditory system: a review and a synthesis. *In* G.L. Rasmussen and W.F. Windle (eds.), Neural Mechanisms of the Auditory and Vestibular Systems, p. 165. Charles C. Thomas, Springfield, Ill.
88. Sousa-Pinto, A. (1973). The structure of the first auditory cortex (AI) in the cat. I. Light microscopic observations on its organization. Arch. Ital. Biol. 111:112.
89. Evans, E.F. (1968). Cortical representation. *In* A.V.S. de Reuck and J. Knight (eds.), Hearing Mechanisms in Vertebrates, p. 272. Churchill, London.
90. Butler, R.A., Diamond, I.T., and Neff, W.D. (1957). Role of auditory cortex in discrimination of changes in frequency. J.Neurophysiol. 20:108.
91. Thompson, R.F. (1960). Function of the auditory cortex of the cat in frequency discrimination. J.Neurophysiol. 23:321.
92. Goldberg, J.M., and Neff, W.D. (1961). Frequency discrimination after ablation of the cortical auditory areas. J.Neurophysiol. 24:119.
93. Diamond, I.T., Goldberg, J.M., and Neff, W.D. (1962). Tonal discrimination after ablation of auditory cortex. J.Neurophysiol. 25:223.
94. Meyer, D.R., and Woolsey, C.N. (1952). Effects of localized cortical destruction upon auditory discriminative conditioning in cat. J.Neurophysiol. 15:149.
95. Penfield, W., and Rasmussen, T. (1950). The cerebral cortex of Man: a clinical study of localization of function, p. 149. Macmillan, New York.
96. Jerger, J., Weikers, N.J., Sharbrough, F.W., and Jerger, S. (1969). Bilateral lesions of the temporal lobe. Acta Otolaryng. Suppl. 258:1.

97. Dobelle, W.H., Mladejovsky, M.G., Stensaas, S.S., and Smith, S.B. (1973). A prosthesis for the deaf based on cortical stimulation. Ann. Otol. 82:445.
98. Powell, T.P.S., and Mountcastle, V.B. (1959). Some aspects of the functional organization of the cortex of the postcentral gyrus of the monkey: a correlation of findings obtained in a single unit analysis with cytoarchitecture. Bull.Johns Hopkins Hosp. 104:211.
99. Hubel, D.H., and Wiesel, T.N. (1962). Receptive fields, binocular interaction and functional architecture in cat's visual cortex. J.Physiol.(Lond.) 160:106.
100. Abeles, M., and Goldstein, M.H. (1970). Functional architecture in cat auditory cortex: columnar organization and organization according to depth. J.Neurophysiol. 33:172.
101. Licklider, J.C.R., and Kryter, K.D. (1942). Frequency-localization in the auditory cortex of the monkey. Fed. Proc. 1:51.
102. Woolsey, C.N., and Walzl, E.M. (1944). Topical projection of the cochlea to the cerebral cortex of the monkey. Amer. J.Med.Sci. 207:685.
103. Stebbins, W.C. (1970). Studies of hearing and hearing loss. *In* W.C. Stebbins (ed.), Animal psychophysics. The design and conduct of sensory experiments, p. 41. Appleton-Century-Crofts, New York.
104. Neff, W.D., and Hind, J.E. (1955). Auditory thresholds of the cat. J.Acoust.Soc. Amer. 27:480.
105. Colonnier, M.L., and Rossignol, S. (1969). Heterogeneity of the cerebral cortex. *In* H. Jasper, A. Ward, and A. Pope (eds.), Basic mechanism of the Epilepsies, p. 29. Little, Brown & Co., Boston.
106. Cragg, B.G. (1971). The fate of axon terminals in visual cortex during trans-synaptic atrophy of the lateral geniculate nucleus. Brain Res. 34:53.
107. Garey, L.J., and Powell, T.P.S. (1971). An experimental study of the termination of the lateral geniculo-cortical pathway in the cat and the monkey. Proc.Roy.Soc. Lond. B. 179:41.
108. Diamond, I.T., Jones, E.G., and Powell, T.P.S. (1968). Interhemispheric fiber connections of the auditory cortex of the cat. Brain Res. 11:177.
109. Diamond, I.T., Jones, E.G., and Powell, T.P.S. (1968). The association connections of the auditory cortex of the cat. Brain Res. 11:560.
110. Diamond, I.T., Jones, E.G., and Powell, T.P.S. (1969). The projection of the auditory cortex upon the diencephalon and brain stem in the cat. Brain Res. 15:305.
111. Greenwood, D.D., and Maruyama, N. (1965). Excitatory and inhibitory response areas of auditory neurons in the cochlear nucleus. J.Neurophysiol. 28:863.
112. Aitkin, L.M., and Dunlop, C.W. (1969). Inhibition in the medial geniculate body of the cat. Exp. Brain Res. 7:68.
113. Brugge, J.F., Dubrovsky, N.A., Aitkin, L.M., and Anderson, D.J. (1969). Sensitivity of single neurons in auditory cortex of cat to binaural tonal stimulation: effects of varying interaural time and intensity. J.Neurophysiol. 32:1005.
114. Boord, R.L. (1968). Ascending projections of the primary cochlear nuclei and nucleus laminaris in the pigeon. J.Comp.Neurol. 133:523.
115. Imig, T.J., Ruggero, M.A., Kitzes, L.M., Javel, E., and Brugge, J.F. (1974). Organization of auditory cortex in the owl monkey (*Aotus trivirgatus*). J.Acoust.Soc.Amer. 56:23.
116. Merzenich, M.M., Kaas, J.H., and Roth, G.L. Auditory cortex in the gray squirrel: tonotopic organization and architectonic fields. Submitted for publication.
117. Geisler, C.D., Rhode, W.S., and Hazelton, D.W. (1969). Responses of inferior colliculus neurons in the cat to binaural acoustic stimuli having wide-band spectra. J.Neurophysiol. 32:960.
118. Merzenich, M.M., and Colwell, S.A. Spatially ordered convergent projection from the auditory thalamus to and from AI in the cat. J.Acoust.Soc.Amer. In press.
119. Licklider, J.C.R. (1965). Basic correlates of the auditory stimulus. *In* S.S. Stevens (ed.), Handbook of Experimental Psychology, p. 1000. Wiley, New York.
120. Butler, R.A. (1974). Does tonotopicity subserve the perceived elevation of a sound? Fed.Proc. 33:1920.

International Review of Physiology
Neurophysiology II, Volume 10
Edited by Robert Porter
 University Park Press Baltimore

7
Control of Locomotion: A Neurophysiological Analysis of the Cat Locomotor System

G. N. ORLOVSKY AND M. L. SHIK
Moscow State University and Institute of Information Transmission Problems, Academy of Sciences, Moscow, USSR

In memory of F. V. Severin

INTRODUCTION

Locomotion, i.e., movement of an animal relative to its environment, is one of the most complicated modes of natural motor activity. Most muscles are involved in the performance of movements, and a considerable part of the nervous system participates in the control of movements. Problems arising while investigating locomotion can be divided conventionally into two groups: 1) problems connected with the activity of the executive motor apparatus (in animals having extremities, muscular activity, gait, equilibrium, etc.); and 2) problems connected with the control of movements. In this review, we consider mainly the second group of problems; the executive motor apparatus being touched on only to the extent required for discussing activity of the nervous system. Many aspects of activity of the motor apparatus in various animals are considered by Gray (41); for the cat they are considered in a number of special studies (32, 37, 38, 89) (cf. also Ref. 45).

The data concerning activity of the nervous centers controlling movements can be obtained in a variety of ways. For example, observations have been made on the work of the executive motor apparatus during natural movements performed by an animal in varying external conditions. Also, data can be obtained while comparing movements of the intact animal with those of the animal deprived of some brain structures. But we think that study of interneuronal connections and study of the activity of individual nerve cells directly during movements can present the most "adequate" data, since the control mechanisms in movement performance are the neuronal mechanisms. Investigation of the neuronal activity during locomotion of invertebrates (insects and crustacea) resulted in many very interesting findings (26, 27, 153).

We have chosen to consider the neurophysiological problems of the cat's locomotion, not because of any specific features of this animal or of its movements, but because, till now, the cat has been the only animal among mammals in which detailed investigation of the activity of various nervous mechanisms has been carried out during locomotion. Such a unique position of the cat is determined, first, by the fact that both anatomical and electrophysiological studies of the cat's nervous system are the most advanced and, second, by the development of a special preparation (mesencephalic cat) in which locomotion can be evoked and controlled by the investigator (137). This preparation presents unique facilities for study of neuronal activity directly during locomotion. The review, therefore, begins with a description of the mesencephalic preparation. The special feature of this review in comparison with others on the same problem (45, 134) is that attention is directed specifically to the neuronal activity occurring during locomotion.

AUTOMATIC SYSTEM FOR THE CONTROL OF LOCOMOTION

Locomotion of the Mesencephalic Cat

The cat can move straightforward on a flat horizontal surface after removal of some higher centers. For example, in acute experiments, a cat can walk almost normally after removal of the forebrain and rostral part of the thalamus, provided the caudal subthalamus is left intact (59). Usually the animal moves slowly, but it can be forced to run by electrical stimulation of a certain region of the subthalamus (the subthalamic "locomotor region," SLR, which is marked by a continuous circle in Figure 1). If, in an acute experiment, the brainstem is sectioned more caudally, i.e., at the rostral border of the superior colliculus (*interrupted line* in Figure 1*A*), this mesencephalic (precollicular post-

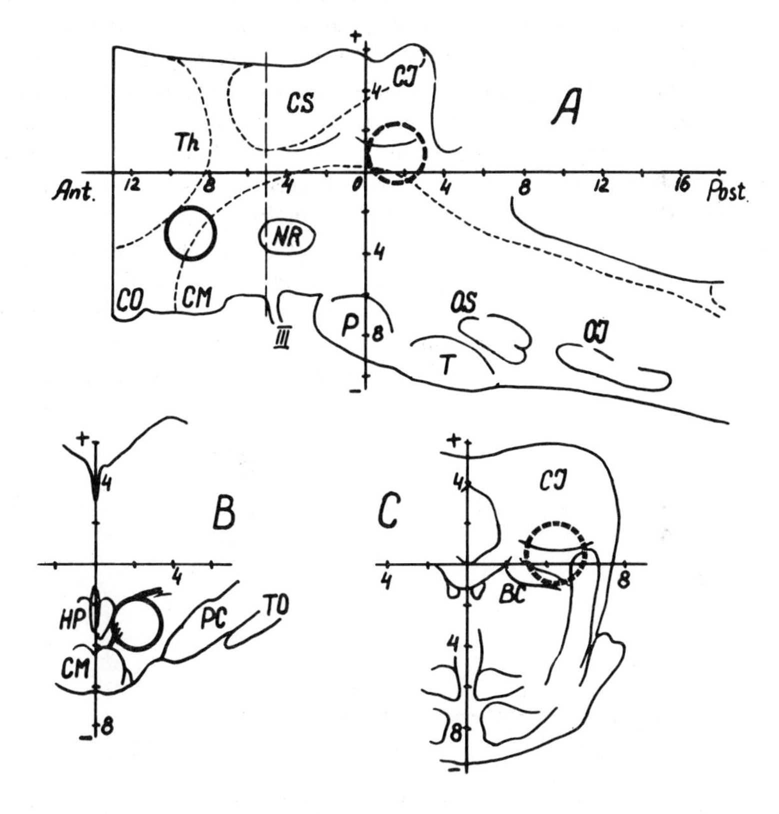

Figure 1. Scheme of the brainstem of the cat (95, 138). *A,* a parasagittal plane (about 2 mm of the midline); *B,* a frontal plane A9 (Horsley-Clarke's coordinates); *C,* a frontal plane P2. The level of section (A13) to obtain the thalamic preparation and that (A5, *dashed line*) to obtain the mesencephalic preparation are shown in *A*. The circles drawn by a *dashed line* show the mesencephalic "locomotor region" (MLR); the circles enclosed by a *solid line* show the subthalamic "locomotor region" (SLR). *BC,* brachium conjunctivum; *CI,* colliculus inferior; *CM,* mammillary bodies; *CO,* chiasma opticus; *CS,* colliculus superior; *HP,* hypothalamus posterior; *NR,* nucleus ruber; *OI,* inferior olive; *OS,* superior olive; *P,* pons; *PC,* pedunculus cerebri; *T,* trapezoid body; *Th,* thalamus; *TO,* tractus opticus; *III,* third cranial nerve.

mammillary) cat exhibits no motor activity. But electrical stimulation of a definite region of the midbrain (the mesencephalic "locomotor region," MLR, which is marked by an *interrupted circle* in Figure 1) evokes walking or running, the animal maintaining equilibrium throughout (137, 138). The preparation is deprived of nociceptive perception; therefore, the skull and vertebral column can be rigidly fixed. The latter possibility is very important in experiments involving recording activity of single neurons. To make external conditions for limb movements more natural, a treadmill with a moving band is placed under the cat so that the animal can walk or run on the band (Figure 2).

Locomotor movements of limbs in the mesencephalic cat are similar to normal ones. With weak mesencephalic MLR stimulation, the cat uses a diagonal gait. In this gait, limbs of the hind girdle are stepping alternately (the *left part* in Figure 3*A*); the same is true for the forelimbs. Each forelimb moves, passing ahead of the diagonal hindlimb; the phase shift is equal to a quarter of the cycle at slow walking, decreasing gradually to zero at higher speeds. With stronger mesencephalic MLR stimulation, muscular activity is higher, and the cat runs faster, accelerating the treadmill band. When stimulation is strong enough, the diagonal gait is sharply replaced by the gallop in which hindlimbs move nearly in-phase. Corresponding gaits were observed in intact animals (14, 41, 93).

Movements in joints and activity of various limb muscles in the mesencephalic cat are also very close to normal ones. It is convenient to divide the whole step cycle into two parts–the stance phase, when the limb touches the ground, and the swing phase (Figure 4). In the stance phase, a group of extensor muscles is active. While extending the limb joints, these muscles counteract the body weight and develop a propulsive force for moving the animal forwards (or, in

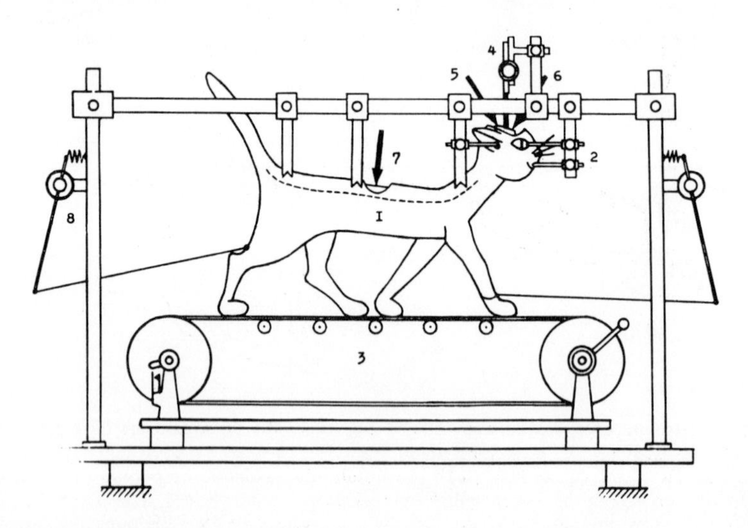

Figure 2. Experimental arrangement (137). The thalamic or mesencephalic cat (*1*) is fixed in a stereotaxic device (*2*) with its limbs on the treadmill (*3*); electrodes (*4–7*) are inserted into the spinal cord and the brain for stimulation and recordings. Limb movements are recorded by a potentiometric transducer (*8*).

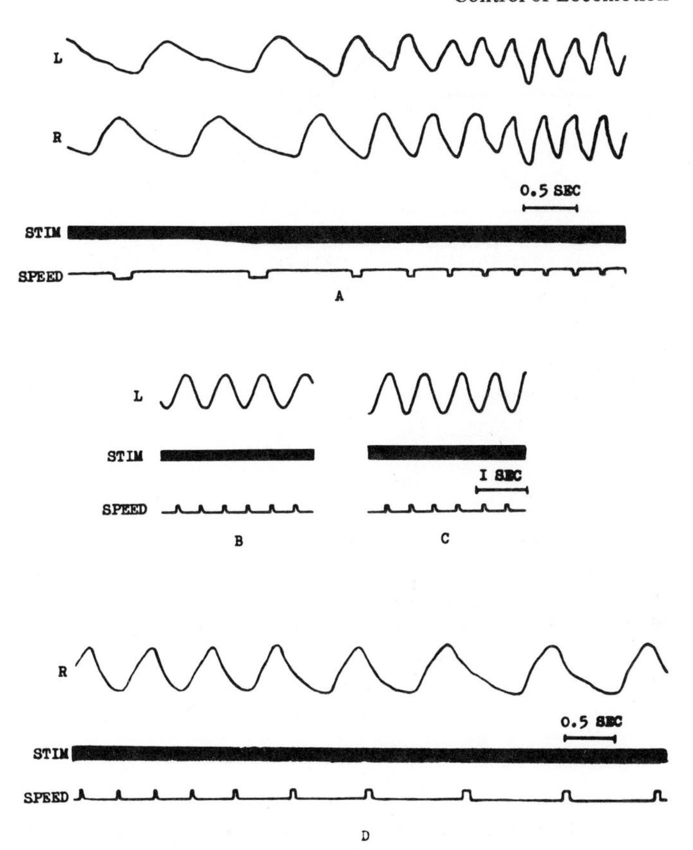

Figure 3. Limb movements under different conditions of locomotion (the mesencephalic cat) (128, 137). *A,* an acceleration of the treadmill band by the animal (increase of the speed of "running") and a transition from a walk to gallop with stronger MLR stimulation. *B* and *C,* effects of increasing of MLR stimulation at a fixed speed of the band. *D,* an effect of the deceleration of the band at constant MLR stimulation. *L* and *R,* movements of the left and right hind limbs (curves deflect upwards when limbs move forwards); *STIM,* MLR stimulation (30 pps); *SPEED,* marks of the band movement, an interval between marks corresponds to 0.5 m.

the experimental conditions shown in Figure 2, accelerate the treadmill band). Not only "pure" extensors but also some two-joint muscles (which extend one joint and flex another) are active in the stance phase. Movements in joints are also schematically shown in Figure 4. The hip joint is extending throughout the stance phase because of contraction of its extensors, and the hip (and the whole limb) is moving backwards relative to the body. Knee and ankle movements are not so simple. During the first half of the stance phase they are flexing to some extent, in spite of the counteraction of the active extensors, that is considered to be a "yield" to the body weight (44, 83, 121). Then, to the very end of the stance phase, all joints are extending because of active extensor contraction and of movement of the animal relative to the supporting surface.

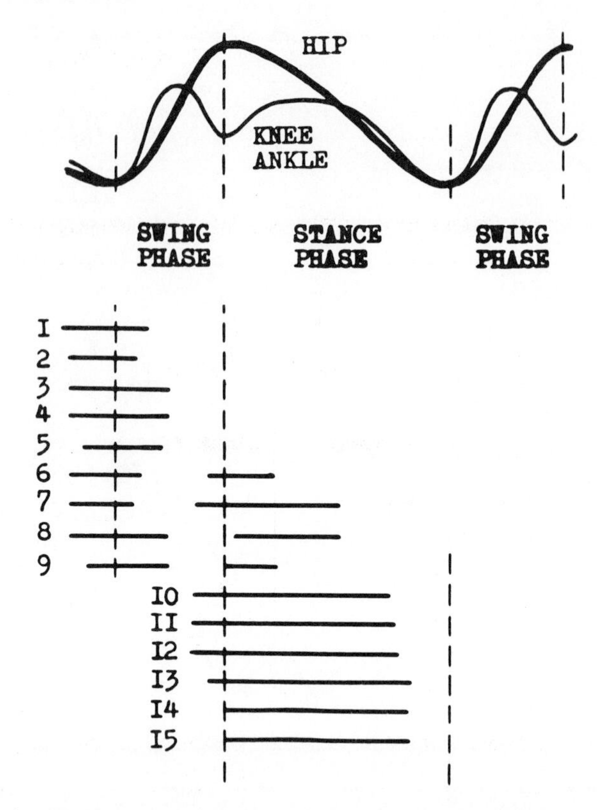

Figure 4. Scheme of the joint movements and muscular activity during locomotion of the mesencephalic cat evoked by MLR stimulation (37). Curves deflect upwards when joints are flexing; muscles illustrated are: *1,* rectus femoris; *2,* tensor fascia latae; *3,* tibialis anterior; *4,* extensor digitorum longus; *5,* iliopsoas; *6,* biceps pars posterior; *7,* gracilis; *8,* sartorius; *9,* semitendinosus; *10,* soleus; *11,* gastrocnemius, plantaris; *12,* vastus; *13,* adductor, semimembranosus; *14,* biceps pars anterior; *15,* gluteus.

When the limb reaches the most caudal position, the swing phase (i.e., protraction of the limb forwards) begins. Both flexors and some two-joint muscles become active at the end of the stance phase which results in flexion of all limb joints after cessation of the extensor activity. The flexion terminates earlier in the knee and ankle joints, and they begin to extend. This extension seems to be determined by a decrease of flexor activity, by onset of weak extensor activity and by inertial forces acting on the distal parts of the limb (32, 37, 83, 121). All limb extensors have been strongly activated a short time (about 20 ms) before the limb touches the ground (32). Therefore, the limb is capable of counteracting and supporting the body weight from the very beginning of the stance phase.

This sequence of activities of various muscles as well as the sequence of movements in various joints is, for the most part, preserved both during weak and sustained (intense) locomotion, as has also been demonstrated in intact animals (32, 38, 121). Main changes are observed not in "pure" flexors and extensors, but in two-joint muscles, which can exhibit one or two bursts of activity during the step cycle, depending on the intensity of locomotion.

Thus, the pattern of muscular activity in stepping is very complex; it can be considered as reciprocal, alternating activity of the flexor and extensor groups only as a rough approximation (83, 131). Indeed, only because of differences in activities of hip flexors and of flexors of other joints can the complex limb movement in the swing phase be realized. Nevertheless, we shall often use a traditionally simplified pattern of muscular activity (i.e., reciprocal, alternating activities of flexors and extensors) while discussing neuronal mechanisms controlling limb movements.

Control of Level of Muscular Activity and of Frequency of Stepping

With stronger mesencephalic locomotor region (MLR) stimulation, both the frequency of stepping and the speed of the treadmill band ("speed of locomotion") increase (Figure 3*A*). It is evident that acceleration of the band is determined by an increase of forces developed by the limb muscles.

Does MLR stimulation influence both these values (the level of muscular activity and the frequency of stepping) in a similar way? To answer this question, MLR stimulation was varied while a speed of the treadmill band was maintained constant (Figure 3, *B* and *C*). In this case, strengthening of MLR stimulation resulted in a considerable increase of muscular activity (both in extensors and in flexors) but the frequency of stepping changed only slightly (128, 137). This result suggested that an increase of the frequency of stepping, which was observed with stronger MLR stimulation when the animal was allowed to accelerate the treadmill band (Figure 3*A*), was determined just by this acceleration. The suggestion was confirmed when MLR stimulation was constant, but the speed of the band was varied by the experimenter (Figure 3*D*). Under these conditions, a deceleration of the band resulted in a lengthening of the step cycle, and an acceleration resulted in a shortening of the step cycle. The frequency of stepping can be changed several times by varying the band speed (77, 128, 137).

Thus one can conclude that MLR stimulation does not directly affect the frequency of stepping. MLR stimulation affects the level of muscular activity, which determines the speed of locomotion; the latter value, in turn, determines the frequency of stepping.

But the level of muscular activity depends not only on MLR stimulation. It can be changed considerably even at constant strength of MLR stimulation and constant speed of the treadmill band. To do this, one may load the limbs by putting a weight on the animal's back or, if the vertebral column is fixed (Figure 2), by raising the treadmill. Loading of the limbs results in considerable activation of the knee and ankle extensors during the stance phase (99).

Therefore, MLR stimulation directly influences the level of muscular activity and, indirectly, the frequency of stepping but it does not determine these values completely, because they also depend on external conditions (loading of limbs, resistance of the moving band, etc.). But by varying MLR stimulation one can obtain the whole range of locomotions (speeds, intensities, and gait) which an intact animal realizes in association with given external conditions.

It seems likely that the nervous system of the intact animal changes a mode (regime) of locomotion by varying also only one "parameter." Some indirect data in favor of this suggestion were obtained in the experiments on intact dogs which ran at various speeds either on the horizontal treadmill band or uphill. The dog was also forced to pull various loads. It was found that the amplitudes of joint movements increased in a similar way with any "hampering" of running (i.e., with increase of the speed, of the uphill angle, or of the load) (109).

The system for the control of locomotion in the mesencephalic cat is concerned with a number of very difficult problems: activation (in strict succession) of a great number of muscles, co-ordination of movements in various limb joints, co-ordination of various limbs, adaptation of movements to external conditions, etc. Nevertheless, control of this system itself is very simple—one has only to regulate MLR stimulation. This is why the system can undoubtedly be called the automatic system for the control of locomotion. Having this system at their disposal, higher brain centers would be released from direct participation in the detailed control and co-ordination of locomotor movements; to establish any regime of locomotion, they need only excite, to a definite level, a certain group of neurons.

Corresponding results were obtained in the study of the fish locomotor system. It was found that electrical stimulation of the midbrain tegmentum resulted in rhythmic movements of the pectoral fins as well as undulatory body movements typical of swimming. These locomotor movements can be regulated through a wide range by varying the strength of stimulation (74). "Controlled" locomotion has also been obtained in some invertebrates. For example, the flight of the locust can be evoked and regulated by stimulation of some nervous structures (154). In the crayfish and lobster, locomotor movements can be evoked by stimulation of single ("command") neurons (27, 153).

In the subsequent parts of this review we discuss organization of the automatic system for the control of locomotion and we suggest that this system includes four separate mechanisms, controlling individual limbs. There is much data indicating that these mechanisms are relatively autonomous and that their co-ordination is achieved by mutual interaction (62, 93, 133, 144). For example, under certain conditions, one can observe different frequencies of stepping in the fore- and hindlimbs in cats (77) and dogs (61, 133) and different step cycle structure of the left and right hindlimbs (77). Also, fluctuations of the phase shifts between movements of various limbs which can be observed in the dog's running were statistically analyzed. Results of this analysis are in agreement with a model in which interaction of autonomous limb centers was proposed (133).

One should not consider locomotion only as rhythmical, coordinated limb movements. Another problem which the automatic system for the control of locomotion solves is maintenance of equilibrium during the movement. The role of various nervous structures (preserved in the mesencephalic cat) in the control of limb movements, in interlimb coordination, and in maintenance of equilibrium is discussed further.

SPINAL MECHANISM OF STEPPING

Role of the Spinal Cord in Locomotion

It is known that the mammalian spinal cord can, under certain conditions, generate stepping movements (36, 121, 129, 131). Of course, preventing the brain from participation in this process leads to some defects. These can be either defects in coordination (i.e., wrong pattern of muscular activity and of joint movements) or limitations of the variety of modes of locomotion. But study of spinal animals (transection of the spinal cord being usually performed at a low thoracic level) shows that defects in hindlimb movement are not very pronounced.

In the chronic spinal dog, stepping movements can be evoked by exteroceptive and proprioceptive stimulation, the hindlimbs stepping alternately as in a walk and a trot (131). But the muscular activity is weak, and the animal has to be kept above the supporting surface. The muscular activity can be intensified by a small dose of strychnine (57). Considerably better stepping can be observed in chronic cats, in which the spinal cord is transected soon after birth (46, 140). Their hindlimbs can bear the body weight. The animals can also vary the frequency of stepping in accordance with the speed of locomotion and they are capable of various gaits (alternating or in-phase stepping).

In acute experiments on cats, stepping can be evoked by an injection of 3,4-dihydroxyphenylalanine (DOPA) or Clonidine (20, 35, 42). Exteroceptive stimulation considerably facilitates this process. In these experiments, the head of the animal is usually fixed while the body is either free or slightly suspended above the treadmill. The speed of the band determines a frequency of stepping and the type of gait; at a high speed the limbs pass from alternating to in-phase stepping (gallop). In the best preparations, limb movements resemble those of intact animals and the muscular pattern is also nearly normal. But in other cases, some lack of coordination is observed, mainly in movements at the ankle joint.

Thus, the main features of stepping movements of the hindlimbs are determined by spinal mechanisms. These mechanisms can generate various rhythms corresponding to various speeds of locomotion; they can activate the limb muscles to various degrees that result in more or less intense stepping; under certain conditions they can even generate various gaits, i.e., various phase relations between the hindlimbs. The fact that it is usually difficult to evoke intense movements indicates that the spinal mechanism is capable of generating intense stepping but that we have not found a fully satisfactory way to activate the mechanism. Indeed, now that "locomotor effects" of DOPA and Clonidine have been discovered, it becomes possible to evoke more intense stepping in spinal preparations.

There are some data indicating that the spinal mechanism which possesses such potent resources for the control of stepping is also used in locomotion of the mesencephalic cat. Short-term electrical stimulation of descending tracts (vestibulospinal, reticulospinal, rubrospinal, or pyramidal) strongly influences the level of muscular activity but does not affect the rhythm of stepping (see "Regulation of Input Signals") (99), which can be explained by a spinal origin of this rhythm. In the mesencephalic decerebellate cat, there is almost no rhythmical activity in descending tracts correlating with limb movements. This also supports an hypothesis of the spinal origin of the locomotor rhythm (see "Activity of Fast-conducting Descending Pathways") (97, 98, 100, 101). As a consequence, for the purposes of this discussion of the activity of some spinal neurons, we shall assume that the activity is determined to a large extent by spinal mechanisms, even though in most cases the data have been obtained not in the spinal but in the mesencephalic or other types of preparations.

Unfortunately, a role for the spinal cord in the control of stepping was studied only for the hindlimbs. One of the reasons is that after transection of the spinal cord at a high cervical level (that is necessary to preserve spinal centers of the forelimbs) regular locomotor movements are difficult to obtain (131).

Central Program of Stepping

A rhythm of stepping is not directly established by the nervous system; it depends on the time interval required for the limb to fulfil the stance phase. This interval, in turn, depends on the speed of the animal's movement relative to the supporting surface (see "Control of Level of Muscular Activity and of Frequency of Stepping"). It seems likely that an animal begins to transfer the limb forwards (i.e., begins the swing phase) at the moment when the limb has reached a definite caudal position, independently of how it got there–rapidly or slowly. In other words, a signal for starting a subsequent phase is termination of the previous one (121).

The spinal cord can "know" about fulfilment of any phase of the movement only by means of afferent signals coming from the limb receptors. Therefore, an afferent inflow has, undoubtedly, a crucial meaning for normal stepping. What would happen if the spinal cord were deprived of any information about limb movements? Could it pass from one phase of the cycle to another (i.e., generate a rhythmical process) or would it "stick" somewhere in the cycle? Brown's experiments show that the deafferented hindlimb of the cat performs stepping movements some dozen seconds after the spinal cord transection (18, 19), i.e., the spinal cord, deprived of the afferent inflow from the limb, is capable of generating the rhythmical process.

Corresponding results can be obtained in other experimental conditions as well. In the mesencephalic cat with de-efferented hindlimbs (ventral roots in the lumbosacral spinal segments being cut), MLR stimulation evokes rhythmical processes in the lumbosacral division of the spinal cord, i.e., rhythmical bursts of activity of the hindlimb motoneurons, without any rhythmical movements of the forelimbs (Figure 5) (104). This rhythmical process has, therefore, an

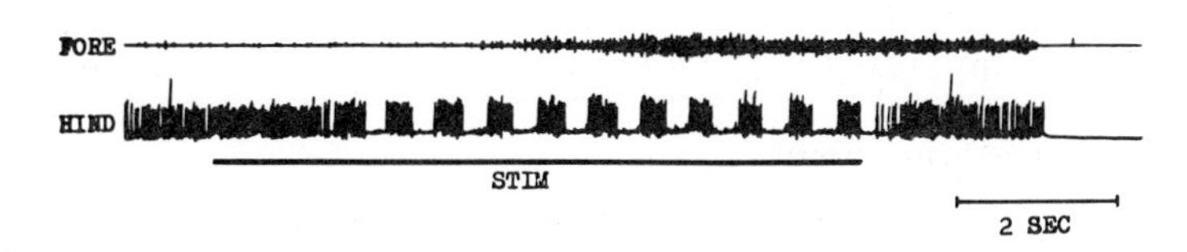

Figure 5. "Fictitious locomotion" of the mesencephalic cat with de-efferented hindlimbs (104). There are presented: EMG of the elbow extensor (*FORE*) and activity of the filament of the SI ventral root (*HIND*); *STIM,* MLR stimulation (30 pps).

intrinsic origin; strengthening of MLR stimulation results in shortening of the cycle from 1–2 to 0.5 s. In the curarized decorticate cat, a rhythmic process (i.e., periodic bursts of activity in the muscle nerves with "locomotor" rhythm) can often be found (114). This process has been named "fictitious locomotion." "Fictitious locomotion" can also be obtained in the curarized rabbit (intact, decerebrate, or spinal); it arises either spontaneously or after an injection of DOPA or 5-hydroxytryptophan (5-HTP) (148, 149). In the curarized spinal cat, "fictitious locomotion" can be evoked by an injection of DOPA or Clonidine combined with electrical stimulation of the dorsal roots. If the spinal cat is not curarized but deafferented, these stimuli evoke actual stepping movements of the hindlimbs (53). These movements are considered to be performed in accordance with *a central program of stepping,* i.e., without any information concerning the effects which the motor commands produce in the executive motor apparatus. In section "Role of the Brainstem and Cerebellum in the Control of Locomotion," data are presented showing that some brain structures are involved in the rhythmical activity during "fictitious locomotion." Therefore, the central program of stepping can differ to some extent in the spinal preparation and, for instance, in the mesencephalic one.

What main defects occur in the movements controlled by the central program only? First, due to an intrinsic origin of the locomotor rhythm in this case, limb movements are not adapted to a real speed of locomotion. Second, in the absence of any information concerning muscle lengths and forces, the limb cannot adapt to various loads. However, some important features of stepping are still preserved after deafferentation; they also can be found in "fictitious locomotion." The central program for stepping produces a nearly normal sequence of muscular activity in the cycle at least for "pure" flexors and extensors, and a normal (in-phase or alternating) relation between the hindlimbs (53, 114, 115, 148, 149).

One should note that consideration of the spinal mechanism for stepping as consisting of two different parts, i.e., the central program and the peripheral feedback which must "improve" movements, would be too simplified an approach to the problem. In many cases an afferent inflow, especially the cyclic one, is very important for the rhythmic generation itself. Stepping movements can be evoked much more easily if deafferentation of the hindlimbs is not complete and one of the dorsal roots is left intact (41, 104). It is possible that afferent signals can trigger (or facilitate) some parts of the central program and, on the contrary, the central program can strongly influence the afferent signals.

Activity of Alpha-Motoneurons

A muscle force is determined by simultaneous activity of many motor units. Extensor muscles are active during most of the stance phase, but corresponding alpha-motoneurons are active during only 1/3–1/2 of this period; only a few of the cells are active throughout the stance phase (128). A decrease of the band speed (at constant MLR stimulation) results in a lengthening of the stance phase and of extensor activity, but values of EMGs usually do not change (128). With a longer stance phase, extensor motoneurons are also active during longer periods.

With stronger MLR stimulation, extensor muscles are activated more intensely. This is achieved by recruiting of new alpha-motoneurons but not by increasing the discharge rates of active units (128). Some changes of discharge rates can be observed only if MLR stimulation (and intensity of locomotion) is near a threshold of the motoneuron. With stronger MLR stimulation, a given motoneuron's discharge rate remains nearly constant (in various neurons it usually ranges from 30 to 50 impulses per second) in spite of more intense locomotion and a considerable increase of EMG activity.

Discharge rates of individual flexor motoneurons during locomotion are also rather constant; an increase of muscle force is achieved mainly by recruiting of new units. Since duration of the swing phase scarcely depends on the intensity of locomotion, duration of activity of flexor motoneurons is also nearly constant.

Unfortunately behavior of motoneurons while muscle activity is increased not by MLR stimulation but by a stretch-reflex (by loading the limbs; see "Control of Level of Muscular Activity and of Frequency of Stepping") has not been studied. Recently it was reported that an increase of the muscle force in the static stretch-reflex is also achieved mainly by recruiting of new units (and by lengthening of activated muscle fibers) but not by increasing discharge rate (52).

Role of Gamma-Motoneurons and Spindle Afferents in the Control of Muscular Activity

Since under certain conditions a deafferented hindlimb is capable of stepping movements (18, 53, 135), an afferent inflow from the limb is not necessary for the activity of its motoneurons. Nevertheless, in normal stepping, limb afferents (especially afferents from the muscle spindles) make a valuable contribution to excitation of alpha-motoneurons. Indeed, when activity of spindle afferents was lowered by novocaine blocking of axons of gamma-motoneurons in a muscle nerve, the EMG of the corresponding extensor muscle decreased considerably (125). The contribution of spindle afferents differs essentially from that of the central sources, because a degree of the muscle activity becomes dependent on the muscle state, i.e., on its length, and velocity of lengthening or shortening. This is important for adaptation of the limb movements to external conditions.

For example, loading of the hindlimbs that leads to a lengthening of the knee and ankle extensors at any point of the stance phase (in comparison with normal stepping) results in a considerable increase of the extensor activity (99).

But such a reaction to lengthening of the muscle (stretch-reflex) is not useful at all phases of the step cycle. Cyclic regulation of the reflex is, in fact, performed by the nervous system in two different ways: 1) by cyclic changes of the excitability of alpha-motoneurons (135); and 2) by cyclic changes of the activity of gamma-motoneurons that lead to corresponding changes of the sensitivity of spindle afferent. Recordings of activity of single spindle afferents during evoked locomotion of the mesencephalic cat (125, 126), during spontaneous locomotion of the decorticate cat (114, 116, 117), and during stepping movements of the spinal cat (46) showed that activity of these afferents was considerably higher in the phase of *excitation* of their "own" muscle, in comparison with the phase of its inactive state, in spite of the fact that in the latter case the muscle can be longer and stretched at higher speed (see the curves of joint angles in Figure 4 reflecting muscle lengths as well).

Figure 6*A* shows activity of a spindle afferent from the ankle extensor when the ankle joint was passively flexed, i.e., the muscle was stretched. During locomotion, a flexion of the ankle joint in the swing phase activates the afferent only slightly, but the afferent was active in the stance phase simultaneously with the ankle extensor (Figure 6*B*); with stronger MLR stimulation (Figure 6*C*) extensor activity increased in parallel with the activity of the afferent. Stretching of the muscle in the swing phase now did not activate the afferent at all. This result can be explained by suggesting that gamma-motoneurons are activated in-phase with alpha-motoneurons of the same muscle, and to a proportional extent. On the other hand, when alpha-motoneurons are not active, corresponding gamma-motoneurons in the above-mentioned experiment (125) brought the spindle behavior nearer to that of "pure" receptors of muscle length. Direct recording of the activity of gamma-motoneurons themselves showed that their discharge rate strongly depended on the phase of the step cycle (125). Thus,

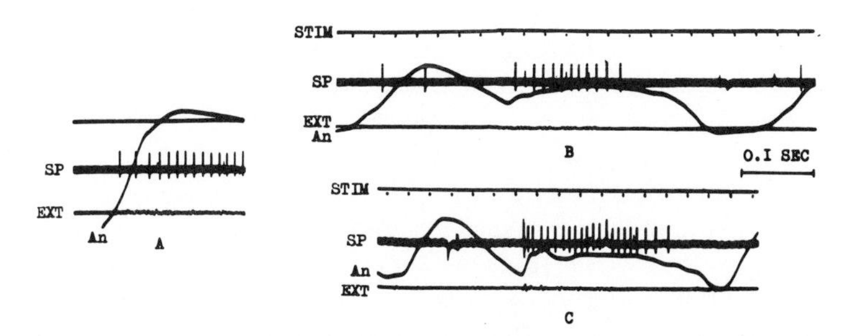

Figure 6. Activity of a spindle afferent from the ankle extensor in the mesencephalic cat (126). *A*, with passive ankle flexion; *B*, during weak locomotion; *C*, during intense locomotion evoked by stronger MLR stimulation. *SP*, activity of the spindle afferent; *An*, the ankle angle (a curve deflects upwards when the ankle is flexing); *EXT*, EMG of the ankle extensor; *STIM*, MLR stimulation (30 pps).

because of the linked activation of gamma- and alpha-motoneurons, the spindle receptors are "switched off" (or their sensitivity is considerably decreased) during the period when the muscle is passively stretched. Had the sensitivity not been decreased, the afferents would react to muscle lengthening and could activate alpha-motoneurons in a "wrong" phase, i.e., when they must be silent and must not counteract the active muscle. The receptors are "switched on" again in the phase when their muscle becomes active. In this phase, activity of the spindle afferents is determined by two main factors: 1) by the discharge rate of corresponding gamma-motoneurons, and 2) by the length (and by the velocity of lengthening or shortening) of the muscle. Since the spindle afferents, especially of group Ia, exert a strong excitatory action on alpha-motoneurons, muscular activity becomes also dependent on these two factors, the second one being important for the adaptation of the muscle activity to external conditions.

"Linked" activation of homonymous alpha- and gamma-motoneurons, that was first postulated on a basis of similarity found in their central connection (40), was also found in some other movements (42, 91, 143). This "linkage" is important not only for an adaptation of the limb to external conditions, but also for a broadening of the range of the muscle force and length regulation (33). Co-activation of alpha- and gamma-motoneurons is performed in accordance with the central program of stepping, since it has also been found in animals with deafferented hindlimbs (114, 118).

Regulation of Input Signals

Various signals come to the spinal cord both from the periphery and from higher centers. Some of them can be important for stepping, others can be useless or even interfere. Therefore, one of the functions of the spinal mechanism of stepping is regulation of "input" signals themselves or of their efficacy. The simplest example is a "tonic" regulation, when the efficacy of the signal is decreased or increased at the beginning of locomotion, and then remains constant. For example, recurrent inhibition of motoneurons is lowered during locomotion (127). This seems to be determined by the inhibition of Renshaw cells which are known to mediate the recurrent inhibition (28, 123). The inhibition of Renshaw cells seems to be "useful" for locomotion since disinhibited alpha-motoneurons can reach higher discharge rates. Also, Renshaw cells are known to suppress a system of reciprocal inhibition (65), and the inhibition of Renshaw cells during locomotion would promote activation of this other system (see below).

More complicated than a "tonic" is a cyclic regulation of input signals. An example was presented (see "Role of Gamma-Motoneurons and Spindle Afferents in the Control of Mucular Activity") showing a very reasonable regulation of the sensitivity of spindle afferents produced by gamma-motoneurons: the sensitivity was decreased in that phase of the cycle in which afferent signals could interfere with the activity of the spinal mechanism. In this case, the regulation of afferent signals is performed directly in receptors, i.e., before the signals arrive in the spinal cord.

Another example demonstrating a regulation of the efficacy of afferent signals was obtained in a study of spinal neurons mediating reciprocal inhibition (34). It is known that alpha-motoneurons are inhibited when an antagonistic muscle is passively stretched (78). This is a result of excitation of group Ia spindle afferents (39, 66). These afferents exert an excitatory action on a special group of spinal neurons which, in turn, inhibit motoneurons of the antagonistic muscle (29, 30, 65, 73). It was found that during locomotion the neurons mediating reciprocal inhibition act like "switches": they respond to afferent impulses in one phase of the locomotor cycle and do not respond in another phase (Figure 7). In the first case, afferent signals can inhibit motoneurons of the antagonistic muscle, but in the second case they cannot. This modulation is organized in a reasonable manner, i.e., afferent signals can inhibit motoneurons of the antagonistic muscle only in that phase, in which the muscle must not be active.

The group of neurons described not only mediates reciprocal inhibition, i.e., they act not only as "switches" in the pathway from Ia afferents to antagonist motoneurons. These neurons, even when deprived of an afferent input, generate impulses in that phase of the cycle when motoneurons of the antagonistic muscle must not be active (Figure 7). Modulation of such a mode seems to be performed by the spinal mechanism in accordance with the central program of stepping (34), and promotes the reciprocal activity of flexor and extensor motoneurons in "fictitious locomotion." In normal locomotion, the effects produced by the afferent signals and by the central program are summed up,

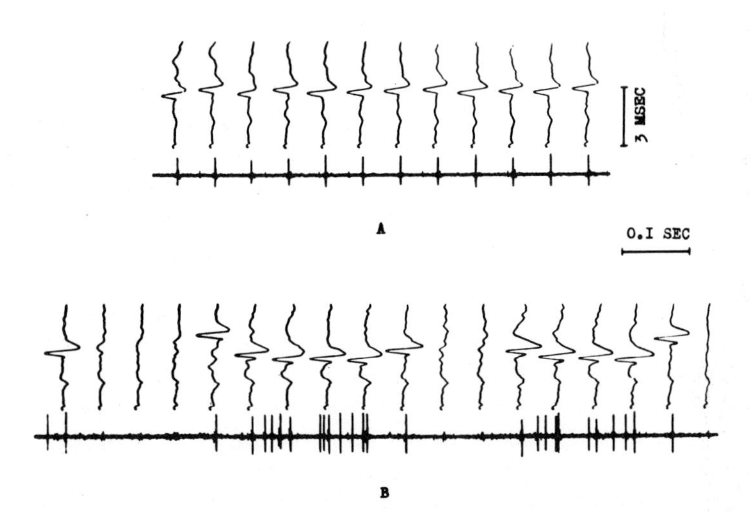

Figure 7. Activity of a neuron mediating the reciprocal inhibition from the knee extensor to its antagonist (the mesencephalic cat, hind limbs de-efferented) (34). The *lower trace* is a continuous recording of the activity of the neuron, while the *upper trace* shows the responses of the neuron to stimulation of the knee extensor nerve with a strength sufficient to excite only group 1a afferents. The recording is performed at rest (*A*) and during "fictitious locomotion" (*B*).

activity of neurons increases, which results in strong inhibition of the antagonist in the appropriate phase of the cycle (34).

The neurons mediating reciprocal inhibition during locomotion are activated in-phase with alpha- and gamma-motoneurons of that muscle from which they receive group Ia afferent input. An hypothesis that these groups of neurons must be co-activated in those movements for which reciprocal activity of the antagonistic muscles is necessary, was first formulated on the basis of similarity found in the central connection of these neurons (64), and has been confirmed during locomotion.

Not only efficacy of afferent signals but also efficacy of supraspinal influences in the spinal cord is regulated during locomotion. This has been demonstrated in experiments with stimulation of various descending tracts during locomotion (99). The following tracts were investigated: 1) the vestibulospinal tract originating from neurons of Deiters' nucleus; in resting conditions this tract exerts an excitatory action on extensors and an inhibitory one on flexors (49); 2) the reticulospinal tract originating from neurons of the medial pontine reticular formation; at rest this tract exerts mainly an excitatory action on flexors and an inhibitory one on extensors (50, 80); 3) the rubrospinal tract originating from neurons of the red nucleus; this tract exerts mainly an excitatory action on flexors (63) and, 4) the pyramidal tract exerting mainly an excitatory action on flexors (79).

Figure 8 shows that stimulation of Deiters' nucleus during locomotion results in a considerable increase of the extensor EMG. But extensor motoneurons respond to the stimulation only in that phase of the cycle when they "must" be active, i.e., in the stance phase, and do not respond in the swing phase. These sharp cyclic changes of the efficacy of supraspinal signals are determined, first of all, by cyclic modulation of the excitability of motoneurons. They are facilitated in the stance phase both by central mechanisms and by the afferent inflow (for example, through Ia afferents; see "Central Program of Stepping" and "Role of Gamma-Motoneurons and Spindle Afferents in the Control of Muscular Activity"). On the contrary, in the swing phase they are inhibited both by the central mechanisms and by the reflex system of reciprocal inhibition.

Corresponding results have been obtained for other descending tracts. Stimulation of the reticulospinal tract resulted in an increase of flexor EMGs and in inhibition of extensors, while stimulation of the red nucleus and of the pyramidal tract led mainly to an increase of flexor EMGs. But in all cases, stimulation of the descending tracts, exerting strong influences on the magnitude of EMGs, did not shift phases of the muscular activity in the cycle, i.e., the muscles switched on and off at a "proper" time. Only very strong stimulation could disturb the rhythmic activity of the spinal mechanism and lead, for example, to the excitation of flexors in the stance phase and to anticipatory protraction of the limb forwards.

Cyclic changes of the efficiency of supraspinal influences can be determined not only by cyclic changes of motoneuron excitability. Indeed, descending tracts affect motoneurons to a large extent polysynaptically, i.e., via interneurons (49, 50, 64, 76, 87). Cyclic changes in interneurons' excitability can also be a reason for cyclic modulation of the efficiency of descending signals. For example, it is known that an inhibitory action of the vestibulospinal tract on flexor motoneurons is mediated, at least partly, by the same neurons which mediate reciprocal inhibition (48). Neurons of this system, which inhibits flexor motoneurons, are known to be active in-phase with the antagonistic, extensor muscle and to be suppressed when the flexor muscle is active (34). This can explain why activation of the vestibulospinal tract does not result in an inhibition of the flexor muscles during locomotion (Figure 8).

Cyclic changes of activity of spinal interneurons can result not only in cyclic changes of the efficiency of input signals, but in more complicated events as well, for example, in "readdressing" of the signals. Thus, stimulation of the foot dorsum performed in the swing phase leads to the activation of flexors, but in the stance phase, it leads to the activation of extensors (45).

Signals Sent to the Brain and to Other Divisions of the Spinal Cord

Signals sent by the spinal cord to the brain can be divided into two classes: 1) signals concerning activity of the neuronal spinal mechanisms; and 2) signals concerning activity of the executive motor apparatus of the limb. Of course, this classification is conventional, especially in the case of the intact animal, since many segmental feedbacks influence the state of the spinal mechanisms and, on the contrary, signals from the limb receptors reflect to some extent motor commands which the limb fulfils.

Signals from the spinal mechanisms and from the limb receptors can be addressed to various recipients. One of the main motor brain centers, which is preserved in the mesencephalic cat after decerebration, is the cerebellum. There are several pathways carrying signals to the cerebellum from the lumbosacral division of the spinal cord (which controls the hindlimbs): the dorsal and ventral spinocerebellar tracts (DSCT and VSCT), as well as the spino-reticulocerebellar and spino-olivocerebellar pathways (110, 112). The first three of these have been studied during locomotion.

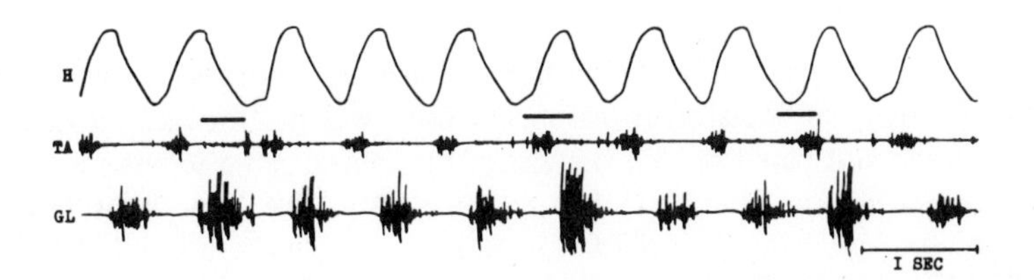

Figure 8. Effects of stimulation of Deiters' nucleus during locomotion of the mesencephalic cat (91). Periods of stimulation (50 pps) are marked by lines. *TA* and *GL*, EMGs of the ankle flexor (tibialis anterior) and extensor (gastrocnemius lateralis) of the ipsilateral (to the nucleus) hindlimb; *H*, a hip angle of this limb (the curve deflects upwards when the hip moves forwards).

Neurons giving rise to DSCT receive an afferent input from various limb receptors located in the skin and joints, but mainly from the muscle spindle receptors (112). In locomotion, these neurons are active in that phase of the cycle, when corresponding receptors are active (7, 10). For example, Figure 9*A* shows activity of a DSCT neuron whose input is derived from 1a afferents from spindles of the ankle extensor. These spindle afferents were active in-phase with the muscle due to the action of gamma-motoneurons (see "Role of Gamma-Motoneurons and Spindle Afferents in the Control of Muscular Activity"). This is why the DSCT neuron was also active in the stance phase, its discharge rate being proportional to the muscle activity. After deafferentation, cyclic modulation of the discharge rate of DSCT neurons disappeared (Figure 9*B*). Thus, DSCT neurons inform the cerebellum mainly about the state of the executive motor apparatus, i.e., about the muscular activity, joint movement, touching the ground, etc.

One should note that DSCT neurons receiving input from group Ia spindle afferents do not directly "measure" any value related to the state of the muscle, but only reflect activity of these afferents. However, activity of alpha-motoneurons strongly depends on the excitatory inflow from Ia spindle afferents. Therefore, by "measuring" activity of Ia spindle afferents, DSCT neurons convey information about one of the main excitatory inputs to motoneurons, this input depending on the muscle length and on the central influences on the spindle receptors.

During locomotion, neurons of VSCT are also active cyclically (Figure 9*C*), but this modulation is not determined directly (or is determined only to a small extent) by afferent signals coming from the moving limb, since the modulation is mainly preserved after deafferentation of the hindlimbs (Figure 9*D*) (8, 9, 11). Therefore, these neurons inform the cerebellum not only about the activity of the executive motor apparatus, but also about processes within the spinal cord, i.e., about cyclic activity of the spinal generator of stepping. At least, they convey signals about a phase of the cycle and the intensity of locomotion, since bursts of their activity are temporally linked with definite phases, and the discharge rate within the bursts is higher at more intense locomotion. It is also possible that they convey more detailed information about the spinal generator or about the state of some other neuronal mechanisms involved in the control of stepping.

An hypothesis that some spinocerebellar pathways may signal information regarding activity of spinal neuronal mechanisms was put forward by Lundberg and Oscarsson and based on the analysis of the central and afferent connections of spinocerebellar neurons (81, 82, 84, 92, 111). The hypothesis has been supported for VSCT in the study of locomotion. Seemingly, not only VSCT but also the spino-reticulocerebellar and spino-olivocerebellar pathways can convey signals about activity of spinal mechanisms (92, 111). Indeed, a rhythmic modulation of neurons of the lateral reticular nucleus (the second-order neurons in the spino-reticulocerebellar pathway) was found in the mesencephalic cat when MLR was stimulated but before stepping movements began (13). Cyclic

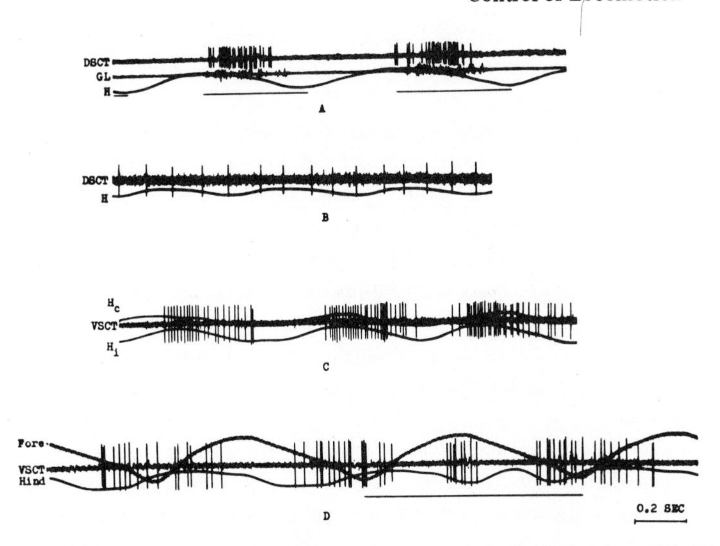

Figure 9. Activity of neurons of the dorsal (*A,B*) and ventral (*C,D*) spinocerebellar tracts during locomotion of the mesencephalic cat with the hind limbs intact (*A,C*) and deafferented (*B,D*) (7–9). A line in *D* shows a period of non-synchronous movements of the fore and hindlimbs. *DSCT* and *VSCT,* activity of corresponding neurons; *GL,* EMG of m. gastrocnemius lateralis; *H*, H_i, and *Hind,* a hip angle of the ipsilateral (to the neuron) limb; H_c, a hip angle of the contralateral limb; *Fore,* movement of the fore limb (curves *H*, H_i, *Hind,* and *Fore* deflect upwards when a limb moves forwards; curve H_c, when a limb moves backwards).

changes of activity of some ascending spinal fibers were also recorded during "fictitious locomotion" of the decorticate cat (114, 119).

Information regarding activity of the spinal generator of stepping can be addressed not only to the brain but also to other divisions of the spinal cord (to the mechanisms controlling other limbs) for interlimb coordination. Indeed, appropriate interlimb relations are observed in the spinal cat during "fictitious locomotion" or during stepping of the deafferented hindlimbs (18, 53, 149). But afferent signals from moving limbs may also participate in the interlimb coordination, since the interlimb reflexes have been found to be facilitated by MLR stimulation (21). Pathways for these reflexes are either propriospinal or spino-bulbospinal (139).

Hypotheses about Arrangement of the Spinal Mechanism of Stepping

In previous sections, some spinal mechanisms responsible for stepping have been discussed. But one of the main problems–origin of the locomotor rhythm–has not been solved yet and presents a wide field for hypotheses. These hypotheses attempt to explain the following experimental findings:

1. In "normal" locomotion, a rhythm of stepping is not determined directly by the central mechanisms but is strongly dependent on the afferent inflow from the moving limb.

2. In the absence of this inflow, a rhythm can be evoked within the spinal cord which, in this case, is completely determined by the central mechanisms.

Unfortunately, it is still unknown whether in these two cases the generation is performed by the *same* neuronal network, or in the latter case (deafferentation or "fictitious locomotion") another system, capable of rhythmical generation without rhythmical external influences, comes into operation.

A Chain-Reflex Hypothesis This hypothesis was briefly mentioned above (see section "Central Program of Stepping"). It claims to explain only "normal" stepping. According to the hypothesis, a signal for starting of the subsequent phase of the step cycle is a fulfilment of the previous phase. For example, protraction of the limb forwards begins only when the limb reaches a definite caudal position.

But a system of proprioceptive reflexes, in the form it exists during resting conditions, is hardly suited for generation of a rhythmic process. So, a reflex activation of the passively stretched antagonist muscle (due to activation of spindle afferents) counteracts the active contraction of the agonist muscle. Similarly, a reflex system of reciprocal inhibition does not promote movements either, since passively stretched antagonist muscles exert an inhibitory action on the actively contracting agonist muscles. However, during locomotion, the system of proprioceptive reflexes is radically changed (see sections "Role of Gamma-Motoneurons and Spindle Afferents in the Control of Muscle Activity" and "Regulation of Input Signals") mainly because of establishment of rigid linkage between activities of homonymous alpha-motoneurons, gamma-motoneurons, and interneurons mediating reciprocal inhibition. Now a reflex activation of the antagonist muscle during its passive stretching is suppressed, as well as an inhibitory action of this muscle on the actively contracting agonist muscle. On the other hand, reflexes from actively contracting muscle contribute to excitation of its own motoneurons and (via the system of reciprocal inhibition) to inhibition of motoneurons of the antagonist muscle.

Unfortunately, it remains unknown how co-activation of motoneurons (alpha and gamma) and interneurons mediating reciprocal inhibition is achieved. But it seems likely that the spinal cord, whose proprioceptive reflexes are reorganized as described above, is capable of generating stepping movements, these movements being adapted to external conditions (speeds and loads). For example, when the limb is moving backwards in the stance phase, extensors are shortening and activity of their spindle afferents is decreasing especially at the end of this phase. This results in a decrease of facilitation of extensor motoneurons and, at the same time, in disinhibition of flexor motoneurons. At a definite caudal position of the limb, flexor alpha-motoneurons become active (note that flexor gamma-motoneurons and interneurons mediating reciprocal inhibition of extensors become active as well), and the swing phase begins.

A real pattern of muscular activity, especially in two-joint muscles, is more complicated than reciprocal activity of two muscular groups which this model can generate (see section "Locomotion of the Mesencephalic Cat"). But two-joint muscles have specific organization of proprioceptive connections that allow one to explain some features of their activity in stepping even provided that the central generator has simple (reciprocal) outputs for flexors and extensors (83).

An Hypothesis Involving Two Half-Centers Brown's hypothesis (18, 19) claims to explain only stepping movements of the deafferented limb, i.e., the origin of a rhythmical process within the spinal cord. According to the hypothesis in its modern form (72, 83), there are in the spinal cord two groups of neurons inhibiting each other. The inhibitory interaction is so strong that when one group ("half-center") is active, activity of the other group is completely suppressed. In addition there are excitatory connections within each half-center, which recruit all the cells into activity if some of them become active. To explain alternating activity of the half-centers, a process similar to fatigue is suggested. One half-center is thought to exert excitatory influences on flexor motoneurons, while the other one acts on extensor motoneurons. Therefore, when a system of two half-centers generates a rhythmical process, flexors and extensors are activated alternately, a phenomenon that is observed (as a first approximation) in stepping.

Some indirect data in favor of Brown's hypothesis were obtained in the study of influences of DOPA on the spinal cord. It was noted above (see section "Role of the Spinal Cord in Locomotion") that an injection of DOPA evokes stepping movements in the spinal cat (20, 42, 53); therefore changes of a state of the spinal cord after DOPA injection can be treated as facilitating generation of stepping. It was shown that, after DOPA injection, stimulation of various nerves of the ipsi- and contralateral hindlimb evokes long-lasting discharges in some spinal neurons (duration of the discharge being comparable with that of the swing and stance phases of the step cycle) (71, 72). According to the pattern of the response to stimulation of various nerves, the neurons can be divided into two reciprocal groups, which might correspond to Brown's half-centers.

Neurons with such a unique behavior are located in the dorsolateral part of the ventral horn. It is in this region that most neurons having rhythmic modulation related to stepping were found during locomotion of the mesencephalic cat with deafferented hindlimbs (105). Therefore, it seems likely that neurons of this region are closely related to the origin of the rhythmical process. But an analysis of the distribution of neurons throughout the step cycle has not revealed two reciprocal groups as required for Brown's model. This negative finding can, however, be explained by suggesting that not all recorded neurons belong to the generator.

In Brown's model, a rhythm of oscillations is determined by intrinsic properties of spinal neurons: the neuron is active for a time, then it is silent, etc. It is evident that with neurons of this kind it is not necessary to have two half-centers. Even one neuron can generate the basic pattern of stepping. If the output of a single "generator" neuron is not "powerful" enough to drive a lot of spinal neurons, many "generator" neurons can operate "in parallel" provided their rhythms are equal to each other or they are synchronized by special interconnections. This modification of the original hypothesis can be named a pacemaker model of the rhythmic generator. A locomotor generator of similar kind has been found in the cockroach (113).

We have discussed two hypotheses about arrangement of the spinal mechanism of stepping. The chain reflex hypothesis satisfactorily explains normal stepping and its strong dependence on external conditions; the hypothesis of two half-centers satisfactorily explains stepping of the deafferented limb and "fictitious locomotion." It is quite possible that a successful combination of these two hypotheses would explain most experimental findings, but till now such a combined model has not been developed.

A Ring Hypothesis This hypothesis (55, 56) claims to explain both normal stepping and stepping of the deafferented limb. It differs from other models with circulating excitation (75, 145) mainly because of the suggestion that there is not one closed chain consisting of successive neurons, but that each cross section of the ring contains many neurons, each neuron being under excitatory influences from several neurons of a preceding ring segment. If a wave of excitation is circulating along the ring, a cycle duration would be determined by the speed of propagation and by the ring length. Each of the ring neurons is active during only a part of the cycle. Flexor and extensor motoneurons receive excitatory inputs from opposite parts of the ring, which result in their reciprocal activity. Two-joint muscles would have more complicated inputs from the ring neurons resulting in a more complicated pattern of cyclic activity. A rhythm of generation can be regulated by varying the speed of propagation around the ring. It is supposed that afferent signals from the moving limb can facilitate the ring neurons and thus affect the rhythm, especially the stance phase duration.

The origin of rhythmical oscillations in this model differs in essence from that in Brown's model since it is determined by activity of a large number of "normal" neurons, and no additional hypotheses suggesting any special properties of neurons are required.

There are no direct experimental data in favor of the ring hypothesis, but it is known that locomotion of a large number of animals (fishes, reptiles, myriapods) is generated by a wave of excitation propagating along the spinal cord (45, 60) or along the chain of ganglia (142), and the speed of propagation can vary. Thus, the main part of the hypothetical ring, i.e., the chain capable of conducting excitation, exists in many animals.

ROLE OF THE BRAINSTEM AND CEREBELLUM IN THE CONTROL OF LOCOMOTION

Activation of the Spinal Mechanism of Stepping

In section "Automatic System of the Control of Locomotion" data were presented in favor of a suggestion that in the cat (and, seemingly, in other animals) there is an automatic system for the control of locomotion. This system controls movements of individual limbs, organizes interlimb coordination, and secures maintenance of equilibrium during movement. How might this system come into operation when the animal intends to walk? In this connection, we shall discuss three questions:

1. Does the intact animal use those nervous structures (for example, MLR) whose electrical stimulation evokes locomotion under experimental conditions?
2. What brain neurons, sending axons to the spinal cord, are responsible for activation of the spinal mechanism of stepping?
3. How is this activation produced?

It was noted (see section "Locomotion of the Mesencephalic Cat") that stimulation of MLR elicits locomotion in the mesencephalic cat. In this preparation, locomotion can also be evoked by stimulation of cortico-brainstem fibers descending in the pyramidal tract (136) and terminating mainly in the pontine, bulbar, and mesencephalic reticular formation (88, 124). Preparations with more rostral section of the brainstem are capable of spontaneous locomotion if a caudal part of the subthalamus is left intact (59). These preparations exhibit bursts of locomotor activity. Between the bursts the animals remain motionless. In such preparations, electrical stimulation of the subthalamic "locomotor region" (SLR) regularly evokes locomotion, while MLR stimulation can only intensify locomotion evoked by SLR stimulation or arising spontaneously (95). SLR stimulation also evokes locomotion in intact and decorticate cats under light anesthesia (54, 151).

In chronic experiments on cats, electrical stimulation of MLR evokes "machine-like" locomotion accompanied by some accessory effects (mewing, flight, micturition, vomiting, etc.). These accessory effects can be eliminated by destruction of either the caudal subthalamus or the centrum medianum-nucleus parafascicularis complex (141). During the evoked locomotion, the animal exhibits perfect space orientation, avoiding collisions with walls and rounding or jumping over obstacles. Bilateral lesions of MLR in chronic experiments lead to a decrease of locomotor activity, but the animals are capable of almost normal walking. On the contrary, bilateral lesion of SLR leads to a striking result—the animal does not walk at all, even with very strong exteroceptive stimulation. Nevertheless, electric stimulation of MLR in such an animal evokes apparently normal locomotion.

These data suggest that SLR and MLR could, in fact, be the "inputs" to the automatic system for the control of locomotion in intact animals, but could have different functional meanings. A role of the cortico-brainstem pathway (stimulation of which can evoke locomotion in the mesencephalic cat) in intact animals is not clear.

How does stimulation of MLR and SLR influence the spinal cord? A study of neurons of the nucleus cuneiformis (which corresponds to MLR) showed that the neurons can easily be activated at those strengths of MLR stimulation which evoke locomotion (132), but there are no direct pathways from MLR and SLR to the spinal cord. Therefore, effects of stimulation of these regions must be mediated by some brainstem structures (94, 138). One of the main candidates for this role is a descending monoaminergic system having at least two subdivisions, i.e., a noradrenergic and a serotoninergic group of neurons. We shall discuss some data concerning the role of these neurons.

It was noted (see section "Role of the Spinal Cord in Locomotion") that an injection of DOPA (3, 4-dihydroxyphenylalanine) evokes stepping movements in the spinal cat (20, 42, 53). DOPA is said to facilitate liberation of noradrenaline from synaptic axonal terminals (2, 3). Since cell bodies of noradrenergic neurons have not been found in the spinal cord (25), the effect of DOPA on the spinal cord is determined by liberation of noradrenaline from the terminals of axons descending to the spinal cord from the brain. In other words, DOPA injection seems to be equivalent to excitation of the brain noradrenergic neurons. An injection of Clonidine also evokes locomotion in the spinal cat (35). This drug is known to excite alpha-adrenoreceptors of spinal neurons (1) which is also equivalent to excitation of the brain noradrenergic neurons.

Locomotor effects can also be produced by an injection of 5-hydroxytryptophan (5-HTP) (149) which promotes liberation of another mediator, serotonin, from the axonal terminals of serotoninergic brainstem spinal neurons (6). Thus, 5-HTP injection can be considered to be equivalent to excitation of the brain serotoninergic neurons.

DOPA and 5-HTP injections exert very prominent effects on the neuronal mechanisms of the spinal cord. It was noted (see section "Hypotheses about Arrangement of the Spinal Mechanism of Stepping") that in the spinal cat reciprocal long-lasting discharges appeared in flexor and extensor motoneurons in response to nerve stimulation (71, 72). Many other effects appear—in the system of presynaptic inhibition (4, 5), and in the activity of gamma-motoneurons (15, 43), for example. Comparison of some of these effects with those produced by MLR stimulation in the mesencephalic immobilized cat showed a striking resemblance (51).

The lumbosacral division of the spinal cord can generate not only stepping, but also some other movements of the hindlimbs: scratching (130), an extensor jerk (131), etc. Generation of a definite movement can be triggered by appropriate afferent influences, especially in spinal animals, but it is natural to suggest that, in intact animals, the spinal cord generates certain types of movements in response to descending commands. So, generation of stepping is triggered by MLR or SLR stimulation. A question arises as to whether the aminergic system is "specialized" for activation of the spinal locomotor mechanisms. Unfortunately, we have no answer to this question. It is also not known what the fundamental mechanism of action of noradrenaline and serotonin in the spinal cord is. It is only known that noradrenaline exerts an inhibitory action on some spinal neurons (31).

How might the monoaminergic neurons of the lower brainstem (sending axons to the spinal cord) be stimulated to evoke locomotion? This question is also without answer; it is only known that MLR and SLR stimulation results in excitation (sometimes monosynaptic) of the reticulospinal pontine and bulbar neurons (96, 97); therefore, there are connections between "locomotor regions" and lower parts of the brainstem. It is possible that some reticulospinal neurons responding to MLR or SLR stimulation belong to the "dorsal reticulospinal system" which exerts influences on the spinal cord similar to those of the monoaminergic system (45).

Activity of Fast-Conducting Descending Pathways

If the monoaminergic slow-conducting descending system of the lower brainstem is considered to have a prominent role in activation of the spinal mechanisms of stepping, what is the role in locomotion of the well known fast-conducting descending systems, i.e., the reticulospinal, vestibulospinal, and rubrospinal? To obtain data concerning this problem, the activity of neurons giving rise to the tracts descending to the lumbosacral spinal cord (i.e., participating in the control of the hindlimbs) was recorded during locomotion of mesencephalic and thalamic cats. An example of the activity of a rubrospinal neuron is given in Figure 10, *A* and *B*. At the beginning of locomotion the neuron becomes more active, and during locomotion its discharge rate is modulated so that it is considerably higher in the swing phase. In preparations having the cerebellum intact, a rhythmic modulation related to stepping was revealed in most reticulospinal (97), vestibulospinal (100), and rubrospinal (101) neurons. Figure 11 shows a mean frequency of different neurons contributing to descending tracts plotted as a function of the position of the corresponding hindlimb (97, 100, 101). The curves define flow of impulses descending through the tracts to the spinal mechanism of stepping. It is seen that modulation in the various tracts has a "reasonable" form. Thus, the vestibulospinal tract exerting excitatory effects on extensors (49) is maximally active just at the moment of the extensor activation. The tracts' exciting flexors, i.e., reticulospinal (originating from neurons of the medial pontine and bulbar reticular formation) and rubrospinal (50, 63, 80), reach a maximal level of activity earlier in comparison with the vestibulospinal one, i.e., in the swing phase when flexors are active.

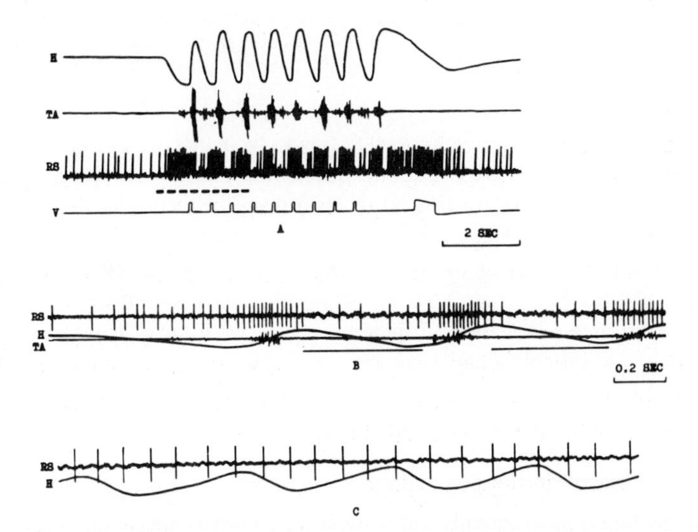

Figure 10. Activity of rubrospinal neurons during locomotion of the thalamic cat with the cerebellum intact (*A,B*) and decerebellate (*C*) (101). The interval marked in *A* by an interrupted line is presented in *B* with the faster sweep. *RS*, activity of the neuron; *H*, a hip angle of the contralateral (to the neuron) limb; *TA*, EMG of the ankle flexor m.tibialis anterior; *V*, marks of the band movement, an interval between marks corresponds to 0.5 m.

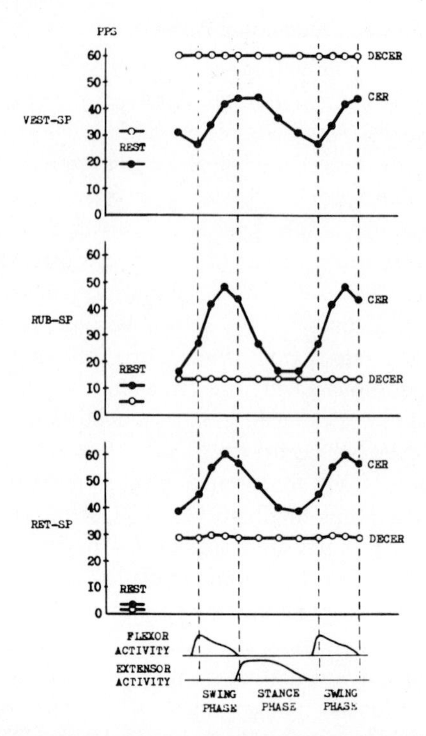

Figure 11. Activity of descending tracts during locomotion. Mean values of the discharge rate of neurons of the vestibulospinal (100), rubrospinal (101), and reticulospinal (97) tracts are plotted as a function of the hind position (ipsilateral for vestibulo- and reticulospinal tracts; contralateral for a rubrospinal one). Curves obtained in cats with the cerebellum intact (*CER*) and in ones decerebellate (*DECER*) are presented. A mean value of the resting discharge rate is also presented (*REST*). Flexor and extensor activity is shown schematically.

After removal of the cerebellum, stepping movements are preserved, but they are poorly coordinated (98). The rhythmic modulation of descending tracts disappears almost completely (98, 100, 101) (see, for example, the rubrospinal neuron in Figure 10*C*). This is also demonstrated in Figure 11 where discharge rates of "mean" neurons of the descending tracts are presented as a function of the limb position. These data strongly support a spinal origin of the locomotor rhythm in the mesencephalic and thalamic cats. Rhythmic changes of activity of rubrospinal and reticulospinal neurons are preserved to a larger extent after cerebellar ablation in the decorticate cats in comparison with the changes in the mesencephalic and thalamic ones (23, 114).

Processing of Signals in the Cerebellum

In mesencephalic and thalamic cats, the cerebellum is the main structure performing modulation of the activity in descending tracts. It performs this modulation in accordance with the signals arriving by various spinocerebellar pathways. The VSCT spino-reticulocerebellar and, probably, spino-olivocerebellar pathways mainly convey information concerning activity of the spinal neuronal mechanism generating stepping, while DSCT also conveys information regarding activity of the executive motor apparatus of the limb (see section "Signals Sent to the Brain and to Other Divisions of the Spinal Cord"). On the basis of these "input" signals, the cerebellum exerts a modulatory influence on the "output" brainstem neurons. As a result, their activity turns out to be dependent on the phase of the step cycle, and this dependence is different in various tracts (Figure 11).

Vestibulospinal neurons of Deiters' nucleus and the reticulospinal neurons are under the influence of the medial (vermal) part of the cerebellum (122). An input to this area is formed by VSCT, spino-reticulocerebellar, and spino-olivocerebellar pathways (86, 111, 112). Taking into consideration the peculiarities of these pathways, one can suggest that modulation of the vestibulospinal and reticulospinal tracts is performed by the cerebellum mainly in accordance with signals coming directly from the spinal mechanism generating stepping. Therefore, this modulation should be preserved, to a large extent, in the absence of afferent feedback from the moving limb. Indeed, a modulation of the reticulospinal neurons has been found during "fictitious locomotion" of the decorticate cat (114). The rhythmical modulation has also been found in neurons of the cerebellar cortex (granular cells, interneurons, and Purkinje cells) during "fictitious locomotion" of the decorticate rabbit (24, 150). Phases of activity of various Purkinje cells within the cycle are different. It is possible that this difference correlates with the different patterns of modulation found in the reticulospinal and vestibulospinal tracts.

Neurons of the red nucleus are under influences from other areas of the cerebellar cortex, i.e., the intermediate part and the paramedian lobule (90, 122). Purkinje cells of these areas affect the red nucleus through the interposed nucleus. An input to these cerebellar areas is formed by DSCT (85, 112), VSCT (22), and spino-reticulocerebellar and spino-olivocerebellar pathways (112). Neurons of the red nucleus were found to be rhythmically modulated during "fictitious locomotion" of the decorticate cat (114). This finding suggests that those tracts which convey information about intraspinal processes (for example, VSCT) exert strong influences on the intermediate part and paramedian lobule of the cerebellum. In addition, experiments carried out on thalamic cats showed that artificial disturbances of movements in the knee and ankle joints during locomotion weakly affect the pattern of modulation of rubrospinal neurons (101). If the DSCT (conveying, to a large extent, information about the activity of the executive motor apparatus of the limb) were to form the main input to those parts of the cerebellar cortex which are responsible for the modulation of rubrospinal neurons, any disturbance of the limb movements should strongly affect rubrospinal neurons. However, the rhythmical modulation of these neurons was interrupted only when the whole limb was arrested. Arrest of the limb resulted in cessation of the rhythmical activity of the spinal generator.

Figure 12 illustrates the "processing" of the input signals in the cerebellum that results in the formation of the signals descending by the rubrospinal tract to

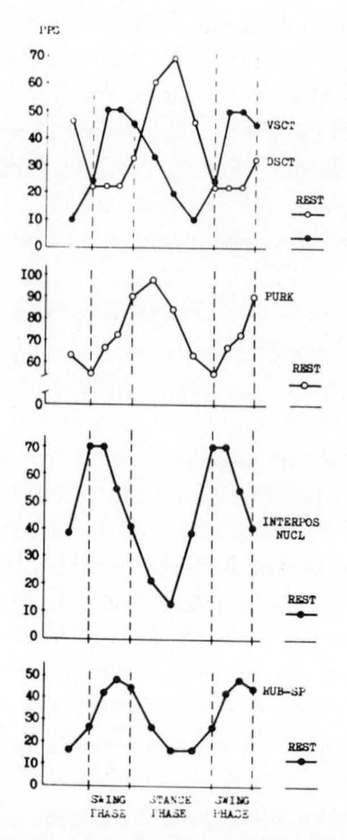

Figure 12. Processing of signals in the cerebellum during locomotion. Mean values of the discharge rate of various neuronal groups are shown as a function of the hindlimb position. There are presented: the dorsal and ventral spinocerebellar tracts projecting to the intermediate part of the cerebellar cortex (7, 8); Purkinje cells (*PURK*) from the hind limb projection zone of this part of the cerebellar cortex (102); neurons from the hind limb projection zone of the interposed cerebellar nucleus (*INTERPOS, NUCL*) (103); and rubrospinal neurons (*RUB-SP*) from the contralateral red nucleus (101).

the spinal mechanisms. Activity of Purkinje cells in the intermediate part of the cerebellum is maximal at the beginning of the stance phase (102). This pattern of modulation seems to be a complex function of various input signals—activity of DSCT having a maximum in the same phase (7), and activity of the fraction of VSCT (projecting to the intermediate part of the cerebellar cortex) having a maximum in the swing phase (8). Influences of other inputs, which have not been studied, cannot by excluded either. Purkinje cells exert inhibitory influences on neurons of the interposed nucleus (68, 69); correspondingly, a curve of the activity in this nucleus shows modulation which is opposite to that of Purkinje cells (102, 103). Neurons of the interposed nucleus, in their turn, excite rubrospinal neurons (147), and the curves of activity in these nuclei are similar, but the rubrospinal one has a small time lag (101, 103).

Hypotheses about Cerebellar Functions in Locomotion

Removal of the cerebellum results in the strongly expressed lack of coordination of locomotor movements both in acute (98) and chronic experiments (152).

This lack of coordination can be correlated with the changes found in activity of the descending tracts after cerebellectomy (Figure 11). First, cerebellectomy results in a considerable change of the mean values (throughout the cycle) of the descending impulse flow in all tracts. The mean discharge rate in the vestibulospinal tract becomes almost twice as high as in normal conditions. This result can be explained by disinhibition of Deiters' neurons since the direct inhibitory action of Purkinje cells on these neurons (67) is now interrupted. On the contrary, the mean discharge rate in the rubro- and reticulospinal tracts becomes 1.5 to 2 times lower. This can be explained by cessation of the excitatory inflow from cerebellar nuclei (70, 147). These changes of the mean descending impulse flow lead to inappropriate relations between flexor and extensor activities; in many cases an extensor rigidity appears (98). Thus, a maintenance of proper relations between descending impulse flow to the flexor and extensor muscular groups can be considered as one of the cerebellar functions in locomotion.

Second, after removal of the cerebellum, the periodic modulation of the activity disappears almost completely in all descending tracts. It is evident that this modulation is not a cause of stepping movements since decerebellate cats are capable of locomotion. But elimination of the modulation means that the supraspinal centers are now deprived of the ability to increase or decrease the activity of certain muscles in certain phases of the step cycle. The data presented above (see section "Processing of Signals in the Cerebellum") show that modulation of the descending tracts is determined, to a large extent, by information regarding activity of the spinal mechanism generating stepping movements. Therefore, by sending the phase-dependent signals to the spinal cord by the descending tracts, the cerebellum participates in generation (or detailing) of the central program of stepping. But it remains unknown why the spinal cord needs the cerebellar assistance during distribution of motor commands to various muscles.

Thus, the experimental data concerning cerebellar function in locomotion suggest that the cerebellum participates in generation of the central pattern of the movement rather than in the processing of afferent information. But in other types of movement its role might be as a center of afferent integration.

It is known that the cerebellum is closely related with the vestibular apparatus (17). Influences of the vestibular receptors on descending tracts (and, therefore, on the muscular activity) are mediated, to a large extent, by the cerebellum (58, 106, 120). Therefore, one could expect that, during locomotion, the cerebellum would participate in the regulation of equilibrium using information from vestibular receptors. But it was found that vestibular reactions of neurons of the vestibulo-, reticulo-, and rubrospinal tracts were prominent only at rest and during weak locomotion. With more intense locomotion (with stronger MLR stimulation) these reactions decreased or disappeared altogether (Figure 13) (107, 108). What could be the functional significance of the weakening of the vestibular responses of descending tracts during locomotion? One might suppose that discharges of vestibular receptors in relation to irregular and accidental head and body movements during rapid running could interfere with the rhythmic activity of descending tracts and thus disturb the stepping

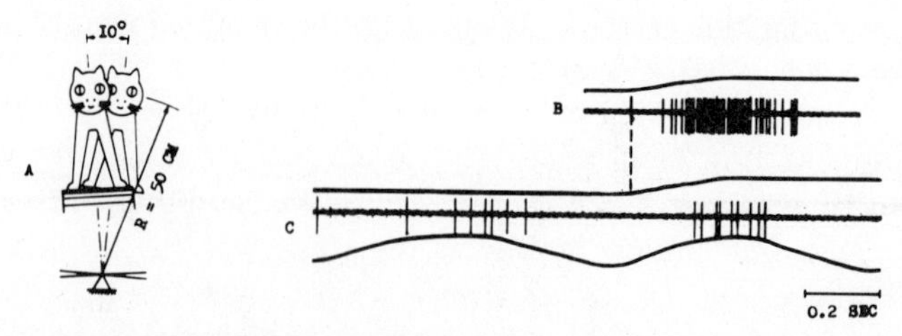

Figure 13. Response of a vestibulospinal neuron to tilt (the mesencephalic cat) (107). *A*, a scheme of the experiment; *B*, a response under resting conditions; *C*, a response during locomotion evoked by MLR stimulation. In *B* and *C* the *upper trace* shows the angle of the tilt (deflection of the curve upwards corresponds to the ipsilateral, to a neuron, tilt); in *C* the *lower trace* shows the hip angle of the ipsilateral limb (deflection of the curve upwards corresponds to the limb movement forwards).

performance. It is possible that a problem of maintaining the equilibrium during rapid running is solved with the help of other afferent systems (proprioceptive and visual), and with participation of the neck reflexes. But mechanisms of the cerebellar participation in this process are unknown.

CONCLUSION

Stimulation of the mesencephalic locomotor region (MLR) in the mesencephalic cat and of the subthalamic locomotor region (SLR) in the thalamic cat results in activation of the automatic system for the control of locomotion. It seems likely that the same system is used in locomotion of intact animals. What are the main events that happen in the nervous system when the automatic system for the control of locomotion comes into operation? First, the mechanisms generating stepping movements of individual limbs are activated in the spinal cord. Second, interconnections between these mechanisms promoting their synchronization and phasing are established. Third, some changes in the state of the brain occur, for example, "switching off" of the vestibular influences on neurons of the brainstem descending tracts. It is possible that in the intact animal some other brain centers (which are absent in the thalamic and mesencephalic cats) are also involved in the automatic system for the control of locomotion. For instance, the pyramidal tract controlling usually fine and non-standard movements of the hand, foot, and fingers could be used in locomotion as a part of the automatic system for the control of the same elements but in accordance with the standard locomotor program.

Simplicity of triggering and regulation of the automatic system for the control of locomotion considerably facilitates operation of those centers which control the motor behavior of the animal.

REFERENCES

1. Andén, N-E., Corrodi, H., Fuxe, K., Hökfelt, B., Ryden, C., and Svensson, T. (1970). Evidence for a central noradrenaline receptor stimulation by Clonidine. Life Sci. 9 Part I:513.

2. Andén, N-E., Jukes, M.G.M., and Lundberg, A. (1964). Spinal reflexes and monoamine liberation. Nature 202:1222.
3. Andén, N-E., Jukes, M.G.M., and Lundberg, A. (1966). The effect of DOPA on the spinal cord. 2. A pharmacological analysis. Acta Physiol. Scand. 67:387.
4. Andén, N-E., Jukes, M.G.M., Lundberg, A., and Vyklicky, L. (1966). The effect of DOPA on the spinal cord. I. Influence on transmission from primary afferents. Acta Physiol. Scand. 67:373.
5. Andén, N-E., Jukes, M.G.M., Lundberg, A., and Vyklicky, L. (1966). The effect of DOPA on the spinal cord. 3. Depolarization evoked in the central terminals of ipsilateral Ia afferents by volleys in the flexor reflex afferents. Acta Physiol. Scand. 68:322.
6. Anderson, E.G., and Shibuya, T. (1966). The effect of 5-hydroxytryptophan and L-tryptophan on spinal synaptic activity. J. Pharmacol. Exp. Ther. 153:352.
7. Arshavsky, YU.I., Berkinblit, M.B., Gel'fand, I.M., Orlovsky, G.N., and Fukson, O.I. (1972). Activity of the neurons of the dorsal spinocerebellar tract during locomotion. Biophysics. 17:506.
8. Arshavsky, YU.I., Berkinblit, M.B., Gel'fand, I.M., Orlovsky, G.N., and Fukson, O.I. (1972). Activity of the neurons of the ventral spinocerebellar tract during locomotion. Biophysics. 17:926.
9. Arshavsky, YU.I., Berkinblit, M.B., Gel'fand, I.M., Orlovsky, G.N., and Fukson, O.I. (1970). Activity of the neurons of the ventral spinocerebellar tract during locomotion of cats with deafferentated hindlimbs. Biophysics 17:1169.
10. Arshavsky, YU.I., Berkinblit, M.B., Fukson, O.I., Gel'fand, I.M., and Orlovsky, G.N. (1972). Recording of neurons of the dorsal spinocerebellar tract during evoked locomotion. Brain Res. 43:272.
11. Arshavsky, YU.I., Berkinblit, M.B., Fukson, O.I., Gel'fand, I.M., and Orlovsky. G.N. (1972). Origin of modulation in neurons of the ventral spinocerebellar tract during locomotion. Brain Res. 43:276.
12. Arshavsky, YU.I., Berkinblit, M.B., Gel'fand, I.M., Orlovsky, G.N., and Fukson, O.I. (1973). Activity of the neurons of the cuneo-cerebellar tract during locomotion. Biophysics 18:132.
13. Arshavsky, YU.I., Berkinblit, M.B., Gel'fand, I.M., Orlovsky, G.N., and Fukson, O.I. (1974). Differences in activity of spinocerebellar tracts during artificial stimulation and locomotion. Proceedings of the Fifth Symposium on Problems in General Physiology, pp. 99–105. Nauka, Leningrad.
14. Arshavsky, YU.I., Kots, YA.M., Orlovsky, G.N., Rodionov, I.M., and Shik, M.L. (1965). Investigation of the biomechanics of running by the dog. Biofizika 10:665. (Eng. transl. pp. 737–746).
15. Bergmans, J., and Grillner, S. (1969). Reciprocal control of spontaneous activity and reflex effects in static and dynamic flexor motoneurons revealed by an injection of DOPA. Acta Physiol. Scand. 77:106.
16. Bergmans, J., Miller, S., and Reitsma, D.J. (1972). Effect of L-DOPA on long ascending propriospinal pathways in the cat. Acta Physiol. Scand. 84:2A.
17. Brodal, A., Pompeiano, O., and Walberg, F. (1962). The Vestibular Nuclei and Their Connections, Anatomy and Functional Correlations. Oliver & Boyd, Edinburgh.
18. Brown, T.G. (1911). The intrinsic factors in the act of progression in the mammal. Proc. Roy. Soc. Lond. B 84:308.
19. Brown, T.G. (1914). On the nature of the fundamental activity of the nervous centers, together with an analysis of the conditioning of rhythmic activity in progression, and a theory of the evolution of function in the nervous system. J. Physiol. (Lond.) 48:18.
20. Budakova, N.N. (1973). [Stepping movements in a spinal cat after DOPA administration.] Fiziol., Zh. im.I.M. Sechenova 59:1190. (In Russian).
21. Budakova, N.N., and Shik, M.L. (1970). Effect of brain-stem stimulation evoking locomotion on ascending reflexes in the mesencephalic cat. Bull. Exp. Biol. Med. 69:17.
22. Burke, R., Lundberg, A., and Weight, F. (1971). Spinal border cell origin of the ventral spinocerebellar tract. Exp. Brain Res. 12:283.
23. Cabelguen, J.M., Millanvoye, M., and Perret, C. (1973). Caractéristiques centrales et efférentes de l'activité rhythmique de type locomoteur chez le Chat décortiqué aigu après ablation du cervele. J. Physiol. (Paris) 67:253A.

24. Coston, J.A., and Viala, G. (1970). Relations entre l'activité de cellules du cortex cérébelleux et les décharges rythmiques locomotrices chez le Lapin. J. Physiol. (Paris) 62 Suppl. 2:264.
25. Dahlstrom, A., and Fuxe, K. (1964). Evidence for the existence of monoamine containing neurons in the CNS. Acta Physiol. Scand. 62 Suppl. 232:3.
26. Davis, W.J. (1973). Neuronal organization and ontogeny in the lobster swimmeret system. *In* R.B. Stein *et al.* (eds.), Control of Posture and Locomotion. pp. 437–455. Plenum Press, New York.
27. Davis, W.D., and Kennedy, D. (1972). Command interneurons controlling swimmeret movements in the lobster. I. Types of effects on motoneurons. J. Neurophysiol. 35:1.
28. Eccles, J.C., Fatt, P., and Koketsu, K. (1954). Cholinergic and inhibitory synapses in a pathway from motor-axon collaterals to motoneurons. J. Physiol. (Lond.) 126:524.
29. Eccles, J.C., Fatt, P., and Landgren, S. (1956). Central pathway for direct inhibitory action of impulses in largest afferent nerve fibers to muscle. J. Neurophysiol. 19:75.
30. Eccles, R.M., and Lundberg, A. (1958). The synaptic linkage of "direct" inhibition. Acta Physiol. Scand. 43:204.
31. Engberg, I., Flatman, J.A., and Kadzielawa, K. (1974). The hyperpolarization of motorneurons by electrophoretically applied amines and other agents. Acta Physiol. Scand. 91:3A.
32. Engberg, I., and Lundberg, A. (1969). An electromyographic analysis of muscular activity in the hindlimb of the cat during unrestrained locomotion. Acta Physiol. Scand. 75:614.
33. Feldman, A.G. (1974). [The control of the muscle length.] Biofizika 19:749. (In Russian.)
34. Feldman, A.G., and Orlovsky, G.N. (1975). Activity of interneurons mediating reciprocal Ia inhibition during locomotion. Brain Res. 84:181.
35. Forssberg, H., and Grillner, S. (1973). The locomotion of the acute spinal cat injected with clonidine i.v. Brain Res. 50:184.
36. Freusberg, A. (1874). Reflexbewegungen beim Hunde. Pflügers Arch. Ges. Physiol. 9:358.
37. Gambaryan, P.P., Orlovsky, G.N., Protopopova, T.G., Severin, F.V., and Shik, M.L. (1971). Work of muscles in various forms of progressive movement of the cat and adaptive changes of the moving organs in the family Felidae. Tr. Zool. Inst. Akad. Nauk, USSR 48:220.
38. Goslow, G.E., Jr., Reinking, R.M., and Stuart, D.G. (1973). The cat step cycle: Hind limb joint angles and muscle lengths during unrestrained locomotion. J. Morphol. 141:1.
39. Granit, R. (1950). Reflex self-regulation of muscle contraction and autogenetic inhibition. J. Neurophysiol. 13:351.
40. Granit, R. (1955). Receptors and Sensory Perception. Yale University Press, New Haven.
41. Gray, J. (1968). Animal Locomotion. Weidenfeld and Nicolson, London.
42. Grillner, S. (1969). Supraspinal and segmental control of static and dynamic γ-motoneurons in the cat. Acta Physiol. Scand. Suppl.327:1.
43. Grillner, S. (1969). The influence of DOPA on the static and the dynamic fusimotor activity to the triceps surae of the spinal cat. Acta Physiol. Scand. 77:490.
44. Grillner, S. (1972). The role of muscle stiffness in meeting the changing postural and locomotor requirements for force development by the ankle extensors. Acta Physiol. Scand. 86:92.
45. Grillner, S. (1975). Locomotion in vertebrates—central mechanisms and reflex interaction. Physiol. Rev. 55:247.
46. Grillner, S. (1973). Locomotion in the spinal cat. *In* R.B. Stein, K.G. Pearson, R.S. Smith, and J.B. Redford (eds.), Control of Posture and Locomotion, pp. 515–535. Plenum Press, New York.
47. Grillner, S. (1974). On the generation of locomotion in the spinal dogfish. Exp. Brain Res. 20:459.
48. Grillner, S., and Hongo, T. (1972). Vestibulospinal effects on motoneurons and interneurons in the lumbosacral cord. *In* A. Brodal and O. Pompeiano (eds.), Basic Aspects of Central Vestibular Mechanisms. Progr. Brain Res. 37:243. Elsvier Publishing Co., Amsterdam.

49. Grillner, S., Hongo, T., and Lund, S. (1970). The vestibulospinal tract. Effects on alpha-motoneurons in the lumbosacral spinal cord in the cat. Exp. Brain Res. 10:94.
50. Grillner, S., and Lund, S. (1968). The origin of a descending pathway with monosynaptic action of flexor motoneurons. Acta Physiol. Scand. 74:274.
51. Grillner, S., and Shik, M.L. (1973). On the descending control of the lumbosacral spinal cord from the "mesencephalic locomotor region." Acta Physiol. Scand. 87:320.
52. Grillner, S., and Udo, M. (1971). Recruitment in the tonic stretch reflex. Acta Physiol. Scand. 81:571.
53. Grillner, S., and Zangger, P. (1974). Locomotor movements generated by the deafferented spinal cord. Acta Physiol. Scand. 91:38A.
54. Grossman, R.G. (1958). Effects of stimulation of non-specific thalamic system on locomotor movements in cat. J. Neurophysiol. 21:85.
55. Gurfinkel, V.S., Kostyuk, P.G., and Shik, M.L. (1973). On some possible modes of descending control of the spinal cord activity in connection with the problem of motor control. *In* Proceedings of Symposial Papers 4th International Biophysics Congress, Moscow.
56. Gurfinkel, V.S., and Shik, M.L. (1973). The control of posture and locomotion. *In* A.A. Gydikov *et al.* (eds.), Motor Control, pp. 217–234. Plenum Press, New York.
57. Hart, B.L. (1971). Facilitation by strychnine of reflex walking in spinal dogs. *In* Physiology and Behavior, Vol. 6, pp. 627–628. Pergamon Press, Elmsford, N.Y.
58. Hiebert, T.G., and Fernandez, C. (1965). Deitersian unit response to tilt. Acta Otolaringol. 60:180.
59. Hinsey, J.C., Ranson, S.W., and McNattin, R.F. (1930). The role of the hypothalamus and the mesencephalon in locomotion. Arch. Neurol. Psychiat. 23:1.
60. Holst, E. Von. (1935). Erregungsbildung und Erregungsleitung im Fischrückenmark. Pflügers Arch. Ges. Physiol. 235:345.
61. Holst, E. Von. (1938). Uber relative Koordination bei Säugern und beim Menschen. Pflügers Arch. Ges. Physiol. 240:44.
62. Holst, E. Von. (1939). Die relative Koordination als Phänomen und als methode zentral nervösen funktions analyse. Ergebn. Physiol.42:228.
63. Hongo, T., Jankowska, E., and Lundberg, A. (1969). The rubrospinal tract. I. Effects on alpha motoneurons innervating hindlimb muscles in cats. Exp. Brain Res. 7:344.
64. Hongo, T., Jankowska, E., and Lundberg, A. (1969). The rubrospinal tract. II. Facilitation of interneuronal transmission in reflex paths to motoneurons. Exp. Brain Res. 7:365.
65. Hultborn, H. (1972). Convergence on interneurons in the reciprocal Ia inhibitory pathway to motoneurons. Acta Physiol. Scand. Suppl.375:1.
66. Hunt, C.C. (1952). The effect of stretch receptors from muscle on the discharge of motoneurons. J. Physiol. (Lond.) 117:359.
67. Ito, M., and Yoshida, M. (1964). The cerebellar evoked monosynaptic inhibition of Deiters neurons. Experientia (Basel) 20:515.
68. Ito, M., Yoshida, M., and Obata, K. (1964). Monosynaptic inhibition of the intracerebellar nuclei induced from the cerebellar cortex. Experientia (Basel) 20:575.
69. Ito, M., Yoshida, M., Obata, K., Kawai, N., and Udo, M. (1970). Inhibitory control of intracerebellar nuclei by the Purkinje cell axons. Exp. Brain Res. 10:64.
70. Ito, M., Udo, M., Mano, N., and Kawai, N. (1970). Synaptic action of the fastigiobulbar impulses upon neurons in the medullary reticular formation and vestibular nuclei. Exp. Brain Res. 11:29.
71. Jankowska, D., Jukes, M.G.M., Lund, S., and Lundberg, A. (1967). The effect of DOPA on the spinal cord. 5. Reciprocal organization of pathways transmitting excitatory action to alpha motoneurons of flexors and extensors. Acta Physiol. Scand. 70:369.
72. Jankowska, E., Jukes, M.G.M., Lund, S., and Lundberg, A. (1967). The effect of DOPA on the spinal cord. 6. Half center organization of interneurons transmitting effects from the flexor reflex afferents. Acta Physiol. Scand. 70:389.
73. Jankowska, E., and Roberts, W.J. (1972). Synaptic action of single interneurons mediating reciprocal Ia inhibition of motoneurons. J. Physiol. (Lond.) 222:623.
74. Kashin, S.M., Feldman, A.G., and Orlovsky, G.N. (1974). Locomotion of fish evoked by electrical stimulation of the brain. Brain Res. 82:41.

75. Kling, U. (1971). Simulation neuronaler impulsrhythmen. Zur Theorie der Netzwerke mit cyclischen Hemmverbindungen. Kybernetik 9:123.
76. Kostyuk, P.G. (1973). [Structure and Function of Descending Systems of the Spinal Cord.] Nauka, Leningrad. (In Russian.)
77. Kulagin, A.S., and Shik, M.L. (1970). Interaction of symmetrical limbs during controlled locomotion. Biophysics 15:171.
78. Liddell, E.G.T., and Sherrington, C. (1924). Reflexes in response to stretch (myotactic reflexes). Proc. Roy. Soc. 96B:212.
79. Lloyd, D.P.C. (1941). The spinal mechanism of the pyramidal system in cats. J. Neurophysiol. 4:525.
80. Lund, S., and Pompeiano, O. (1968). Monosynaptic excitation of alpha motoneurons from supraspinal structures in the cat. Acta Physiol. Scand. 73:1.
81. Lundberg, A. (1959). Integrative significance of patterns of connections made by muscle afferents in the spinal cord. Symp. XXI Int. Physiol. Congr., Buenos Aires, pp. 100–105.
82. Lundberg, A. (1966). Integration in the reflex pathway. *In* R. Granit (ed.), Muscular Afferents and Motor Control, pp. 275–304. Almquist and Wiksell, Stockholm.
83. Lundberg, A. (1969). Reflex control of stepping. The Nansen Memorial Lecture V, pp. 1–42. Universitetsforlaget, Oslo.
84. Lundberg, A. (1971). Function of the ventral spinocerebellar tract–a new hypothesis. Exp. Brain Res. 12:317.
85. Lundberg, A., and Oscarsson, O. (1960). Functional organization of the dorsal spinocerebellar tract in the cat. VII. Identification of units by antidromic activation from the cerebellar cortex with recognition of the five functional subdivisions. Acta Physiol. Scand. 50:356.
86. Lundberg, A. and Oscarsson, O. (1962). Functional organization of the ventral spinocerebellar tract in the cat. IV. Identification of units by antidromic activation from the cerebellar cortex. Acta Physiol. Scand. 54:252.
87. Lundberg, A., and Voorhoeve, P. (1962). Effects from the pyramidal tract on spinal reflex arcs. Acta Physiol. Scand. 56:201.
88. Magni, F., and Willis, W.D. (1964). Cortical control of brain stem reticular neurons. Arch. Ital. Biol. 102:418.
89. Manter, J.T. (1938). The dynamics of quadrupedal walking. J. Exp. Biol. 15:522.
90. Massion, J. (1967). The mammalian red nucleus. Physiol. Rev. 47:383.
91. Matthews, P.B.C. (1972). Mammalian muscle receptors and their central actions. Edward Arnold (Publishers) Ltd., London.
92. Miller, S., and Oscarsson, O. (1970). Termination and functional organization of spino-olivocerebellar paths. *In* W.S. Fields and W.D. Willis (eds.), The Cerebellum in Health and Disease, pp. 172–200. Warren H. Green, St. Louis.
93. Miller, S., and Van Der Burg, J. (1973). The function of long propriospinal pathways in the co-ordination of quadrupedal stepping in the cat. *In* R.B. Stein *et al.* (eds.), Control of Posture and Locomotion, pp. 561–577. Plenum Press, New York.
94. Orlovsky, G.N. (1969). Electrical activity in brainstem and descending paths in controlled locomotion. Sechenov Physiol. J. of USSR. 55:437. (In Russian).
95. Orlovsky, G.N. (1969). [Spontaneous and induced locomotion of the thalamic cat.] Biofizika, 14:1095. (Eng. transl. pp. 1154–1162.)
96. Orlovsky, G.N. (1970). Connexions of the reticulo-spinal neurons with the "locomotor sections" of the brain stem. Biophysics 15:178.
97. Orlovsky, G.N. (1970). Work of the reticulo-spinal neurons during locomotion. Biophysics 15:761.
98. Orlovsky, G.N. (1970). Influence of the cerebellum on the reticulospinal neurons during locomotion. Biophysics 15:928.
99. Orlovsky, G.N. (1972). The effect of different descending systems of flexor and extensor activity during locomotion. Brain Res. 40:359.
100. Orlovsky, G.N. (1972). Activity of vestibulospinal neurons during locomotion. Brain Res. 46:85.
101. Orlovsky, G.N. (1972). Activity of rubrospinal neurons during locomotion. Brain Res. 46:99.
102. Orlovsky, G.N. (1972). Work of the Purkinje cells during locomotion. Biophysics 17:935.

103. Orlovsky, G.N. (1972). Work of the neurons of cerebellar nuclei during locomotion. Biophysics 17:1177.
104. Orlovsky, G.N., and Feldman, A.G. (1972). [On the role of afferent activity in generation of stepping movements.] Neirofisiologia 4:401. (Eng. abst. p. 409)
105. Orlovsky, G.N., and Feldman, A.G. (1972). [Classification of lumbrosacral neurons according to their discharge patterns during evoked locomotion.] Neirofisiologia 4:410. (Eng. abst. p. 417)
106. Orlovsky, G.N., and Pavlova, G.A. (1972). [Vestibular responses of neurons of different descending pathways in cats with intact cerebellum and in decerebellated ones.] Neirofisiologia 4:303. (Eng. abst. p. 316).
107. Orlovsky, G.N., and Pavlova, G.A. (1972). [Vestibular responses in neurons of descending pathways during locomotion.] Neirofisiologia 4:311. (Eng. abst. p. 316.)
108. Orlovsky, G.N., and Pavlova, G.A. (1972). Response of Deiters' neurons to tilt during locomotion. Brain Res. 42:212.
109. Orlovsky, G.N., Severin, F.V., and Shik, M.L. (1966). [Effect of speed and load on co-ordination of movements during running of the dog.] Biofizika 11:364. (Eng. transl. pp. 414–417.)
110. Oscarsson, O. (1965). Functional organization of the spino- and cuneo-cerebellar tracts. Physiol. Rev. 45:495.
111. Oscarsson, O. (1967). Functional significance of information channels from the spinal cord to the cerebellum. *In* Neurophysiological Basis of Normal and Abnormal Motor Activities, M.D. Yahr and D.P. Purpura (eds.), pp. 93–117, 3rd Symposium of the Parkinson's Disease Information and Research Center of Columbia Univ. Raven Press, New York.
112. Oscarsson, O. (1973). Functional organization of spinocerebellar paths. *In* A. Iggo (ed.), Handbook of Sensory Physiology, Vol. II, pp. 339–380. Springer-Verlag, Berlin.
113. Pearson, K.G., Fourtner, C.R., and Wong, R.K. (1973). Nervous control of walking in the cockroach. *In* R.B. Stein *et al.* (eds.), Control of Posture and Locomotion, pp. 495–514. Plenum Press, New York.
114. Perret, C. (1973). Analyse des mécanismes d'une activité de type locomoteur chez le chat. These de Doct. Sci., Paris, CNRS AL 8342.
115. Perret, C. (1974). Activités efférentes et générateur de rythme locomoteur chez le chat. J. Physiol. (Paris) 69:284A.
116. Perret, C., and Berthoz, A. (1973). Evidence of static and dynamic fusiomotor actions on the spindle response to sinusoidal stretch during locomotor activity in the cat. Exp. Brain Res. 18:178.
117. Perret, C., and Buser, C. (1972). Static and dynamic fusiomotor activity during locomotor movements in the cat. Brain Res. 40:165.
118. Perret, C., Cabelguen, J.M., and Millanvoye, M. (1972). Caractéristiques d'un rythme de type locomoteur chez le chat spinal aigu. J. Physiol. (Paris) 65:473A.
119. Perret, C., Millanvoye, M., and Cabelguen, J.M. (1972). Messages spinaux ascendants pendant une locomotion fictive chez le chat curarisé. J. Physiol. (Paris) 65:153A.
120. Peterson, B.W. (1970). Distribution of neural responses to tilting within vestibular nuclei of the cat. J. Neurophysiol. 33:750.
121. Philippson, M. (1905). L'autonomie et la centralisation dans le système nerveux des animaux. Trav. Lab. Physiol. Inst. Solvay (Bruxelles) 7:1.
122. Pompeiano, O (1967). Functional organization of the cerebellar projections to the spinal cord. *In* C.A. Fox and R.S. Snider (eds.), The Cerebellum. Prog. Brain Res. 25:282–321.
123. Renshaw, B. (1941). Influence of the discharge of motoneurons upon excitation of neighbouring motoneurons. J. Neurophysiol. 4:167.
124. Rossi, G.F., and Brodal, A. (1956). Corticofugal fibers to the brain stem reticular formation. An experimental study in the cat. J. Anat. (Lond.) 90:42.
125. Severin, F.V. (1970). The role of the gamma motor system in the activation of the extensor alpha motor neurons during controlled locomotion. Biophysics 15:1138.
126. Severin, F.V., Orlovsky, G.N., and Shik, M.L. (1967). [Work of the muscle receptors during controlled locomotion.] Biofizika 12:502. (Eng. transl. pp. 575–586.
127. Severin, F.V., Orlovsky, G.N., and Shik, M.L. (1968). Recurrent influences on work of single motoneurons during controlled locomotion. Bull. Exp. Biol. Med. 66:713.
128. Severin, F.V., Shik, M.L., and Orlovsky, G.N. (1967). [Work of the muscles and single

motoneurons during controlled locomotion.] Biofizika 12:660. (Eng. transl. pp. 762–772.)

129. Sherrington, C.S. (1906). The integrative action of the nervous system. Yale University Press, New Haven.
130. Sherrington, C.S. (1906). Observations on the scratch-reflex in the spinal dog. J. Physiol. (Lond.) 34:1.
131. Sherrington, C.S. (1910). Flexion-reflex of the limb, crossed extension reflex, and reflex stepping and standing. J. Physiol. (Lond.) 40:28.
132. Shik, M.L., and Yagodnitsyn, A.S. (1973). [Study of the neuronal connections in the dorsocaudal midbrain tegmentum of the cat by the method of microstimulation.] Neirofisiologia 5:593. (In Russian).
133. Shik, M.L., and Orlovsky, G.N. (1965). [Co-ordination of the limbs during running of the dog.] Biofizika 10:1037. (Eng. transl. pp. 1148–1159.)
134. Shik, M.L., and Orlovsky, G.N. Neurophysiology of the locomotor automatism. Physiol. Rev. In press.
135. Shik, M.L., Orlovsky, G.N., and Severin, F.V. [Organization of locomotor synergism.] Biofizika 11:879. (Eng. transl. pp. 1011–1019)
136. Shik, M.L., Orlovsky, G.N., and Severin, F.V. (1968). [Locomotion of the mesencephalic cat elicited by stimulation of the pyramids.] Biofizika 13:127. (Eng. transl. pp. 143–152.)
137. Shik, M.L., Severin, F.V., and Orlovsky, G.N. (1966). [Control of walking and running by means of electrical stimulation of the midbrain.] Biofizika 11:659. (Eng. transl. pp. 756–765.
138. Shik, M.L., Severin, F.V., and Orlovsky, G.N. (1967). [Brain stem structures responsible for evoked locomotion.] Fiziol. Zh. im.I.M. Sechenova 53:1125.
139. Shimamura, M., and Livingston, R.B. (1963). Longitudinal conduction systems serving spinal and brain stem co-ordination. J. Neurophysiol. 26:258.
140. Shurrager, P.S. (1955). Walking in spinal kittens and puppies. *In* W.F. Windle (ed.), Regeneration in the Central Nervous System, pp. 208–218. Charles C. Thomas Publishers, Springfield.
141. Sirota, M.G., and Shik, M.L. (1973). [The cat locomotion during midbrain stimulation.] Sechenov Physiol. J. of USSR 59:1314. (In Russian).
142. Smolyaninov, V.V., and Karpovich, A.L. (1975). Cinematics of metachronal locomotion. I. Configurations. Biofizika 20:527. (In Russian.)
143. Stein, R.B. (1974). Peripheral control of movement. Physiol. Rev. 54:215.
144. Stuart, D.G., Withey, T.P., Wetzel, M.C., and Goslow, G.E., Jr. (1973). Time constraints for inter-limb co-ordination in the cat during unrestrained locomotion. *In* R.B. Stein *et al.* (eds.), Control of Posture and Locomotion, pp. 537–560. Plenum Press, New York.
145. Szekely, G. (1968). Development of limb movements; embryological, physiological and model studies. *In* G.E.W. Wolstenholme and M. O'Connor (eds.), A Ciba Foundation Symposium: Growth of the Nervous System, pp. 77–93. J. & A. Churchill Ltd., London.
146. Szekely, G., Czéh, and Vörös, G. (1969). The activity pattern of limb muscles in freely moving normal and deafferented newts. Exp. Brain Res. 9:53.
147. Tsukahara, N., Toyama, K., and Kosaka, K. (1964). Intracellularly recorded responses of red nucleus neurons during antidromic and arthodromic activation. Experientia (Basel) 20:632.
148. Viala, D., and Buser, P. (1969). The effects of DOPA and 5-HTP on rhythmic efferent discharges in hind limb nerves in the rabbit. Brain Res. 12:437.
149. Viala, D., and Buser, P. (1971). Modalités d'obtention de rythmes locomoteurs chez le lapin spinal par traitements pharmacologiques (DOPA, 5 HTP, d'amphétamine). Brain Res. 35:151.
150. Viala, D., Coston, A., and Buser, P. (1970). Participation de cellules du cortex cérébelleux aux rythmes "locomoteurs" chez le lapin curarisé, en absence d'informations somatiques liées au mouvement. C.R. Acad. Sci. Paris 271:688.
151. Waller, W.H. (1940). Progression movements elicited by subthalamic stimulation. J. Neurophysiol. 3:300.
152. Wang, G.H., and Welker, W. (1961). Behavior of decerebellate cats with or without neocortex in chronic state. Fed. Proc. 20:332. Abst.

153. Wiersma, C.A.G., and Ikeda, K. (1964). Interneurons commanding swimmeret movements in the crayfish, Procambarus clarki (Girard). Comp. Biochem. Physiol. 12:509.
154. Wilson, D.M. (1961). The central nervous control of flight in a locust. J. Exp. Biol. 38:471.
155. Wilson, D.M. (1964). The origin of the flight-motor command in grasshoppers. *In* R.F. Reiss (ed.), Neuronal Theory and Modelling, pp. 331–345. Stanford University Press, Stanford.
156. Wilson, D.M. (1972). Genetic and sensory mechanisms for locomotion and orientation in animals. Amer. Sci. 60:358.

Index